Statistical Methods in Analytical Chemistry

CHEMICAL ANALYSIS

A SERIES OF MONOGRAPHS ON
ANALYTICAL CHEMISTRY AND ITS APPLICATIONS

Editor

J. D. WINEFORDNER

VOLUME 153

A WILEY-INTERSCIENCE PUBLICATION

JOHN WILEY & SONS, INC.

New York / Chichester / Weinheim / Brisbane / Singapore / Toronto

Statistical Methods in Analytical Chemistry

Second Edition

PETER C. MEIER

CILAG A.G.
(A Johnson & Johnson Company)
Schaffhausen, Switzerland

RICHARD E. ZÜND

TERANOL A.G.
(A Hoffmann-LaRoche Company)
Visp, Switzerland

A WILEY-INTERSCIENCE PUBLICATION

JOHN WILEY & SONS, INC.

New York / Chichester / Weinheim / Brisbane / Singapore / Toronto

Published simultaneously in Canada.

For ordering and customer service, call 1-800-CALL-WILEY.

Library of Congress Cataloging-in-Publication Data
Meier, Peter C., 1945–
    Statistical methods in analytical chemistry / Peter C. Meier, Richard E. Zünd. — 2nd ed.
        p.   cm. — (Chemical analysis ; v. 153)
    "A Wiley-Interscience publication."
    Includes bibliographical references and index.
    ISBN 0-471-29363-6 (cloth : alk. paper)
    1. Chemistry, Analytic—Statistical methods.   I. Zünd, Richard E.
II. Title.   III. Series.
QD75.4.S8M45   2000
543′.007′2—dc21   99-25291
                                                                              CIP

Printed in the United States of America.

10 9 8 7 6 5 4 3 2

To our wives, Therese and Edith, respectively, who granted us the privilege of "book" time, and spurred us on when our motivation flagged.

To our children, Lukas and Irène, respectively, and Sabrina and Simona, who finally have their fathers back.

CONTENTS

PREFACE

This book focuses on statistical data evaluation, but does so in a fashion that integrates the question—plan—experiment—result—interpretation—answer cycle by offering a multitude of real-life examples and numerical simulations to show what information can, or cannot, be extracted from a given data set. This perspective covers both the daily experience of the lab supervisor and the worries of the project manager. Only the bare minimum of theory is presented, but is extensively referenced to educational articles in easily accessible journals.

The context of this work, at least superficially, is quality control in the chemical and pharmaceutical industries. The general principles apply to any form of (chemical) analysis, however, whether in an industrial setting or not. Other readers need only to replace some phrases, such as "Health Authority" with "discriminating customer" or "official requirements" with "market expectations," to bridge the gap. The specifically chemical or pharmaceutical nomenclature is either explained or then sufficiently circumscribed so that the essentials can be understood by students of other disciplines.

The quality and reliability of generated data is either central to the work of a variety of operators, professionals, or managers, or is simply taken for granted. This book offers insights for all of them, whether they are mainly interested in applying statistics (cf. worked examples) or in getting a feeling for the connections and consequences (cf. the criminalistic examples). Some of the appended programs are strictly production-oriented (cf. Histo, Similar, Data, etc.), while others illustrate an idea (cf. Pedigree, SimCal, OOS-Risk, etc.).

When the first edition was being prepared in the late 1980s, both authors worked out of cubicles tucked into the corner of an analytical laboratory and were still very much engaged in hands-on detail work. In the intervening years, responsibilities grew, and the bigger the offices got, the larger became the distance from the work bench. Diminishing immediacy of experience may be something to bemoan, but compensation comes in the form of a wider view, i.e., how the origin and quality of the samples tie in with the product's history and the company's policies and interests.

Life at the project and/or line manager level sharpens awareness that "quality" is something that is not declared, but designed into the product and the manufacturing process. Quality is an asset, something that needs management attention, particularly in large, multinational organizations. Laboratory instrumentation is largely computerized these days, a fact that certainly fosters standardization and method transfer across continents. The computational power makes child's play of many an intricate procedure of yesteryear, and the excellent report-writing features generate marvels of GMP-compliant documentation (GMP = Good Manufacturing Practices). Taken at face value, one could gain the impression that analytical chemistry is easy, and results are inevitably reliable and not worthy of introspection. This history is reflected in the statistically oriented chemical literature: 10–15 years ago, basic math and its computer-implementation were at the forefront; today's literature seeks ways to mine huge, multidimensional data sets. That numbers might be tainted by artifacts of nonideal chemistry or human imperfection is gradually being acknowledged; the more complex the algorithms, though, the more difficult it becomes to recognize, track, and convincingly discuss the ramifications. This is reason enough to ask for upfront quality checks using simple statistical tools before the individual numbers disappear in large data banks.

In a (laboratory) world increasingly dominated by specialization, the vendor knows what makes the instrument tick, the technician runs the samples, and the statistician crunches numbers. The all-arounder who is aware of how these elements interact, unfortunately, is an endangered species.

Health authorities have laid down a framework of regulations ("GMPs" in the pharmaceutical industry) that covers the basics and the most error-prone steps of the development and manufacturing process, for instance, analytical method validation. The interaction of elements is more difficult to legislate the higher the degree of intended integration, say, at the method, the laboratory, the factory levels, or at the sample, the batch, and the project perspectives. This second edition places even greater emphasis on these aspects and shows how to detect and interpret errors.

PETER C. MEIER

Schaffhausen, Switzerland

RICHARD E. ZÜND

Visp, Switzerland

PREFACE, First Edition

Both authors are analytical chemists. Our cooperation dates back to those happy days we spent getting educated and later instructing undergraduates and PhD candidates in Prof. W. Simon's laboratory at the Swiss Federal Institute of Technology in Zürich (ETH-Z). Interests ranged far beyond the mere mechanics of running and maintaining instruments. Designing experiments and interpreting the results in a wider context were primary motives, and the advent of computerized instrumentation added further dimensions. Masses of data awaiting efficient and thorough analysis on the one hand, and introductory courses in statistics slanted toward pure mathematics on the other, drove us to the autodidactic acquisition of the necessary tools. Mastery was slow in coming because texts geared to chemistry were rare, such important techniques as linear regression were relegated to the "advanced topics" page, and idiosyncratic nomenclatures confused the issues.

Having been through despiriting experiences, we happily accepted, at the suggestion of Dr. Simon, an offer to submit a manuscript. We were guided in this present enterprise by the wish to combine the cookbook approach with the timely use of PCs and programmable calculators. Furthermore, the when-and-how of tests would be explained in both simple and complex examples of the type a chemist understands. Because many analysts are involved in quality control work, we felt that the consequences statistics have for the accept/reject decision would have to be spelled out. The formalization that the analyst's habitual quest for high-quality results has undergone—the keywords being GMP and ISO 9000—is increasingly forcing the use of statistics.

PETER C. MEIER

Schaffhausen, Switzerland

RICHARD E. ZÜND

Visp, Switzerland
September 1992

CHEMICAL ANALYSIS

A SERIES OF MONOGRAPHS ON ANALYTICAL CHEMISTRY AND ITS APPLICATIONS

J. D. WINEFORDNER, *Series Editor*

Vol. 1. **The Analytical Chemistry of Industrial Poisons, Hazards, and Solvents.** *Second Edition.* By the late Morris B. Jacobs

Vol. 2. **Chromatographic Adsorption Analysis.** By Harold H. Strain (*out of print*)

Vol. 3. **Photometric Determination of Traces of Metals.**
Fourth Edition
Part I: General Aspects. By E. B. Sandell and Hiroshi Onishi
Part IIA: Individual Metals, Aluminum to Lithium. By Hiroshi Onishi
Part IIB: Individual Metals, Magnesium to Zirconium. By Hiroshi Onishi

Vol. 4. **Organic Reagents Used in Gravimetric and Volumetric Analysis.** By John F. Flagg (*out of print*)

Vol. 5. **Aquametry: A Treatise on Methods for the Determination of Water.** *Second Edition (in three parts).* By John Mitchell, Jr. and Donald Milton Smith

Vol. 6. **Analysis of Insecticides and Acaricides.** By Francis A. Gunther and Roger C. Blinn (*out of print*)

Vol. 7. **Chemical Analysis of Industrial Solvents.** By the late Morris B. Jacobs and Leopold Schetlan

Vol. 8. **Colorimetric Determination of Nonmetals.** *Second Edition.* Edited by the late David F. Boltz and James A. Howell

Vol. 9. **Analytical Chemistry of Titanium Metals and Compounds.** By Maurice Codell

INTRODUCTION

Modern instrumental analysis is an outgrowth of the technological advances made in physics and electronics since the middle of this century. Statistics have been with us somewhat longer, but were impractical until the advent of powerful electronic data processing equipment in the late 1960s and early 1970s, and even then remained bottled up in the central computer department.

Chemistry may be a forbidding environment for many nonchemists: there are few rules that link basic physics with the observable world, and typical molecules sport so many degrees of freedom that predictions of any kind inevitably involve gross simplifications. So, analytical chemistry thrives on very reproducible measurements that just scratch the phenomenological surface and are only indirectly linked to whatever one should determine. A case in point: what is perceived as off-white color in a bulk powder can be due to any form of weak absorption in the VIS(ible) range ($\lambda \approx 400–800$ nm), but typically just one wavelength is monitored.

For these reasons, the application of statistics in an analytical setting will first demand chemical experience, full appreciation of what happens between start of sampling[§] and the instrument's dumping numbers on the screen, and an understanding of which theories might apply, before one can even think of crunching numbers. This book was written to tie together these aspects, to demonstrate how every-day problems can be solved, and how quality is recognized and poor practices are exposed.

Analytical chemistry can be viewed from two perspectives: the insider sees the subject as a science in its own right, where applied physics, math, and chemistry join hands to make measurements happen in a reliable and representative way; the outsider might see the service maid that without further effort yields accurate results that will bring glory to some higher project.

The first perspective, taken here, revolves around calibration, finding reasons for numbers that are remarkable or out of line in some way, and validation. The examples given in this book are straight from the world of routine quality control and the workhorse instruments found there: gas chromatography (GC), high-pressure liquid chromatography (HPLC), acidity (pH) meters, and the like. Whether we like it or not, this represents analytical "ground truth." The employed statistical techniques will be of the simpler type. No statistical theory can straighten out slips in manufacturing

1

or sampling. Those lucky enough to have access to high-profile, one-of-a-kind instruments in special settings are a minority that is overrepresented in the literature, which conveys the wrong picture.

The second perspective might be that of the leader of some large project where chemical analyses are just a side issue, where sample numbers are large and chemical niceties might be completely swamped by, say, biological variability; here a statistician will be necessary to make sense of the results in the context of a very complex model. Chemistry is a bit harder to relate to than many other industries in that the measured quantities are often abstract, invisible, and only indirectly linked to what one wants to control.

This book, written by two passionate analysts, treats the application of statistics to analytical chemistry[1,2] in a very practical manner. A minimum of tools is explained and then applied to everyday, that is, complex situations. The examples should be illuminating to both beginners and specialists from other fields in their quest to evaluate data and make decisions.

What are the circumstances of this decision process? The scheme in Table 1 might serve as an illustration. Every step is fraught with uncertainties and is subject to artifacts.[32] Sampling,[3,6] a weak point for various reasons, is the hinge on which everything else depends: carelessness in this area makes the best intentions meaningless. Lack of maintenance and instrument calibration endangers the relevance of the decisions made. Note that statistics are only part of the picture; the decision process has to be viewed as a coherent whole; a decision can only be passed by taking into account the complex interrelations among the chemical species being consumed and formed, and the legal, economic, scientific, and environmental characteristics of the analytical process.

The sensor only incompletely maps physicochemical reality into an electrical signal that is then subject to signal-conditioning hardware, and the instrument again transforms the individual measurements into time averages, signal areas,[17] or other user-oriented information. Just filtering out noise, strictly speaking, already constitutes an adulteration. The instrument configuration (hard- and software) is to be regarded as just one element that the analyst has to put to use without exceeding its inherent limitations (not necessarily the design specifications quoted in the prospectus!).[15]

Thus, one can be far from the ideal world often assumed by statisticians: tidy models, theoretical distribution functions, and independent, essentially uncorrupted measured values with just a bit of measurement noise superimposed. Furthermore, because of the costs associated with obtaining and analyzing samples, small sample numbers are the rule. On the other hand, linear ranges upwards of 1 : 100 and relative standard deviations of usually 2% and less compensate for the lack of data points.

The authors have found the tests and procedures described in Chapters 1

Table 1   The evaluation of raw data is embedded in a process

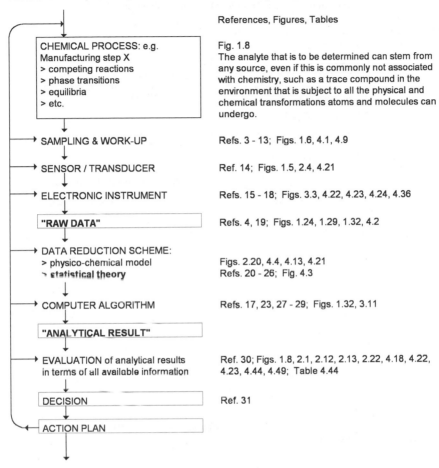

	References, Figures, Tables
CHEMICAL PROCESS: e.g. Manufacturing step X > competing reactions > phase transitions > equilibria > etc.	Fig. 1.8 The analyte that is to be determined can stem from any source, even if this is commonly not associated with chemistry, such as a trace compound in the environment that is subject to all the physical and chemical transformations atoms and molecules can undergo.
SAMPLING & WORK-UP	Refs. 3 - 13; Figs. 1.6, 4.1, 4.9
SENSOR / TRANSDUCER	Ref. 14; Figs. 1.5, 2.4, 4.21
ELECTRONIC INSTRUMENT	Refs. 15 - 18; Figs. 3.3, 4.22, 4.23, 4.24, 4.36
"RAW DATA"	Refs. 4, 19; Figs. 1.24, 1.29, 1.32, 4.2
DATA REDUCTION SCHEME: > physico-chemical model > statistical theory	Figs. 2.20, 4.4, 4.13, 4.21 Refs. 20 - 26; Fig. 4.3
COMPUTER ALGORITHM	Refs. 17, 23, 27 - 29; Figs. 1.32, 3.11
"ANALYTICAL RESULT"	
EVALUATION of analytical results in terms of all available information	Ref. 30; Figs. 1.8, 2.1, 2.12, 2.13, 2.22, 4.18, 4.22, 4.23, 4.44, 4.49; Table 4.44
DECISION	Ref. 31
ACTION PLAN	

to 3 to be the most useful for the constellation of "a few precise measurements of law-abiding parameters" prevalent in analytical chemistry, but this does not disqualify other perspectives and procedures. For many situations routinely encountered several solutions of varying theoretical rigor are available. A case in point is linear regression, where the assumption of error-free abscissa values is often violated. Is one to propagate formally more correct approaches, such as the maximum likelihood theory, or is a weighted, or even an unweighted least-squares regression sufficient? The exact numerical solutions found by these three models will differ: any practical consequences thereof must be reviewed on a case-by-case basis.

Table 2   The steps of recognizing a problem, proposing a solution, checking it for robust operation, and documenting the procedure and results under GMP are nested operations

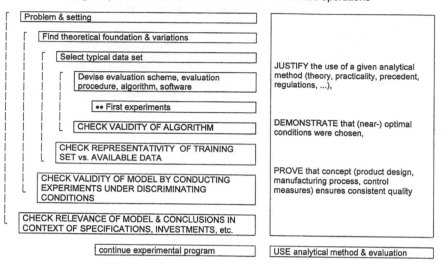

Problem & setting
Find theoretical foundation & variations
Select typical data set
Devise evaluation scheme, evaluation procedure, algorithm, software
•• First experiments
CHECK VALIDITY OF ALGORITHM
CHECK REPRESENTATIVITY OF TRAINING SET vs. AVAILABLE DATA
CHECK VALIDITY OF MODEL BY CONDUCTING EXPERIMENTS UNDER DISCRIMINATING CONDITIONS
CHECK RELEVANCE OF MODEL & CONCLUSIONS IN CONTEXT OF SPECIFICATIONS, INVESTMENTS, etc.
continue experimental program

JUSTIFY the use of a given analytical method (theory, practicality, precedent, regulations, ...),

DEMONSTRATE that (near-) optimal conditions were chosen,

PROVE that concept (product design, manufacturing process, control measures) ensures consistent quality

USE analytical method & evaluation

The choice of subjects, the detail in which they are presented, and the opinions implicitly rendered, of course, reflect the author experiences and outlook. In particular, the GMP aspect had been built in from the beginning, but is expanded in this second edition: the basic rules applicable to the laboratory have been relocated to Chapter 3 and are presented in a more systematic manner (see Table 2) and many additional hints were included. To put things into a broader perspective and alert the analyst to the many factors that might affect his samples before they even hit the lab bench or could influence his evaluation, Section 4.38 was added. It lists many, but by far not all of the obstacles that line the road from the heady atmosphere of the project-launch meeting to when the final judgment is in. Because the GMP philosophy does not always permeate all levels of hierarchy to the same degree, this table, by necessity, also contains elements indicative of managerial style, work habits, and organizational structure, besides pedestrian details like keeping calibration standards in stock.

Some of the VisualBasic programs that come with the book offer approaches to problem-solving or visualization that may not be found elsewhere. Many VB programs and Excel sheets were crafted with a didactical twist: to make the influence of random noise and the bias due to the occasional error apparent. Details are found in Section 5.3.

Many figures illustrate abstract concepts; heavy use is made of numerical simulation to evade the textbook style "constructed" examples that, due to reduction to the bare essentials, answer one simple question, but do not tie into the reader's perceived reality of messy numbers. In the past, many texts assumed little more than a pencil and an adding machine, and so propagated involved schemes for designing experiments to ease the number-crunching load. There are only three worked examples here that make use of integer numbers to ease calculations (see the first three numerical examples in Chapter 1); no algebraic or numerical shortcuts are taken or calculational schemes presented to evade divisions or roots, as was so common in the recent past.

Terminology was chosen to reflect recent guides[33] or, in the case of statistical symbols, common usage.[34]

There are innumerable references that cover theory, and still many more that provide practical applications of statistics to chemistry in general and analytical chemistry in particular. Articles from *Analytical Chemistry* were chosen as far as possible to provide world-wide availability. Where necessary, articles in English that appeared in *Analytica Chimica Acta*, *Analyst*, or *Fresenius Zeitschrift für Analytische Chemie* were cited.

There are a number of authorative articles the reader is urged to study that amplify on issues central to analytical understanding.[35–38]

THE CONCEPT BEHIND THIS BOOK

Background

Textbooks and courses in general statistics are easily accessible to students of chemistry, physics, biology, and related sciences. Some of the more or less explicitly stated assumptions that one often comes across are the following:

- A large population of objects is available from which samples can be pulled.
- Measurements are easy and cheap, that is a large number of measurements is available, either as many repeats on a small number of samples or as single determinations on a large number of independent samples.
- The appropriate theoretical distribution (Gaussian, Poisson, etc.) is known with certainty.
- The governing variables are accurately known, are independent of each other (orthogonal) and span wide value ranges.
- Simple mathematical models apply.
- Sample§ collection and work-up artifacts do not exist.
- Documentation is accurate, timely, and consistent.

- The investigated (chemical) moiety is pure, the major signal (observable) is being investigated.
- Few, if any, minor signals are in evidence for which the signal-to-noise ratio is good and which can be assigned to known chemical entities that are available in quantities sufficiently large to allow for a complete physicochemical characterization.
- The measured quantities accurately represent the system under investigation.
- Statistics are used to prove the appropriateness of the chosen model.
- Nonstatistical decision criteria do not exist.

Professional communities can be very diverse in their thinking:

- Many natural scientists think in terms of measured units like concentration (mg/ml, moles/liter, etc.), and disregard the issue of probabilities.
- In medical circles it is usual to cite the probability of treatment A being better than B.
- A coefficient of determination is known to lab supervisors to be $r^2 > 0.99$ for any worthwhile calibration.
- Instrument makers nonetheless provide this less-than-useful "information," but hardly anybody recognizes r^2 as the outflow of the wide calibration range, the linear concentration-to-signal transfer function, and the excellent repeatability.
- Mathematicians bask in proofs piled on other proofs, each proof being sold as a "practical application of theory."
- Statisticians advise "look for a simpler problem" when confronted with the complexity and "messiness" of practical chemistry.
- Chemists are frustrated when they learn that their problem is mathematically intractable. All sides have to recognize that the other's mental landscape is "valid and different" and that a workable decision necessitates concessions. The chemist (or other natural scientist) will have to frame questions appropriately and might have to do some experiments in a less than straightforward manner; the statistician will have to avoid overly rigorous assumptions.

Fields of Application for Analytical Chemistry

A somewhat simplistic description of reality would classify analytical practice as follows (see Table 3):

Table 3 Types of problems encountered

PROBLEM	
↓	↓
TYPE **One-of-a-kind solution & implementation (e.g. research)**	**Optimized design for routine use (e.g. GMP-type conditions)**
↓	↓
CRITERIA (1) Non-standard process or situation that is amenable to an experiment, or (2) Many low-cost samples and/or a cheap analytical method, and/or (4) Low legal or financial risks	Standardized production process with clear success criteria, high costs & risks, tough specifications, and (usually) few data points
↓	↓
METHOD Any analytical methodology that yields numbers that can somehow be linked to the quantity of interest (e.g. find complexation constants, determine optimal composition, description of one aspect of a multi-facetted problem)	Separation and quantitation of species, use of linear portion of response function, simple statistical tools, results can be interpreted in terms of physico-chemical concepts; binding specification limits, and the obligation to investigate every deviation
↓	↓
EVALUATION Employ a statistician or an investigator who is knowledgable in sophisticated statistical tools; do not impose rules on evaluation team, allow them to concentrate on interesting aspects (EXPLORATIVE DATA ANALYSIS); after the brainstorming, bring in the process and analytics experts for a round-table discussion: CAN THE POSTULATED EFFECTS BE EXPLAINED BY THE INVOLVED CHEMISTRY, BY ARTIFACTS, OR HAS SOMETHING NEW BEEN OBSERVED?	Employ lab staff that will get the work done according to the imposed regulations; scientifically interesting investigations should be the exception, rather than the rule, because the process has been fully investigated and is under control. USE PRE-DETERMINED CRITERIA AND TOOLS FOR RAPID EVALUATION; FULLY DOCUMENT THE RESULTS AND OBTAIN A DECISION.

1. Research use of individual methods or instruments in an academic or basic research environment, with interest centered around obtaining facts and relationships, where specific conditions exist as concerns precision, number of measurements, models, etc. that force the use of particular and/or highly sophisticated statistical techniques.

2. Research use of analytical results in the framework of a nonanalytical setting, such as a governmental investigation into the spread of pollution; here, a strict protocol might exist for the collection of samples (number, locations, time, etc.) and the interpretation of results, as provided by various consultants (biologists, regulators, lawyers, statisticians, etc.); the analytical laboratory would only play the role of a black box that transforms chemistry into numbers; in the perspective of the laboratory worker, calibration, validation, quality control, and interpolation are the foremost problems. Once the reliability and plausibility of the numbers is established, the statisticians take over.

3. Quality control (QC) in connection with manufacturing operations is

probably the most widespread, if mundane, application of analytical (chemical) determinations. The keywords are precision, accuracy, reliability, specifications, costs, and manageability of operations.[31] Since management takes it for granted that a product can be tested for compliance with specifications, and often only sees QC as a cost factor that does not add value (an obsolete notion), the lab staff is left to its own devices when it comes to statistical support.

Statistical Particulars of Analytical Chemistry

Much of today's instrumentation in principle allows for the rapid acquisition of vast amounts of information on a particular sample.[§] However, the instruments and the highly trained staff needed to run them are expensive. Often, samples are not cheap either; this is particularly true if they have to be pulled to confirm the quality of production lots. [See (A).] Then each point on the graph represents a four-, five-, or even six-digit investment in materials and manpower. Insisting on doubling the number of samples N to increase statistical power could easily bankrupt the company. For further factors, see Section 4.38.

A manufacturing process yields a product that is usually characterized by anywhere from one to as many as two dozen specifications, each in general calling for a separate analytical method. For each of these parameters a distribution of values will be observed if the process is carried out sufficiently often. Since the process will change over the years (raw materials, equipment train, synthesis fine-tuning, etc.), as will the analytical methods (better selectivity, lower limit of detection, new technologies), the overall distribution of values must be assigned to an assortment of subpopulations that will not necessarily match in all points. These subpopulations might intrinsically have narrow distributions for any given parameter, but what is observed is often much wider because several layers of effects contribute to statistical variance through insufficient sampling for reasons of time, money, and convenience:

(A) The number of batches produced under a given set of conditions (each batch can cost millions)

(B) The number of points in space or time within one such batch A that needs to be tested (spacial inhomogeneity due to viscosity, temperature gradients, etc.); temporal inhomogeneity due to process start-up and shut-down.

(C) The number of repeat samples§ pulled in one location/time coordinate *B*

(D) The number of sample work-ups conducted on any one sample *C*, and

(E) The number of repeat determinations performed on any one worked-up sample *D*.

Note the following points:

- The number of samples is often restricted for cost reasons (*A*, *B*),
- Careless or misguided sample pulling (*B*, *C*) and/or work-up (*D*) can easily skew the concentration-signal relationship,
- The dynamic range of the instrument can be overwhelmed, leading to signal distortions and/or poor signal-to-noise ratios for the observed moiety (*E*),
- Specificity is often inadequate (i.e., the instrument insufficiently suppresses the signal for the chemical species not being investigated);
- Duplicate determinations (*D* or *D* + *E*) can be justified, but more repeats usually do not improve the interpretation dramatically,
- The instruments generally yield very precise measurements (*E*, CV ≪ 5%),
- For impurity signals in high-quality chemicals, the digital resolution of the instrument may encroach on the repeatability figure-of-merit,
- The appropriate theoretical distribution (*A*, *B*, *C*) can only be guessed at because the high price and/or time loss attached to each result precludes achievement of the large N necessary to distinguish between rival models,
- By the time the analytical result is in, so many selective/non-linear pro-

§*Note*: A chemical sample is a quantity of material that represents physicochemical reality. Each of the samples/processes *A* ⋯ *E* given in the text introduces its own distribution function (Cf. Fig. 1.8.) by repeated measurements *E* on each of several work-ups *D* done on each of several samples *C* pulled from many locations/time points *B* on the number of batches *A* available for investigation. The perceived distribution function (compound data *A* ⋯ *E* at level *E*) may or may not be indicative of the distribution function one is trying to study (e.g., level *A*); a statistician's "sample", however, is a number in a data set which here corresponds to the numerical result of one physical measurement conducted on a chemically processed volume of product/water/soil/ ... /tissue. The unifying thought behind the nomenclature "sample" is that it supposedly accurately represents the population it is drawn from. For the statistician, that means, figuratively speaking, a pixel in a picture (both are in the same plane); for the chemist, the pixel and the chemical picture are separated by a number of veils, each one further blurring the scene one would perceive if it were not there.

cess steps $(B \ldots E)$ have modified the probability density function that is to be probed, that the best-guess assumption of a Gaussian distribution for (A, B, C) may be the only viable approach,

- After the dominant independent variables have been brought under control, many small and poorly characterized ones remain that limit further improvement in modeling the response surface; when going to full-scale production, control of "experimental" conditions drops behind what is possible in laboratory-scale work (e.g., temperature gradients across vessels), but this is where, in the long term, the "real" data is acquired,

- Chemistry abounds with examples of complex interactions among the many compounds found in a simple synthesis step,

- Sample collection and work-up artifacts (D) exist, as do impurities and problems with the workers (experience, motivation, turnover, deadlines, and suboptimal training), all of which impact the quality of the obtained results,

- The measured quantities frequently are related to tracers that only indirectly mirror the behavior of a hard-to-quantitate compound;

- The investigated species or physical parameter may be a convenient handle on an otherwise intangible concept such as "luster," "color," or "tinge,"

- Because physicochemical cause-and-effect models are the basis of all measurements, statistics are used to optimize, validate, and calibrate the analytical method, and then interpolate the obtained measurements; the models tend to be very simple (i.e., linear) in the concentration interval used,

- Particularly if the industry is government regulated (i.e., pharmaceuticals), but also if the supply contract with the customer stipulates numerical specification limits for a variety of quality indicators, the compliance question is legal in nature (rules are set for the method, the number of samples and repeat determinations); the analyst can then only improve precision by honing his/her skills,

- Nonstatistical decision criteria are the norm because specification limits are frequently prescribed (i.e., 95 to 105% of nominal) and the quality of previous deliveries or competitor's warranty raises expectations beyond what statistical common sense might suggest.

Selection of Topics

Since the focus of this book is the use of statistics in practical situations in everyday work, one-of-a-kind demonstrations are avoided, even if the mathematics is spectacular. Thus, the reader will be confronted with each of the following items:

- A brief repetition of the calculation of the mean, the standard deviation, and their confidence limits,
- Hints on how to present results, and limits of interpretation,
- The digital resolution of equipment and the limited numerical accuracy of calculators/programs,
- An explanation of why the normal distribution will have to be used as a model, even though the adherence to ND (or other forms) cannot be demonstrated under typical conditions,
- Comparisons between data sets (t test, multiple range test, F test, simple ANOVA),
- Linear regression with emphasis on the use as a calibration/interpolation tool.

Because the number of data points is low, many of the statistical techniques that are today being discussed in the literature cannot be used. While this is true for the vast majority of control work that is being done in industrial labs, where acceptability and ruggedness of an evaluation scheme are major concerns, this need not be so in R&D situations or exploratory or optimization work, where statisticians could well be involved. For products going to clinical trials or the market, the liability question automatically enforces the tried-and-true sort of solution that can at least be made palatable to lawyers on account of the reams of precedents, even if they do not understand the math involved.

For many, this book will at least offer a glimpse of the nonidealities the average analyst faces every day, of which statistics is just a small part, and the decisions for which we analysts have to take responsibility.

Software

A series of programs is provided that illustrates the statistical techniques that are discussed. The data files that are provided for experimentation in part reflect the examples that are worked in the book, and in part are different. There is a particular data file for each program that illustrates the application. (See Section 5.4.)

Because the general tone is educational, principles are highlighted. The programs can be used to actually work with rather large sets of experimental data, but may fail if too much is demanded of them in terms of speed, data volume, or options.

Liability: The authors have applied their half-century of programming experience to design clean user-interfaces, to get their math straight, and test the resulting applications in all kinds of circumstances. Colleagues were enlisted for "testing", but these programs were not validated in the strictest sense of the word. Given the complexity of today's operating systems, we do not claim they are foolproof. The authors should not be held responsible for any decisions based on output from these programs.

Each program includes the necessary algorithms to generate t-, p-, z-, F-, or χ^2-values (relative errors below 1%; for details, see ⟨Display Accuracy⟩ options in program CALCVAL. Therefore, table-look-up is eliminated.

The source code is now in VisualBasic (it used to be in GW-BASIC, later in QBASIC); the files are provided in compiled form, together with a structured menu. Some Excel files are included in the XLS directory.

UNIVARIATE DATA

The title implies that in this first chapter techniques are dealt with that are useful when the observer concentrates on a single aspect of a chemical system, and repeatedly measures the chosen characteristic. This is a natural approach, first because the treatment of one-dimensional data is definitely easier than that of multidimensional data, and second, because a useful solution to a problem can very often be arrived at in this manner.

A scientist's credo might be "One measurement is no measurement." Thus, take a few measurements and divine the truth! This is an invitation for discussions, worse yet, even disputes among scientists. Science thrives on hypotheses that are either disproven or left to stand; in the natural sciences that essentially means experiments are re-run. Any insufficiency of a model results in a refinement of the existing theory; it is rare that a theory completely fails (the nineteenth-century luminiferous ether theory of electromagnetic waves was one such, and cold fusion was a more shortlived case).

Reproducibility of experiments indicates whether measurements are reliable or not; under GMP regulations this is used in the systems suitability and the method validation settings.

A set of representative data is considered to contain a determinate and a stochastic component. The *determinate* part of a signal is the expected or average outcome. The human eye is good at extracting the average trend of a signal from all the noise superimposed on it; the arithmetic mean is the corresponding statistical technique. The *stochastic* part is what is commonly called noise, that is, the difference between the individual measurement and the average that is wholly determined by chance; this random element comprises both the sign and the size of the deviation. The width of the jittery track the recorder pen traces around the perceived average is commonly obtained by calculating the standard deviation on a continuous series of individual measurements.

1.1 MEAN AND STANDARD DEVIATION

This section treats the calculation of the mean, the standard deviation, and the standard deviation of the mean without recourse to the underlying theory.

It is intended as a quick introduction under the tacit assumption of normally distributed values.

The simplest and most frequent question is "What is the typical value that best represents these measurements, and how reliable is it?"[39]

1.1.1   The Most Probable Value

Given that the assumption of normally distributed data (see Section 1.2.1) is valid, several useful and uncomplicated methods are available for finding the most probable value and its confidence interval, and for comparing such results.

When only a few measurements of a given property are available, and especially if an asymmetry is involved, the median is often more appropriate than the mean. The *median*, x_m, is defined as the value that bisects the set of n ordered observations, that is,

- If n is odd, $(n-1)/2$ observations are smaller than the median, and the next higher value is reported as the median.

Example 1: For $n = 9$ and $x() = 4, 5, 5, 6, 7, 8, 8, 9, 9 \rightarrow x_m = 7.0$ and $\mathbf{x}_{mean} = 6.78$.

- If n is even, the average of the middle two observations is reported.

Example 2: For $n = 6$ and $x() = 2, 3, 4, 5, 6, 6 \rightarrow x_m = 4.5$ and $\mathbf{x}_{mean} = 4.33$.

The most useful characteristic of the median is the small influence exerted on it by extreme values, that is, its robust nature. The median can thus serve as a check on the calculated mean.

The *mean*, $\mathbf{x}_{mean}$, can be shown to be the best estimate of the true value μ; it is calculated as the arithmetic mean of n observations:

$$\mathbf{x}_{mean} = \frac{1}{n} \Sigma (x_i) \tag{1.1}$$

where "Σ" means "obtain the arithmetic sum of all values x_i, with $i = 1 \ldots n$"

Example 3: If the extreme value "15" is added to the data set $x() = 2, 3, 5, 5, 6, 6, 7$, the median changes from $x_m = 5.0$ to 5.5, while the mean changes from $\mathbf{x}_{mean} = 4.8571$ to 6.125. (See Figure 1.1.)

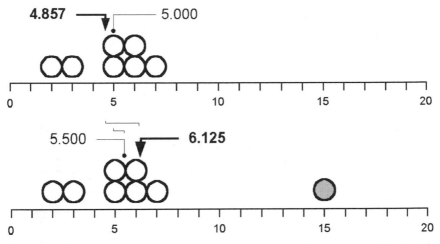

Figure 1.1. The median x_m and the average are given for a set of observations. This figure is a simple form of a histogram; see Section 1.8.1. An additional measurement at $x = 15$ would shift x_{mean} much more than x_{median}.

Notice that by the inclusion of x_8, the mean is much more strongly influenced than the median. The value of such comparisons lies in the automatic processing of large numbers of small data sets, in order to pick out the suspicious ones for manual inspection. (See also the next Section.)

1.1.2   The Dispersion

The reliability of a mean is judged by the distribution of the individual measurements about the mean. There are two generally used measures of the spread (the scatter) of a set of observations, namely the range R and the standard deviation s_x.

The *range*, R, is the difference between the largest and the smallest observation:

$$R = x_{max} - x_{min} \qquad (1.2)$$

Precisely because of this definition, the range is very strongly influenced by extreme values. Typically, for a given sample size n, the average range $R(n)$ will come to a certain expected (and tabulated) multiple of the true standard deviation. In Figure 1.2 the ranges R obtained for 390 simulations are depicted. It is apparent that the larger the sample size n, the more likely the occurrence of extreme values: for $n = 4$ the two extremes are expected to be

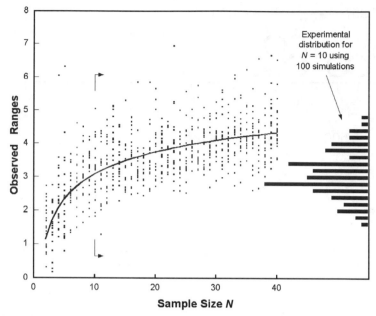

Figure 1.2. The range $R(n)$ for size of sample n, with $n = 2 \ldots 40$ (left). The line gives the tabulated values.[34] The range R is given as $y = R/s_x$ in units of the experimental standard deviation. A total of 8190 normally distributed values with mean 0 and standard deviation 1 was simulated. (See Section 3.5.5.) The righthand figure gives the distribution of ranges found after simulating 100 sets of $n = 10$ normally distributed values.

around ± one standard deviation apart, and the other two values somewhere in between. At $n = 40$, the range is expected to be twice as large.

There is no alternative to using the full range if there are only few observations available and no plausible theoretical description of the distribution. With n larger than, say, 9, the concept of *quantiles* (or percentiles) can be used to buffer the calculated range against chance results: for $n = 10$, throwing out the highest and the lowest observations leaves $n' = 8$; the corresponding range is termed the 10–90% range (difference between the 10th and the 90th percentiles). This process of eliminating the extremes can be repeated to yield the 20–80% range, etc. A very useful application is in the graphical presentation of data, for instance by drawing a box around the central two-thirds of all observations and extensions to mark the overall range. (See Fig. 1.3.)

The advantage of this technique is that no assumptions need to be made about the type of distribution underlying the data and that many sets of observations can be visually compared without being distracted by the individual numbers. For further details, see Section 1.8. Note that the fraction of all

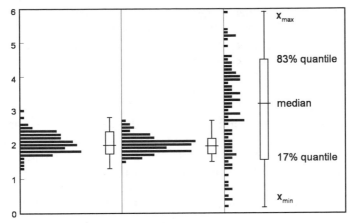

Figure 1.3. The use of quantiles for displaying data. Three distributions of 100 events each are shown in histogram (left) and in quantile (right) form. The reason for choosing the 17th and the 83rd quantiles is explained in the text.

observations bounded by the 17th and the 83rd quantiles encompasses two-thirds of all values, which is very close to the 68.3% expected in the interval $\pm 1 \cdot s_x$ around the mean.

The *standard deviation*, s_x, is the most commonly used measure of dispersion. Theoretically, the parent population from which the n observations are drawn must meet the criteria set down for the normal distribution (see Section 1.2.1); in practice, the requirements are not as stringent, because the standard deviation is a relatively robust statistic. The almost universal implementation of the standard deviation algorithm in calculators and program packages certainly increases the danger of its misapplication, but this is counterbalanced by the observation that the consistent use of a somewhat inappropriate statistic can also lead to the right conclusions.

The standard deviation, s_x, is by definition the square root of the variance, V_x,

$$S_{xx} = \Sigma(x_i - x_{mean})^2 = \Sigma(r_i)^2 \tag{1.3a}$$

$$S_{xx} = \Sigma(x_i^2) - (\Sigma x_i)^2/n \tag{1.3b}$$

$$V_x = S_{xx}/(n-1) \tag{1.3c}$$

$$s_x = \sqrt{V_x} \tag{1.3d}$$

where $f = (n-1)$ is the number of degrees of freedom, by virtue of the fact that, given x_{mean} and $n-1$ freely assigned x_i values, the nth value is fixed.

S_{xx} is the sum of squares of the residuals, r_i, that are obtained when the average value x_{mean} is subtracted from each observation x_i. x_{mean} is the best estimate for the true mean μ. When discussing theoretical concepts or when the standard deviation is precisely known, a small Greek sigma, σ, is used; in all other cases, the estimate s_x appears instead.

Example 4: For a data set $x() = 99.85$, 100.36, 99.75, 99.42, and 100.07 one finds $\Sigma(x_i) = 499.45$, $\Sigma(x_i^2) = 49890.5559$; S_{xx} according to Eq. (1.3b) is thus $49890.5559 - 49890.0605 = 0.4954$. Here the five significant digits "49890" are unnecessarily carried along with the effect that the precision, which is limited by the computer's word length, is compromised. The average x_{mean} is found as 99.89 and the standard deviation s_x as ±0.3519 (via Eqs. (1.3c) and (1.3d); see also Table 1.1).

The calculation via Eq. (1.3b) is the one implemented in most calculators[27,40] because no x_i values need to be stored. The disadvantage inherent in this approach lies in the danger of digit truncation (cf. error propagation, Section 3.3), as is demonstrated earlier. This can be avoided by subtracting a constant from each observed x_i, $x_i' = x_i - c$, so that fewer significant digits result before doing the calculations according to Eq. (1.3b); this constant could be chosen to be $c = 99.00$ in the previous example.

Example 5: The exact volume of a 100 ml graduated flask is to be determined by five times filling it to the mark, weighing the contents to the nearest 0.1 mg, correcting for the density to transform grams to milliliters, and averaging; the density-corrected values are 99.8536, 99.8632, 99.8587, 99.8518, and 99.8531 ml (see data file VOLUME.dat). For the purpose of demonstration, this task is solved using 32-bit VisualBasic 5.0, GW-BASIC and seven different models of calculators; the results are given in Table 1.1. Of course it would be more appropriate to round all volumes to four or at most five significant digits because a difference in filling height of 0.2 mm, which might just be discernible, amounts to a volume difference of $0.02 \cdot (0.8)^2 \cdot \pi/4 \approx 0.01$ ml (inside diameter of the flask's neck: 8 mm), but for the sake of the argument, the full accuracy theoretically available in the standard density tables is used. In all cases the correct mean 99.85608 was found. The digits given in *italics* deviate from the correct value given in **bold** numbers. Drawing the root by hand shows that the VB5 double-precision mode result is actually a bit less accurate than the GW-BASIC one for this particular numerical constellation! This difference is irrelevant for practical applications because the sum of squares S_{xx} and the product $4 \cdot s_x \cdot s_x$ differ by only

Table 1.1. Reliability of calculated standard deviations

Calculator Model	Internal Digits	Digits Displayed	Standard Deviation as Displayed	Case
TI-95	?	13	0.004767074574621	a
			0.004767284342265	b
HP-71B	15	12	0.00476728434226	a
HP-32S	15	11	0.004766235412	a
			0.004767284342	b, c, d
			0.0047672843423	d'
HP-11 and	?	10	0.004878012	a
41C			0.004766235	b
			0.004767284	c, d, e
			0.004767284342	c', d', e'
HP-55	12	10	0.00500000000	a
			0.0047670751	b
			0.0047672841	c
TI-30D	10	8	0.0104403	a
			0.004878	b
			0.0047662	c
			no result	d
			0.0047672843	d'
GW-BASIC*		16	0.004767281342264767	f
		16	0.004767284262925386	g
VB 5.0		15	0.00476728434226759	f
		15	0.00476728426292539	g
Excel 97		15	0.00476728388778602	a
			0.00476728434386556	b
			0.00476728434224702	c
			0.00476728434226458	d
			0.00476728434226447	e
			0.00476728434226759	f
hand calculation		16	0.004767284342264472	—

*) an older version of BASIC that was available on disc operating systems (at least up to DOS 3.11; QBASIC replaced it at least up to DOS 6.21); the mathematical algorithms used in Visual Basic appear to be very similar but not identical, as can be seen from the difference in the last three digits of the double-precision results; the single-precision result is …8446… instead of …8426…

Case	Amount c Subtracted	Digits Typed in	Number of Significant Digits
a	0.0000	99.8536	6
b	90.0000	9.8536	5
c	99.0000	0.8536	4
d	99.8000	0.0536	3
e	99.8500	0.0036	2 (3)
f	99.85608	−0.00248	double-precision mode, Eq. (1.3a) see program MSD
g	99.85608	−0.00248	single-precision mode, Eq. (1.3a) see program MSD

1 in the 15th digit. The square of the result for the TI-95 SD differs from the correct value by about $1 \cdot 10^{-16}\%$, while the corresponding difference for the VB5 SD is about three orders of magnitude better. The best result achieved by Excel uses case "e", with a deviation $(s_x - \sigma)/\sigma$ on the order of $7 \cdot 10^{-16}$. As is pointed out in Section 1.7.2, only the first one or two nonzero digits (rounded) are to be reported (e.g., "0.005" or "0.0048"); all available digits are printed here to demonstrate the limitations inherent in the employed algorithms. The number of significant digits carried along internally (where available) and the those displayed are given in columns 2 and 3. The cases show how, by way of example, the first data point was typed in. A prime $(')$ indicates a multiplication by 10 000 after the subtraction, so as to eliminate the decimal point. The HP-71 displays the last three digits "226" either if cases a′, b′, etc. apply, or if the output SDEV is multipled by 1000. The TI-95 has a feature that allows 13 significant places to be displayed. The TI-30D fails in one case and displays "negative difference"; because of the restricted word length, and hence accuracy and number of displayable digits, this calculator should only be used to check the grocery bill and not for scientific work. The difference between program MSD in single- and double-precision mode (effected by redefining all the variables in the source program) is quite evident: either 7 or 14 significant digits; because the intermediate results are accessible, the fault can be unequivocally assigned to the SQR-function. All told, the user of calculators and software should be aware that the tools at his disposal might not be up to the envisaged task if improperly employed. The particular models of calculators cited are probably no longer available, but the information remains valid because a successful "math package" (the code developed to solve the mathematical functions) designed for a given word length will continue to be used in later models.

The *relative standard deviation* RSD (also known as the *coefficient of variation*, c.o.v., or CV), which is frequently used to compare reproducibilities, is calculated as

$$RSD = 100 \cdot s_x / \mathbf{x}_{mean} \qquad (1.4)$$

and is given as a percentage.

For reasons that will not be detailed here the standard deviation of the mean is found as

$$\mathbf{s}_{x,\,mean} = \sqrt{V_x}/\sqrt{n} = \sqrt{V_x/n} = s_x/\sqrt{n} \qquad (1.5)$$

The difference between s_x and $s_{x,\,mean}$ is crucial: while the first describes

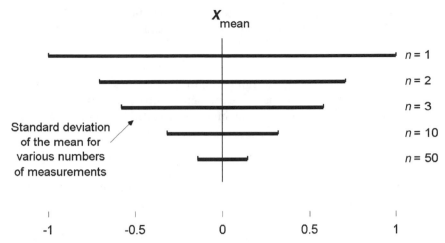

Figure 1.4. The standard deviation of the mean, $s_{x,\,mean}$, converges toward zero for a large number of measurements n. (Cf. Fig. 1.18.)

the population as such and tends with increasing n toward a positive constant, the latter describes the quality of the determination of the population mean, and tends toward zero.

1.1.3   Independency of Measurements

A basic requirement, in order that the above results "mean" and "standard deviation" are truly representative of the sampled population, is that the individual measurements should be independent of each other. Two general cases must be distinguished:

1. Samples are taken for classical off-line processing, e.g., a 10 ml aliquot is withdrawn from a reaction vessel every hour and measurements are conducted thereupon.

2. The sensor is immersed in the reaction medium and continuously transmits values.

In case (1) the different samples must be individually prepared. In the strictest interpretation of this rule, every factor that could conceivably contribute to the result needs to be checked for bias, i.e., solvents, reagents, calibrations, and instruments. That this is impractical is immediately apparent, especially because many potential influences are eliminated by careful experimental design, and because the experienced analytical chemist can often identify the major influences beforehand. Three examples will illustrate the point:

- In UV-spectroscopy the weighing and dilution steps usually introduce more error than does the measurement itself and thus the wish to obtain a replicate measurement involves a second weighing and dilution sequence.

- In contrast, in HPLC assays the chromatographic separation and the integration of the resulting analyte peak normally are just as or even more error-prone than is the preparation of the solutions; here it would be acceptable to simply reinject the same sample solution in order to obtain a quasi-independent measurement. Two independent weighings and duplicate injection for each solution is a commonly applied rule.

- In flame photometry, signal drift and lamp flicker require that one or a few unknowns be bracketed by calibrations. Here, independent measurements on the same solutions means repeating the whole calibration and measurement cycle.

In other words, those factors and operations that contribute the most toward the total variance (see additivity of variances, next section) need to be individually repeated for two measurements on the same sample to be independent. Provided the two samples are taken with a sufficiently long delay between them, they can be regarded as giving independent information on the examined system.

In case 2 the independent variables are time or distance of movement of the sensor. Repeat measurements should only be taken after the sensor has had enough time to adjust to new conditions. Thus if a continuous record of measurements is available (strip chart recorder or digitized readings), an independent measurement constitutes the average over a given time span (hatched bar) at least five time constants τ after the last such average. The time spans from which measurements are drawn for averaging may not overlap. The time constant is determined by provoking a step-response. If the conditions of independency are met as far as the analytical equipment and procedures are concerned, one must still differentiate between rapidly and slowly relaxing chemistry. (See Table 1.2 and Fig. 1.5.)

Taking measurements on a tighter raster or at shorter time intervals increases the workload, improves the plausibility of the results, but does not add any new knowledge about the system under investigation.

Table 1.2. Information Gained from Combinations of Chemical Systems and Analytical Instrumentation

Chemical System	Instrument and Analytical Method		
	Slow	Intermediate	Fast
Slow	Result is not representative of process phase, e.g. multi-step extraction while reaction is ongoing	Determinations that largely reflect repeatability; process can be controlled on the basis of the full instrument response	Ideal case to test repeatability; many mutually independent determinations
Intermediate	↑	Result is not representative of process phase, decisions have to be based on instrument's initial, not full response	Many determinations that largely reflect repeatability; process can be controlled on the basis of the full instrument response
Fast	Situations to be avoided: instrument's response time τ dominates →		Flood of data that needs to be processed before it can be correlated with process parameters

1.1.4 Reproducibility and Repeatibility

Both measures refer to the random error introduced every time a given property of a sample is measured. The distinction between the two must be defined for the specific problem at hand. Examples for continuous (Fig. 1.6) and discrete (Fig. 1.7) records are presented.

Repeatability is most commonly defined as the standard deviation obtained using a given standard operating procedure (SOP) in connection with a particular sample, and repeatedly measuring a parameter in the same laboratory, on the same hardware, and by the same technician in a short period of time. Thus, boundary conditions are as controlled as possible; the standard deviation so obtained could only be improved upon by changing the agreed-upon analytical method (column type, eluent, instrument, integration parameters, etc.). System suitability tests are required under GMP settings to determine whether the hardware is up to the task.

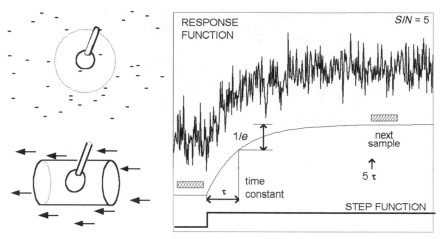

Figure 1.5. Under stagnant conditions a sensor will sample a volume, that is, the average response for that volume is obtained. A sensor in a current yields an average reading over time and cross-section. The observed signal S over time t is the convolute of the local concentration with the sensor's sampling volume and time constant. Two measurements are only then independent when they are separated at least by five time constants and/or a multiple of the sampling volume's diameter. At the left, the sampled volumes are depicted. At the right, a typical signal-*versus*-time record (e.g., strip-chart recorder trace) and the system response to a step change in concentration are shown. Tau (τ) is the time constant defined by an approximately 63.2% change $(1 - 1/e) = 0.63212$, with $e = 2.71828\ldots$. The hatched bars indicate valid averages taken at least $5 \cdot \tau$ after the last disturbance.

Reproducibility is understood to be the standard deviation obtained for the same SOP over a longer period of time. This time frame, along with other particulars, has to be defined. For example, similar but not identical HPLC configurations might be involved, as well as a group of laboratory technicians working in shifts; the working standard and key reagents might have been replaced, and seasonal/diurnal temperature and/or humidity excursions could have taken their toll. The only thing that one has to be careful to really exclude is batch-to-batch variation in the sample. This problem can be circumvented by stashing away enough of a typical (and hopefully stable) batch, so as to be able to run a sample during every analysis campaign; incidentally, this doubles as a form of a system suitability test, cf. Sections 1.8.4 and 3.2.

In mathematical terms, using the additivity of variances rule,

$$V_{\text{reprod}} = V_{\text{repeat}} + V_{\text{temp}} + V_{\text{operator}} + V_{\text{chemicals}} + V_{\text{work-up}} + V_{\text{population}} + \ldots \quad (1.6)$$

Each of these variances is the square of the corresponding standard deviation and describes the effect of one factor on the uncertainty of the result.

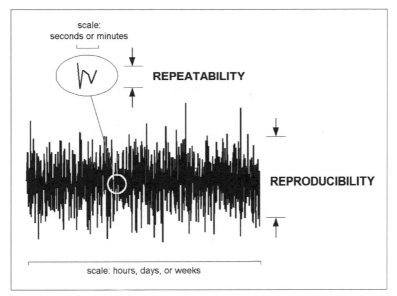

Figure 1.6. Repeatability and reproducibility are defined using historical data. The length of the time interval over which the parameter is reviewed is critical: the shorter it is, the better defined the experimental boundary conditions tend to be; the repeatability sets the limit on what could potentially be attained, the reproducibility defines what is attained in practice using a given set of instrumentation and SOPs.

1.1.5   Reporting the Results

As indicated in Section 1.7.2, the standard deviations determined for the small sets of observations typical for analytical chemistry are trustworthy only to one or two significant digits.

Example 6: Thus, for $x()$: 1.93, 1.92, 2.02, 1.97, 1.98, 1.96, and 1.90, an ordinary pocket calculator will yield

$$x_{mean} = 1.954285714 \qquad s_x = 0.040766469$$

The second significant digit in s_x (underlined) corresponds to the third decimal place of x_{mean}. In reporting this result, one should round as follows:

$$x_{mean} = 1.954 \pm 0.041 \qquad (n = 7)$$

Depending on the circumstances, it might even be advisable to round to one digit less, i.e.,

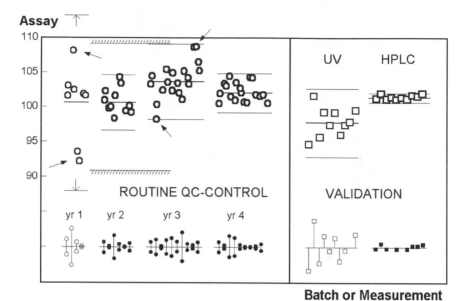

Figure 1.7. Reproducibility and repeatability. For a cream the assay data for the active principle is shown for retrospective surveys (left) and validation runs (right). This particular product is produced in about 20 batches a year. At the end of every year, product release analysis data for a number of randomly picked batches is reviewed for an overall picture of the performance of the laboratory. In four successive years, 30 batches (circles) were investigated; the repeat determinations are given by simple (UV) and bold (HPLC) and the respective mean x_{mean} and $CL(x)$ are indicated by horizontal lines; the $CL(x_{mean})$ are given by the symbols bars. For definitions of CL see Section 1.3. The residuals for the double determinations are shown below (dots). The following conclusions can be drawn: (a) All data are within the $\pm9.1\%$ specifications (hatched bars), because otherwise the releases would not have been granted; (b) The mean of the third group is higher than the others ($p < 0.025$, 95% CL being shown); (c) Four pairs of data points are marked with arrows; because the individual points within a pair give typical residuals, either one of three artifact-causing mechanisms must be investigated: (1) over- or under-dosing during production, (2) inhomogeneity, and (3) errors of calibration. Points 1 and 2 can be cleared up by taking more samples and checking the production records; point 3 is a typical problem found in routine testing laboratories (deadlines, motivation). This is a reason why Good Manufacturing Practices (GMP) regulations mandate that reagent or calibration solutions be marked with the date of production, the shelf life, and the signature of the technician, in order that gross mistakes are avoided and such questions can retrospectively be cleared. In the right panel, validation data for an outdated photometrical method (squares) and the HPLC method (bold squares) are compared. HPLC is obviously much more reliable. The HPLC-residuals in the righthand panel (repeatability, same technician, day, and batch) should be compared with those in the lefthand panel (reproducibility: several technicians, different days and batches) to gain a feeling for the difference between a research and a routine lab.

$x_{mean} = 1.95 \pm 0.04$ $(n = 7)$ or

$x_{mean} = 1.95 \pm 2.1\%$ RSD $(n = 7)$ RSD : relative standard deviation

Notice that a result of this type, in order to be interpretable, must comprise three numbers: the mean, the (relative) standard deviation, and the number of measurements that went into the calculation. All calculations are done using the full precision available, and only the final result is rounded to an appropriate precision. The calculator must be able to handle ≥ 4 significant digits in the standard deviation. (See file SYS_SUITAB.xls.)

1.1.6 Interpreting the Results

The inevitability of systematic and random errors in the measurement process, somewhat loosely circumscribed by "drift" and "noise," means that x_{mean} and s_x can only be approximations to the true values. Thus the results found in the preceding section can be viewed under three different perspectives:

1. Does the found mean x_{mean} correspond to expectations? The expected value $E(x)$ written as μ (Greek mu), is either a theoretical value, or an experimental average underpinned by so many measurements that one is very certain of its numerical value. The question can be answered by the t-test explained in Section 1.5.2. A rough assessment is obtained by checking to see whether μ and x_{mean} are separated by more than $2 \cdot s_x$ or not: if the difference Δx is larger, x_{mean} is probably not a good estimate for μ.

2. Does the found standard deviation, s_x, correspond to expectations? The expected value $E(s_x)$ is σ (Greek sigma), again either a theoretical value or an experimental average. This question is answered by the F-test explained in Section 1.7.1. Proving s_x to be different from σ is not easily accomplished, especially if n is small.

3. Is the mean x_{mean} significant? The answer is the same as for question (1), but with $\mu = 0$. If the values $(x_{mean} - 2 \cdot s_x)$ and $(x_{mean} + 2 \cdot s_x)$ bracket zero, it is improbable that μ differs from zero.

The standard deviation as defined relates to the repeatability of measurements on the same sample. When many samples are taken from a large population, "sampling variability" and "population variability" terms have to be added to Eq. (1.6) and the interpretation will reflect this.

For analytical applications it is important to realize that three distributions are involved, namely one that describes the measurement process, one that brings in the sampling error, and another that characterizes the sampled population. In a *Gedankenexperiment* the difference between the population variability (which does not necessarily follow a symmetrical distribution function, cf. Statistical Particulars in the Introduction), and the errors associated with the measurement process (repeatability, reproducibility, both usually normally distributed) is explored.

In chemical operations (see Fig. 1.8) a synthesis step is governed by a large number of variables, such as the concentration ratios of the reactants, temperature profiles in time and space, presence of trace impurities, etc. The

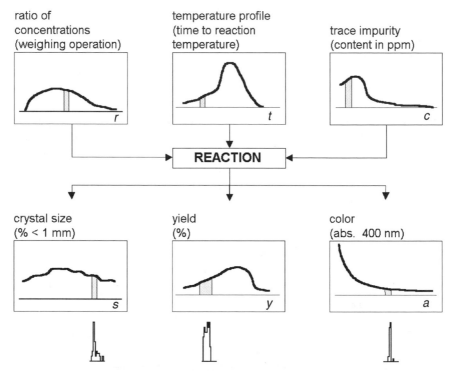

Figure 1.8. Schematic frequency distributions for some independent (reaction input or control) resp. dependent (reaction output) variables to show how non-Gaussian distributions can obtain for a large population of reactions (i.e., all batches of one product in 5 years), while approximate normal distributions are found for repeat measurements on one single batch. For example, the gray areas correspond to the process parameters for a given run, while the histograms give the distribution of repeat determinations on one (several) sample(s) from this run. Because of the huge costs associated with individual production batches, the number of data points measured under closely controlled conditions, i.e., validation runs, is miniscule. Distributions must be estimated from historical data, which typically suffers from ever-changing parameter combinations, such as reagent batches, operators, impurity profiles, etc.

outcome of a single synthesis operation (one member of the sampled population) will yield a set of characteristic results, such as yield, size-distribution of crystals, or purity. If the synthesis is redone many times, it is improbable that the governing variables will assume exactly the same values every time. The small variations encountered in temperature profiles, for example, will lead to a variation in, say, the yield that might well follow a skewed distribution. Repetition of the analyses on one single sample will follow a normal distribution, however. If the same synthesis is carried out on several, most likely slightly different equipment trains, then one might have to contend with several versions of Figure 1.8. Thus, even if a synthesis step is run once a day, it will take years until a sufficient number of points is accumulated to determine the distribution function parameters to any level of accuracy. By that time, it is likely that new operators have been trained, the synthesis has been modified, or key equipment has been replaced.

1.2   DISTRIBUTIONS AND THE PROBLEM OF SMALL NUMBERS

If a large number of repeat observations on one and the same sample are plotted, most fall within a narrow interval around the mean, and a decreasing number is found further out. The familiar term *bell curve* is appropriate. (See Fig. 1.9.)

1.2.1   The Normal Distribution

It would be of obvious interest to have a theoretically underpinned function that describes the observed frequency distribution shown in Fig. 1.9. A number of such distributions (symmetrical or skewed) are described in the statistical literature in full mathematical detail; apart from the normal- and the t-distributions, none is used in analytical chemistry except under very special circumstances, e.g. the Poisson and the binomial distributions. Instrumental methods of analysis that have *Poisson*-distributed noise are optical and mass spectroscopy, for instance. For an introduction to parameter estimation under conditions of linked mean and variance, see Ref. 41.

For a long time it was widely believed that experimental measurements accurately conformed to the normal distribution. On the whole this is a pretty fair approximation, perhaps arrived at by uncritical extrapolation from a few well-documented cases. It is known that real distributions are wider than the normal one; t-distributions for 4 to 9 degrees of freedom (see Section 1.2.2) are said to closely fit actual data.[20]

Does this mean that one should abandon the normal distribution? As will be shown in Sections 1.8.1 through 1.8.3 the practicing analyst rarely gets

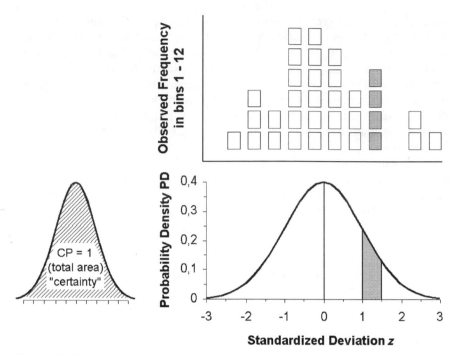

Figure 1.9. A large number of repeat measurements x_i are plotted according to the number of observations per x-interval. A bell-shaped distribution can be discerned. The corresponding probability densities PD are plotted as a curve *versus* the z-value. The probability that an observation is made in the shaded zone is equal to the zone's area relative to the area under the whole curve.

together enough data points to convincingly demonstrate adherence to one or the other distribution model. So, for all practical purposes, the normal distribution remains a viable alternative to unwieldy but "better" models.

For general use, the normal distribution has a number of distinct advantages over other distributions. Some of the more important advantages are as follows:

- Efficiency.

- Lack of bias.

- Wide acceptance.

- Incorporation into many programs and tests.

Its characteristics are described in detail in Fig. 1.10. For practical purposes, distributions over less than 50–100 measurements must be regarded as belonging to the normal distribution class, even if small deviations are observed, because the contrary cannot be proven. The only real exceptions consist in (1) manifest asymmetry, and (2) the a priori knowledge that another model applies. For example, if the outcome of an observation can only be of the type "0" or "1", a binomial distribution must be used. Since nearly all types of measurements in analytical chemistry belong to the class yielding continuous values, however, it is a defensible approach to assume a normal distribution. If results are obtained in digitized form, the Gaussian approximation is valid only if the true standard deviation is at least three to five times greater than the digitizer resolution.

The normal or Gaussian distribution a bell-shaped frequency profile defined by the function

$$PD = \frac{1}{\sigma \cdot \sqrt{2 \cdot \pi}} \cdot \exp\left(\frac{-1}{2} \cdot \left(\frac{x - \mu}{\sigma}\right)^2\right) \qquad (1.7)$$

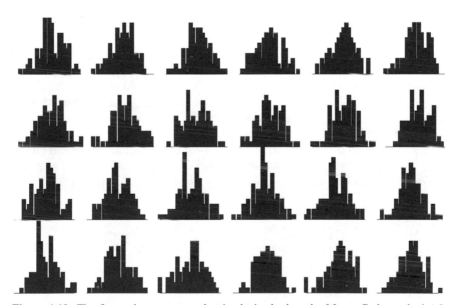

Figure 1.10. The figure demonstrates what is obtained when the Monte Carlo method (cf. Section 3.5.5) is used to simulate normally distributed values: each histogram (cf. Section 1.8.1) summarizes 100 "measurements"; obviously, many do not even come close to what one expects under the label "bell curve."

where μ is the true average, as deduced from theory or through a very large number of measurements, σ is the true standard deviation, as deduced from theory or through a very large number of measurements, x is the observed value, PD is the probability density as a function of x, that is, the expected frequency of observation at x.

Since it is impractical to tabulate PD(x) for various combinations of μ and σ, the normal distribution is usually presented in a normalized form where $\mu = 0$ and $\sigma = 1$, that is

$$PD = 0.39894 \cdot \exp(-z^2/2) \qquad (1.8)$$

where $z = (x-\mu)/\sigma$; this state of affairs is abbreviated "ND(0, 1)", as opposed to "ND(μ, σ^2)". Because of the symmetry inherent in PD $= f(z)$, the ND(0, 1) tables are only given for positive z-values usually over the range $z = 0$ ... 4, with entries for 0.05 or smaller increments of z.

The corresponding statistical table is known as the probability density table; a few entries are given for identification purposes in Figure 1.11.

When many observations are made on the same sample, and these are plotted in histogram form the bell-shaped curve becomes apparent for n larger than about 100. Five such distributions calculated according to the Monte Carlo method (see Section 3.5.5) for $n = 100, 300, 1000, 3000,$ and 10 000 are shown in Fig. 1.12; a scaling factor was introduced to yield the same total area per distribution. The z-axis scale is $-4 ... 4 \cdot$ sigma, resp. $C = 80$ classes (bins), that is, each bin is $\sigma/10$ wide. A rule of thumb for plotting histograms suggests $C = \sqrt{n}$ classes (bins); that would mean about 8–12, 15–20, 30–35, 50–60, respectively 100 bins. The number C is often chosen so as to obtain convenient boundaries, such as whole numbers. A constant bin width was chosen here for illustrative purposes; thus, the left two figures do not represent the optimum in graphical presentation: one could either fuse 5 to 10 adjacent bins into one, e.g., bins 1–10, 11–20, etc, and plot the average, or then one could plot a moving average. (Cf. Section 3.6.)

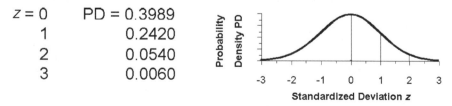

$z = 0$	PD = 0.3989
1	0.2420
2	0.0540
3	0.0060

Figure 1.11. The probability density of the normal distribution. Because of the symmetry often only the right half is tabulated.

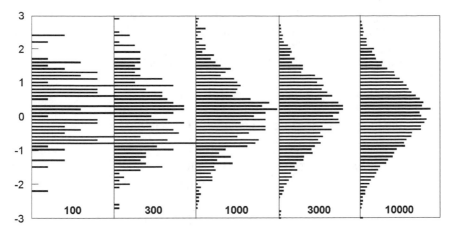

Figure 1.12. Simulated normal distributions for $n = 100$ to 10 000 events. For details see text.

Since one is only rarely interested in the density at a precise point on the z-axis, the cumulative probability (cumulative frequency) tables are more important: in effect, the integral from $-\infty$ to $+z$ over the probability density function for various $z \geq 0$ is tabulated; again a few entries are given in Fig. 1.13.

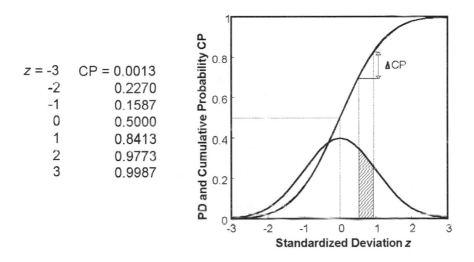

$z = -3$	$CP = 0.0013$
-2	0.2270
-1	0.1587
0	0.5000
1	0.8413
2	0.9773
3	0.9987

Figure 1.13. The cumulative probability of the normal distribution. The hatched area corresponds to the difference ΔCP in the CP plot.

The integral function is symmetrical about the coordinate ($z = 0$, $CP = 0.5000$); for this reason only the right half is tabulated, the other values being obtained by subtraction from 1.000.

Example 7: For $z = -2$, $CP = 1 - 0.97725 = 0.02275$.

Some authors adopt other formats, for instance,

$$\text{the integral } z = 0 \text{ to } +z \text{ is given with } CP = 0.0000 \text{ to } 0.4987$$
$$(\text{at } z = 3), \text{ or then}$$
$$\text{the integral } z = -z \text{ to } +z \text{ is given with } CP = 0.0000 \text{ to } 0.9973$$
$$(\text{at } z = 3; 1 - 2 \cdot 0.00135 = 0.9973)$$

In lieu of normal distribution tables, fairly accurate approximations to the entries can be made by using the following equations:

The cumulative probability table can be presented in two forms, namely

$$1 - CP = P(z) \tag{1.9a}$$
$$z = P'(\text{lgt}(1 - CP)) \tag{1.9b}$$

where $(1 - CP)$ is the area under the curve between $+z$ and $+\infty$.

P and P' are functions that involve polynomials of order 6. The coefficients and measures of accuracy are given in the Appendix 5.1.1. Both functions are used in sample programs in Chapter 5.

1.2.2   Student's *t*-Distribution

The normal distribution is the limiting case ($n = \infty$) for the Student's *t*-distribution. Why a new distribution? The reason is simply as follows: If the number of observations n becomes small, the mean's confidence interval $CI(x_{\text{mean}})$ can no longer be ignored. The same is true for the uncertainty associated with the calculated standard deviation s_x. What is sought, in effect, is a modification of the normal distribution that provides for a normally distributed x_{mean} (instead of a fixed μ) and a variance V_x following a χ^2-distribution (instead of a fixed σ^2). This can be visualized as follows: pick two values $\mu + \Delta\mu$ and $\sigma + \Delta\sigma$, and calculate the normal distribution according to Eq. (1.7). Repeat the procedure many times with different deviations $\Delta\mu$ and $\Delta\sigma$ (cf. algorithm, Section 3.5.5). Add the calculated distributions; this results in a new distribution that is similar in form to the normal one, but only wider and lower in height for the same area. The Student's *t* value is used exactly as is the variable z in the normal distribution (see Table 1.3 and Fig. 1.14.)

Table 1.3. Critical Student's t-Factors for the One- and Two-Sided Cases for Three Values of the Error Probability p and 7 Degrees of Freedom f

	Two-sided			One-sided		
p=	0.1	0.05	0.01	0.1	0.05	0.01
f = 1	6.314	12.706	63.66	3.078	6.314	31.821
2	2.920	4.303	9.925	1.886	2.920	6.965
3	2.353	3.182	5.841	1.638	2.353	4.541
4	2.132	2.776	4.604	1.533	2.132	3.747
5	2.015	2.571	4.032	1.476	2.015	3.365
10	1.812	2.228	3.169	1.372	1.812	2.764
20	1.725	2.086	2.845	1.325	1.725	2.528
∞	1.645	1.960	2.576	1.282	1.645	2.326

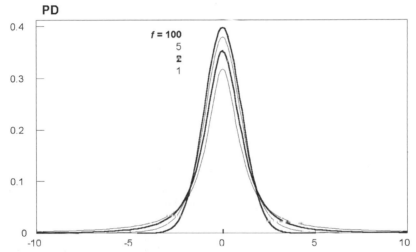

Figure 1.14. The probability density functions for several t-distributions (f = 1, 2, 5, resp. 100) are shown. The t-distribution for f = 100 already very closely matches a normal distribution.

1.3   CONFIDENCE LIMITS

If a result is quoted as having an uncertainty of ±1 standard deviation, an equivalent statement would be "the 68.3% confidence limits are given by $x_{mean} \pm 1 \cdot s_x$," the reason being that the area under a normal distribution curve between $z = -1.0$ to $z = 1.0$ is 0.683. Now, confidence limits on the 68% level are not very useful for decision making because in one-third of all cases, on average, values outside these limits would be found. What is sought is a confidence level that represents a reasonable compromise between these narrow limits and wide limits:

- Wide limits: the statement "the result is within limits" would carry a very low risk of being wrong; the limits would be so far apart as to be meaningless.

Example 8: $\pm 3.5 \cdot s_x$: The probability of error is 0.047%, and the confidence level is 99.953% for $n = 2$.

- Narrow limits: any statement based on a statistical test would be wrong very often, a fact which would certainly not augment the analyst's credibility. Alternatively, the statement would rest on such a large number of repeat measurements that the result would be extremely expensive and perhaps out of date.

Example 9: $\pm 0.5 \cdot s_x$: The probability of error and the confidence level are, respectively

$$
\begin{array}{ll}
61.7\% & 38.3\% \text{ for } n = 2 \\
11.4\% & 88.6\% \text{ for } n = 10 \\
0.6\% & 99.4\% \text{ for } n = 30
\end{array}
$$

Depending on the risks involved, one would like to choose a higher or lower confidence level; as with the many measures of length in use up to the 19th century—nearly every principality defined its own "mile"—confusion would ensue. Standardization is reflected in the confidence levels commonly listed in statistical tables: 90, 95, 98, 99, 99.5, ... %. There is no hard-and-fast rule for choosing a certain confidence level, but one has to take into account such things as the accuracy and precision of the analytical methods, the price of each analysis, time and sample constraints, etc. A balance has to be struck between making it easy and hard to prove a hypothesis. Making it too easy (very narrow limits on H_0) means that no statement will ever hold up and average noise levels would hold sway over the disposition of expensive production runs; making it too hard (very wide limits on H_0) means that one side—manufacturing or QC, depending on how the question is phrased—is always right. A fair compromise has turned out to be the 95% level, i.e. one in 20 tests will suggest a deviation (too high or too low) where none is expected. Rerunning one test out of 20 to see whether a real or a statistical outlier had been observed is an acceptable price to pay. In effect, the 95% confidence level comes close to being an agreed-upon standard. Because confidence limits and the number of measurements n are closely linked, see Figs. 1.18 and 1.26, opting for a higher confidence level, such as 99.9% sharply increases the workload necessary to prove a hypothesis. While this may not be all that difficult if a method with a RSD of ±0.1% were available, in trace analysis, where the RSD is often around ±20% (or more), series of seven or more replicates would be needed just to reduce the confidence limits to ±100% of the estimate. The effect is illustrated in Figure 1.20 and Section 1.6.

Assuming for the moment that a large number of measurements went into a determination of a mean x_{mean} and a standard deviation s_x, what is the width of the 95% confidence interval, what are the 95% confidence limits?

A table of *cumulative probabilities* (CP) lists an area of 0.975002 for z = 1.96, that is 0.025 (2.5%) of the total area under the curve is found between +1.96 standard deviations and +∞. Because of the symmetry of the normal distribution function, the same applies for negative z-values. Together p = 2 · 0.025 = 0.05 of the area, read "probability of observation," is outside the 95% *confidence limits* (outside the 95% *confidence interval* of $-1.96 \cdot s_x$ ... + 1.96 · s_x). The answer to the preceding questions is thus

$$95\% \text{ confidence limits CL}(x): \mathbf{x}_{mean} \pm z \cdot s_x \qquad (1.10a)$$

$$95\% \text{ confidence interval CI}(x): 2 \cdot z \cdot s_x \text{ centered on } \mathbf{x}_{mean} \qquad (1.10b)$$

With z = 1.96 ≈ 2 for the 95% confidence level, this is the explanation for the often-heart term "± two sigma" about some mean.

Unless otherwise stated, the expressions of CL() and Cl() are forthwith assumed to relate to the 95% confidence level.

In everyday analytical work it is improbable that a large number of repeat measurements is performed; most likely one has to make do with less than 20 replications of any determination. No matter which statistical standards are adhered to, such numbers are considered to be "small", and hence, the law of large numbers, that is the normal distribution, does not strictly apply. The t-distributions will have to be used; the plural derives from the fact that the probability density functions vary systematically with the number of degrees of freedom, f. (Cf. Figs. 1.14 through 1.16.)

In connection with the preceding problem, one looks for the list "two-tailed (sym.) Student's t-factors for p = 0.05"; sample values are given for identification in Table 1.3, i.e., 12.7, 4.30, etc.:

In Section 5.1.2, an algorithm is presented that permits one to approximate the t-tables with sufficient accuracy for everyday use.

1.3.1   Confidence Limits of the Distribution

After having characterized a distribution by using n repeat measurements and calculating x_{mean} and s_x, an additional measurement will be found within

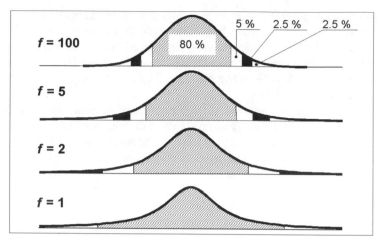

Figure 1.15. Student's *t*-distributions for 1 (bottom), 2, 5, and 100 (top) degrees of freedom *f*. The hatched area between the innermost marks is in all cases 80% of the total area under the respective curve. The other marks designate the points at which the area reaches 90, resp. 95% of the total area. This shows how the *t*-factor varies with *f*. The *t*-distribution for *f* = 100 already very closely matches the normal distribution. The normal distribution, which corresponds to $t(f = \infty)$, does not depend on *f*.

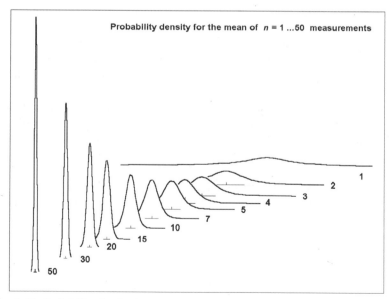

Figure 1.16. Probability density for a mean x_{mean} with $f = n-1 = 1 \ldots 50$ degrees of freedom. The areas under the curves are equal. If points demarking, say, 95% of the area (cf. Fig. 1.15) were connected, Fig. 1.17 (right) would result.

the following limits 19 out of 20 times on average (compare with Eq. 1.10!):

$$95\% \text{ confidence limits } CL(x): \mathbf{x}_{mean} \pm t \cdot s_x \qquad (1.11a)$$
$$95\% \text{ confidence interval } CI(x): 2 \cdot t \cdot s_x \text{ centered on } \mathbf{x}_{mean} \qquad (1.11b)$$

For large n the confidence interval for the distribution converges toward the $\mathbf{x}_{mean} \pm 1.96 \cdot s_x$ range familiar from the normal distribution, cf. Fig. 1.17 (left).

1.3.2   Confidence Limits of the Mean

If, instead of the distribution as such, the calculated mean $\mathbf{x}_{mean}$ is to be qualified:

$$95\% \text{ confidence limits } CL(\mathbf{x}_{mean}): \pm t \cdot s_x / \sqrt{n} \qquad (1.12a)$$
$$95\% \text{ confidence interval } CI(\mathbf{x}_{mean}): 2 \cdot t \cdot s_x / \sqrt{n} \text{ centered on } \mathbf{x}_{mean} \qquad (1.12b)$$

It is apparent that the confidence interval for the mean rapidly converges toward very small values for increasing n, because both $t(f)$ and $1/\sqrt{n}$ become smaller.

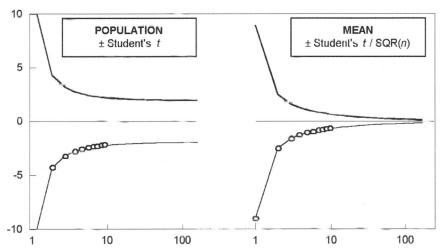

Figure 1.17. The 95% confidence intervals for x and $\mathbf{x}_{mean}$ are depicted. The curves were plotted using the approximations given in Section 5.1.2; the f-axis was logarithmically transformed for a better overview. Note that solid curves are plotted as if the number of degrees of freedom could assume any positive value; this was done to show the trend; f is always a positive integer. The ordinates are scaled in units of the standard deviation.

Example 10: For $p = 0.05$, $x_{mean} = 10$, $s_x = 1$, and different n, Table 1.4 and Fig. 1.18 present CL that are found.

Table 1.4. Confidence Limits for the Population and the Mean

n	t(f)	CL(x)	CI(x)	CL(x_{mean})	CI(x_{mean})
2	12.71	−2.71 ... 22.71	25.42	1.01 ... 18.99	17.97
3	4.303	5.70 ... 14.30	8.61	7.52 ... 12.48	4.97
4	3.182	6.82 ... 13.18	6.36	8.41 ... 11.59	3.18
5	2.776	7.22 ... 12.78	5.55	8.76 ... 11.24	2.48
6	2.571	7.43 ... 12.57	5.14	8.95 ... 11.05	2.10
10	2.262	7.74 ... 12.26	4.52	9.28 ... 10.72	1.43
100	1.984	8.02 ... 11.98	3.97	9.80 ... 10.20	0.40

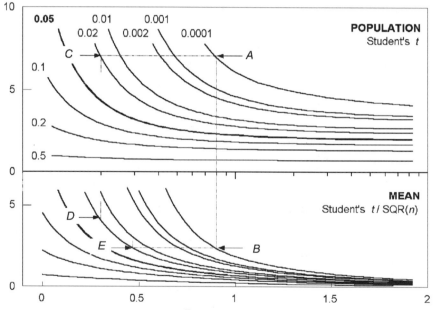

Figure 1.18. The Student's t resp. $t/\sqrt{n}$ for various confidence levels are plotted; the curves for $p = 0.05$ are enhanced. The other curves are for $p = 0.5$ (bottom), 0.2, 0.1, 0.02, 0.01, 0.002, 0.001, and 0.0001 (top). By plotting a horizontal, the number of measurements necessary to obtain the same confidence intervals for different confidence levels can be estimated. While it takes $n = 9$ measurements ($f = 8$) for a t-value of 7.12 and $p = 0.0001$, just $n' = 3$ ($f = 2$) will give the same limits on the population for $p = 0.02$ (line $A \rightarrow C$). For the CL on the mean, in order to obtain the same $t/\sqrt{n}$ for $p = 0.02$ as for $p = 0.0001$, it will take $n = 4$ measurements (line $B \rightarrow E$); note the difference between points D and E!

1.4   THE SIMULATION OF A SERIES OF MEASUREMENTS

Simulation by means of the digital computer has become an extremely useful technique (see Section 3.7) that goes far beyond classical interpolation/extrapolation. The reasons for this are fourfold:

- Very complex systems of equations can be handled; this allows interactions to be studied that elude those who simplify equations to make them manageable at the paper and pencil level.[42]

- Fast iterative root-finding algorithms do away with the necessity of algebraically solving for "buried" variables, an undertaking that often does not yield closed solutions (a solution is "closed" when the equation has the form $x = f(a, b, c, \ldots)$ and "x" does not appear in the function f).

- Nonlinear and discontinuous equations can be easily implemented, e.g., to simulate the effects of a temperature-limiting device or a digital voltmeter.[17]

- Not only deterministic aspects can be modeled, but random ones as well (see Refs. 5, 34 and Section 3.5.5).

This important technique is introduced at this elementary level to demonstrate characteristics of the confidence level concept that would otherwise remain unrecognized. Two models are necessary, one for the deterministic, the other for the stochastic aspects.

As an example, the following very general situation is to be modeled: A physicochemical sensor in contact with an equilibrated chemical system is used to measure the concentration of an analyte.

The measurement has noise superimposed on it, so that the analyst decides to repeat the measurement process several times, and to evaluate the mean and its confidence limits after every determination. (*Note*: This *modus operandi* is forbidden under GMP; the necessary number of measurements and the evaluation scheme must be laid down before the experiments are done.) The simulation is carried out according to the scheme depicted in Fig. 1.19. The computer program that corresponds to the scheme principally contains all of the simulation elements; however, some simplifications can be introduced:

- For the present purposes the deterministic function generator yields a constant signal $x = 0$, which means the summation output is identical with that of the noise generator.

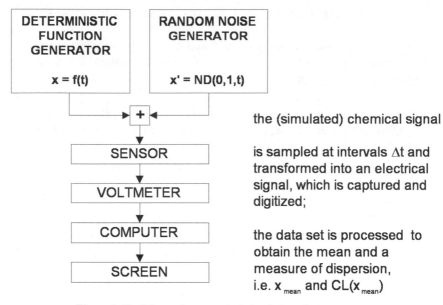

Figure 1.19. Scheme for numerical simulation of measurements.

- The sensor is taken to be of the linear type, i.e., it transduces the incoming chemical information into electrical output according to the equation el.signal = constant + slope · (chemical signal); without loss of clarity, the constant can be set to zero and the slope to 1.00.

- One noise generator in parallel to the chemical function generator suffices for the present purposes; if electrical noise in the sensor electronics is to be separately simulated, a second noise generator in parallel to the sensor and a summation point between the sensor and the volt meter would become necessary. The noise is assumed to be normally distributed with $\mu = 0$ and $\sigma = 1$.

- The computer model does nothing but evaluate the incoming "assay values" in terms of Eqs. (1.1) and (1.12a).

The output of the simulation will be displayed as in Fig. 1.20 or by program CONVERGE.exe; the common abscissa is the sample number i. The ordinates are in signal units; the top window (panel A) shows the individual measurements as points; the bottom window (panel C) shows how the derived standard deviation converges toward its expected value, $E(s_x) = 1.00$; in the middle window (panel B) the mean and the $CL(x_{mean})$ are shown to rapidly, although erratically, converge toward the expected value 0 ± 0. Equa-

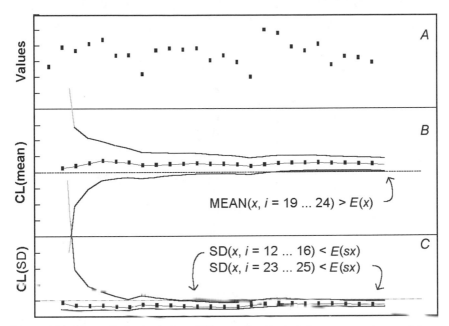

Figure 1.20. Monte Carlo simulation of 25 normally distributed measurements: raw data are depicted in panel A, the derived means $x_{mean} \pm CL(x_{mean})$ in B, and the standard deviation s_x $\pm CL(s_x)$ in C. Notice that the mean and/or the standard deviation can be statistically different from the expected values, for instance in the range $23 \leq n \leq 25$ in this example. The ordinates are scaled in units of 1σ.

tions (1.1), (1.3), (1.5), and (1.12a) (middle window), and (1.3), (1.5), and (1.42) (bottom window) were used.

Figure 1.21 was obtained using Excel file CONV.xls, and shows 8 successive "measurements" per group. It is definitely possible that the $CI(x_{mean})$ does not include the expected value: The first few points suggest a satisfyingly small scatter (Case B). In this particular simulation, this is due to the operation of "pure chance" as defined in the Monte Carlo algorithm. However, inadequate instrument configurations, poor instrument maintenance, improper procedures, or a knowledge of what one is looking for can lead to similar observations. Analysts and managers may (subconsciously) fall prey to the latter, psychological trap, because without a rigid plan they are enticed to act selectively, either by stopping an experiment at the "right" time or by replacing apparent "outliers" by more well-behaved repeat results. Visual-Basic program CONVERGE and Excel sheet CONV allow experimenting with various combinations of p and n.

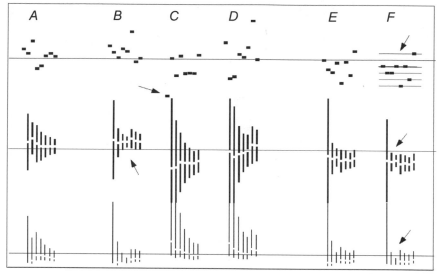

Figure 1.21. Monte Carlo simulation of six groups of eight normally distributed measurements each: raw data are depicted as x_i vs. i (top); the mean (gaps) and its upper and lower confidence limits (full lines, middle); the confidence limits $CL(s_x)$ of the standard deviation converge toward $\sigma = 1$ (bottom, Eq. 1.42). The vertical divisions are in units of $1 \pm \sigma$. The CL are clipped to $\pm 5\sigma$ resp. $0 \ldots 5\sigma$ for better overview. Case A shows the expected behavior, that is for every increase in n the $CL(x_{mean})$ bracket $\mu = 0$ and the $CL(s_x)$ bracket $\sigma = 1$. Cases B, C, and D illustrate the rather frequent occurrence of the CL not bracketing either μ and/or σ, cf. Case B @ $n = 5$. In Case C the low initial value (arrow!) makes x_{mean} low and s_x high from the beginning. In Case D the 7th measurement makes both CI @ $n = 7$ widen relative to the $n = 6$ situation. Case F depicts what happens when the same measurements as in Case E are clipped by the DVM.

1.5   TESTING FOR DEVIATIONS

The comparison of two results is a problem often encountered by the analyst. Intuitively, two classes of problems can be distinguished:

1. A systematic difference is found, supported by indirect evidence that from experience precludes any explanation other than "effect observed." This case does not necessarily call for a statistical evaluation, but an example will nonetheless be provided: in the elemental analysis of organic chemicals (CHN analysis) reproducibilities of 0.2 to 0.3% are routine (for a mean of 38.4 wt-% C, for example, this gives a true value within the bounds 38.0 ... 38.8 wt-% for 95% probability). It is not out of the ordinary that traces of the solvent used in the

last synthesis step remain in the product. So, a certain pattern of carbon, hydrogen, resp. nitrogen deviations from the theoretical element-by-element percentage profile will be indicative of such a situation, whereas any single element comparison, e.g., C(exp) with C(theor), would result in a rejection of the solvent-contamination hypothesis. In practice, varying solvent concentrations would be assumed, and for each one the theoretical elemental composition would be calculated, until the best fit with the experimental observations is found. (See χ^2-test.)

2. A measurement technique such as titration is employed that provides a single result that, on repetition, scatters somewhat around the expected value. If the difference between expected and observed value is so large that a deviation must be suspected, and no other evidence such as gross operator error or instrument malfunction is available to reject this notion, a statistical test is applied. (Note: under GMP, a deviant result may be rejected if and when there is sufficient *documented* evidence of such an error.)

If a statistical test is envisioned, some preparative work is called for; *Every statistical test is based on*

a model,
a confidence level, and
a set of hypotheses.

These three prerequisites of statistical testing must be established and justified before any testing or interpreting is done; since it is good practice to document all work performed, these apparent "details" are best set down in the experimental plan or the relevant SOP, unless one desires that the investigator have unusual freedom to influence the conclusions by choosing the three elements to suit his needs (exploratory data analysis is an exception discussed in Section 3.4). *A point that cannot be stressed enough is that statistics provides a way of quantitizing hidden information and organizing otherwise unmanageable amounts of data in a manner accepted and understood by all parties involved. The outcome of a statistical test is never a "hard fact," but always a statement to the effect that a certain interpretation has a probability of x% or less of not correctly representing the truth.* For lack of a more convincing *model*, the *t*-distribution is usually accepted as a description of measurement variability, the normal distribution being the limiting case, and a *confidence level* in the 95–99% range is more or less tacitly assumed to fairly balance the risks involved. If a difference is found, the wording of the result depends on the con-

fidence level, namely "the difference is 'significant' (95%) or 'highly significant' (99%)." The setting up and testing of *hypotheses* is the subject of Section
1.9. What hypotheses are there?

The "null" Hypothesis H_0

Given that the measured content for a certain product has been within 2%
of the theoretical amount over the past, say, 12 batches, the expectation of a
further result conforming with previous ones constitutes the so-called "null"
hypothesis, H_0, i.e. "no deviation" is said to be observed.

The Alternate Hypothesis H_1

Since it is conceivable that some slight change in a process might lead
to a "different" content, a mental note is made of this by stating that if the
new result differs from the old one, the alternate hypothesis H_1 applies. The
difference between H_0 and H_1 might be due to $\mu_A \neq \mu_B$ and/or $\sigma_A \neq \sigma_B$.
The first possibility is explored in the following section, the second one will
be dealt with in Section 1.7.1.

The situation of H_0 and H_1 differing solely in the true averages A and B,
is summed up in Fig. 1.22.

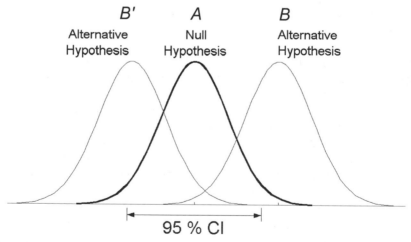

Figure 1.22. The null and the alternate hypotheses H_0 resp. H_1. The normal distribution probability curves show the expected spread of results. Since the alternate distribution $ND(\mu_B, \sigma^2)$
might be shifted toward higher or lower values, two alternative hypotheses H_1 and H_1' are
given. Compare with program HYPOTHESIS. Measurement B is clearly larger than A, whereas
B' is just inside the lower $CL(A)$.

Assuming one wants to be certain that the risk of falsely declaring a good batch B to be different from the previous one A is less than 5%, the symmetrical 95% confidence limits are added to A (see Fig. 1.22): any value B in the shaded area results in the judgment "B different from A, H_1 accepted, H_0 rejected", whereas any result, however suspect, in the unshaded area elicits the comment "no difference between A and B detectable, H_1 rejected, H_0 retained". *Note that the expression "A is identical to* B*" is not used; by statistical means only deviations can be demonstrated, and similarities must be inferred from their absence.*

1.5.1 Examining Two Series of Measurements

Testing requires that a certain protocol be adhered to; this protocol is set forth here in a "how-to" format of questions and instructions. The actual calculations are shown on pp. 48–65.

Data. Two series of measurements are made, such as n repetitive determinations of an analyte concentration in the same sample by two different methods: $x_{11}, x_{12}, \ldots x_{1n}, x_{21}, x_{22}, \ldots x_{2n}$.

Question 1. Are the group variances V_1 and V_2 indistinguishable?

Perform the F-test (Section 1.7.1; most authors find the F-test to be a prerequisite for the t-test). If no significant effect is found (H_0 retained), the two sample variances may be pooled to decrease the uncertainty associated with V through inclusion of a higher number of measurements: Of the models given in Table 1.10, cases b1 or b3, as appropriate, both give the same degree of freedom f, but different variances. However, if a significant difference between V_1 and V_2 is found (H_1 accepted), they may, according to some authors, be pooled with different models yielding the same variance, but different degrees of freedom f (case c).

Question 2. Are the two means $x_{\mathrm{mean}, 1}$ and $x_{\mathrm{mean}, 2}$ distinguishable?

If H_0 holds for question 1, all authors agree that a t-test can be performed. If H_1 is true, opinions diverge as to the propriety of such a t-test. In practice, a t-test will be performed: If the outcome is clear, that is t is much different from t_c, the differences between the models are negligible; if t is close to t_c, more tests should be performed to tighten the confidence limits, or judgment should be suspended until other evidence becomes available.

Obviously, the t-test also involves its own set of null and alternative hypotheses; these are also designated H_0 and H_1, but must not be confused with the hypotheses associated with the F-test.

1.5.2   The t-Test

The most widely used test is that for detecting a deviation of a test object from a standard by comparison of the means, the so-called t-test. Note that before a t-test is decided upon, the confidence level must be declared and a decision made about whether a one- or a two-sided test is to be performed. For details, see shortly. Three levels of complexity, a, b, and c, and subcases are distinguishable. (The necessary equations are assembled in Table 1.10 and are all included in program TTEST.)

a   The standard is a precisely known mean, a theoretical average μ, or a preordained value, such as a specification limit.

b   The test sample and the standard were measured using methods that yield indistinguishable standard deviations s_1 resp. s_2 (cf. F-test, Section 1.7.1).

c   The standard deviations s_1 resp. s_2 are different.

In Case a, only the standard deviation estimated from the experimental data for the test sample is needed, which is then used to normalize the difference $x_{mean} - \mu$. The quotient difference/s_x, the so-called Student's t, is compared with the critical t_c for a chosen confidence level and $f = n - 1$. (Use Eq. 1.13.)

Example 11: $x_{mean} = 12.79$, $s_x = 1.67$, $n = 7$, $s_{x,mean} = 0.63$, $\mu = 14.00$, and the chosen confidence level is 95%.

Table 1.5. Calculation of a Student's t-Factor

$t = |12.79 - 14.00|/0.63 = 1.92$	Table 1.3
$t_c(6, p = 0.05) = 2.45$	two-sided test
$t_c(6, p/2 = 0.05) = 1.94$	one-sided test

Interpretation: If the alternate hypothesis had been stated as "$H_1 : x_{mean}$ is different from μ," a two-sided test is applied with 2.5% probability being provided for each possibility "x_{mean} smaller than μ" resp. "x_{mean} larger than μ". Because 1.92 is smaller than 2.45, the test criterion is not exceeded, so H_1 is rejected. On the other hand, if it was known beforehand that x_{mean} can only be smaller than μ, the one-sided test is conducted under the alternate hypothesis "$H_1 : x_{mean}$ smaller than μ"; in this case the result is close, with 1.92 almost exceeding 1.94.

In *Cases b and c* the standard deviation s_d of the average difference, $d =$

$x_{mean, 1} - x_{mean, 2}$, and the number of degrees of freedom must be calculated
    Case b must be divided into three subcases (see the equations in Table 1.10):

Table 1.6. Applications of the t-Test

b1:	$n_1 = n_2$	• one test, repeatedly performed on each of two samples, • one test, performed once on every sample from two series of samples, or • two tests, each repeatedly performed on the same sample, 1 ... n, n + 1 ... 2n	see Eqn. (1.18) through (1.20)
b2:	$n_1 = n_2$	• two tests being performed pairwise on each of n samples	see Eqn. (1.24)
b3:	$n_1 \neq n_2$	• same situation as b1, but with different n: 1 ... n_1, $n_1 + 1$ ... $n_1 + n_2$	see Eqn. (1.14) through (1.17)

Subcase b1: This case is encountered, for example, when batch records from different production campaigns are compared and the same number of samples was analyzed in each campaign. (*Note*: under GMP, trend analysis has to be performed regularly to stop a process from slowly, over many batches, drifting into a situation where each parameter on its own is within specifications, but collectively there is the risk of sudden, global "loss of control." "Catastrophe theory" has gained a foothold in physical and biological literature to describe such situations; cf. Section 4.14.)

The variance V_d of the average difference is calculated as $(V_1 + V_2)/n$ (see Eq. 1.18–19) with $f = 2n - 2$ degrees of freedom. s_d, the square root of V_d, is used in the calculation of the Student's t-statistic $t = d/s_d$.

Example 12: $x_{mean, 1} = 17.4$, $s_1 = 1.30$, $x_{mean, 2} = 19.5$, $s_2 = 1.15$, $n_1 = n_2 = 8$, p = 0.05.
The standard deviations are not significantly different (*F*-test, Section 1.7.1); the standard deviation of the mean difference is $s_d = 0.61$ and $t = |17.4 - 19.5|/0.61 = 3.42$; $f - 14$. Since the critical t value is 2.145, the two means, $x_{mean, 1}$ and $x_{mean, 2}$, can be distinguished on the 95% confidence level.

Subcase b2: This case, called the "paired t-test", is often done when two test procedures, such as methods *A* and *B*, are applied to the same samples, for instance when validating a proposed procedure with respect to the accepted one. In practicular, an official content uniformity[43] assay might prescribe a photometric measurement (extract the active principle from a tablet and measure the absorbance at a particular wavelength); a new HPLC procedure, which would be advantageous because it would be selective and the extraction step could be replaced by a simple dissolution/filtration step, is proposed as a replacement for the photometric method. In both methods the raw result (absorbance resp. peak area) is first converted to concentration units to make the values comparable. Tests are conducted on 10 tablets. The following statistics are calculated:

$$d_i = x_{PM,i} - y_{HPLC,i} \text{ for } i = 1 \dots 10.$$

The average d and s_d are calculated according to Eqs. (1.1) resp. (1.5). $t = d/s_d$ is compared to the tabulated t-value for $f = n - 1$.

Example 13: Consider the following data: $x() = 1.73, 1.70, 1.53, 1.78, 1.71$; $y() = 1.61, 1.58, 1.41, 1.64, 1.58$. One finds $x_{mean} = 1.690$, $s_x = 0.0946$, $y_{mean} = 1.564$, and $s_y = 0.0896$: the F-value is 1.11, certainly not higher than the $F_c = 3.2$ critical value for $p = 0.05$; the two standard deviations are recognized as being indistinguishable.

- If, despite the fact that the x and the y values are paired, a t-test was applied to the difference ($x_{mean} - y_{mean}$) using the standard deviations $s_d = 0.0583$ according to Eq. (1.18), a t-value of 2.16 would be found, which is smaller than the critical $t_c(2n-2) = 2.306$. The two series thus could not be distinguished. The overall mean, found as 1.63, would be advanced.

- An important aspect, that of pairing, had been ignored in the preceding example; if it is now taken into account (*proviso*: $x(i)$ and $y(i)$ are related as in "i^{th} sample pulled from i^{th} powder mixture and subjected to both PM and HPLC analysis, $i = 1 \dots n$", etc.), a standard deviation of the differences of $s_d = 0.0089$ is found according to Eq. (1.3). With the average difference being $d = 0.126$, this amounts to a relative standard deviation of about 7.1%, or a $t_d = 14.1$, which is clearly much more significant than what was erroneously found according to case $b1$, viz., $t_c(4, 0.05) = 2.77$.

The example demonstrates that all relevant information must be used; ignoring the fact that the PM and HPLC measurements for $i = 1 \dots 5$ are paired results in a loss of information. The paired data should under all circumstances be plotted (Youden plot, Fig. 2.1, and Fig. 1.23) to avoid a pitfall: it must be borne in mind that the paired t-test yields insights only for the particular (additive) model examined (see Section 1.5.6), that is, for the assumption that the regression line for the correlation PM vs. HPLC has a slope $b = 1$, and an intercept $a_{PM, HPLC}$ that is either indistinguishable from zero (H_0), or is significantly different (H_1). If there is any systematic, nonadditive difference between the methods (e.g., interferences, slope $b \neq 1$, nonlinearity, etc.), a regression might be more appropriate (Section 2.2.1). Indeed, in the previous example, the data are highly correlated. ($s_{res} = \pm 0.0083$, $r^2 = 0.994$; see Chapter 2.) Using case $b1$, the overall variance is used without taking into account the correlation between x and y; for this reason the standard deviation of the difference $x_{mean} - y_{mean}$ according to case $b1$ is much larger than the residual standard deviation (± 0.0583 vs. ± 0.0083), or the standard deviation of the mean difference ± 0.0089. A practical example is given in Section 4.14.

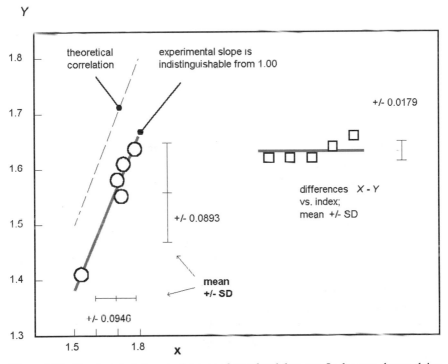

Figure 1.23. Standard deviations in the case of correlated data sets. In the case shown, doing the calculations without taking the correlation into consideration results in the SDs given at left; SD_y appears much larger than expected. When this is recognized, the SD for the difference $|x-y|$ or for $|y-x|$ is calculated (at right) and is seen to conform much better with experience.

The same is true if another situation is considered: if in a batch process a sample is taken before and after the operation under scrutiny, say, impurity elimination by recrystallization, and both samples are subjected to the same test method, the results from, say, 10 batch processes can be analyzed pairwise. If the investigated operation has a strictly additive effect on the measured parameter, this will be seen in the t-test; in all other cases both the difference Δx_{mean} and the standard deviation s_x will be affected.

Subcase b3: When n_1 and n_2 are not equal the degrees of freedom are calculated as $f = n_1 + n_2 - 2$ for the variance of the difference. Up to this point a random pick of statistics textbooks[34,44-49] shows agreement among the authors. The pooled variance is given in Eq. (1.14) where the numerator is the sum of the squares of the residuals, taken relative to $x_{mean,1}$ or $x_{mean,2}$, as appropriate. Some authors simplify the equation by dropping the "-1" in "$n - 1$," under the assumption $n \gg 1$ (Eq. 1.16), something that might have been appropriate to do in the precomputer era in order to simplify the equations and lessen the calculational burden.

In order to get the variance V_d of the difference of the means d, a way must be found to multiply the pooled variance V_p by a number akin to $1/n$, as in Eq. (1.22). A formula is proposed, namely $V_d = V_p \cdot (1/n_1 + 1/n_2)$. Other authors take the sum of the variances of the means, $V_d = V_1/n_1 + V_2/n_2$ (Eq. 1.15).

It is evident that there is no simple, universally agreed-upon formula for solving this problem. For practical applications, then, if t is much different from t_c, any one of the above equations will do, and if t is close to t_c, or if high stakes are involved in a decision, more experiments (to achieve equality $n_1 = n_2$) might be the best recourse. (See subcase $b1$.)

Case c: Subcase b3 is an indication of the difficulties associated with testing hypotheses. There is no theory available if $s_1 \neq s_2$; testing for differences under these premises is deemed improper by some,[44] while others[46,47,49] propose a simple equation for the variance of the difference [see Eq. (1.21)] and a very complicated one for f [see Eq. (1.22)], where V_d is as before and $Q_i = (V_i/n_i)^2/(n_i \pm 1)$, $i = 1, 2$.

The ambiguity over the sign in the denominator is due to the fact that both versions, $(n_i - 1)$[46] and $(n_i + 1)$,[47,49] are found in the literature ("+" resp. "$-$" behind the equation number (1.22) indicate the sign used in the denominator). Equation (1.22$-$) appears to be correct; Eq. (1.22+) might have arisen from transcription errors.

Some numerical examples will illustrate the discussed cases. Means and standard deviations are given with superfluous significant figures to allow recalculation and comparison.

Case b1:

Table 1.7. Case b1

Data	$x_{mean, 1} = 101.26$	$s_{x, 1} = 7.328$	$n_1 = 7$	Interpretation:
	$x_{mean, 2} = 109.73$	$s_{x, 2} = 4.674$	$n_2 = 7$	
F-test		$F = 2.46$	$F_c = 4.29$	H_0
t-tests	Eq. (1.18)	$t = 2.58$		H_1
	Eq. (1.19)	$t = 2.58$		H_1
	Eq. (1.20), f = 12		$t_c = 2.18$	

Comment: The two standard deviations are similar enough to pass the F-test, and the data are then pooled. The difference $x_{mean, 1} - x_{mean, 2}$ is significant. Eq. (1.14) gives the same results; if Eq. (1.16) had been applied, $t = 2.387$ would have been found.

Case b3:

Table 1.8. Case b3

Data	$x_{mean, 1} = 101.26$	$s_{x, 1} = 7.328$	$n_1 = 6$	Interpretation:
	$x_{mean, 2} = 109.73$	$s_{x, 2} = 4.674$	$n_2 = 8$	
F-test		$F = 2.46$	$F_c = 3.97$	H_0
t-tests	Eq. (1.14)	$t = 2.65$		H_1
	Eq. (1.15)	$t = 2.48$		H_1
	Eq. (1.16)	$t = 2.44$		H_1
	Eq. (1.17), f = 12		$t_c = 2.18$	

Comment: The same as for case *b*1, except that n_1 and n_2 are not the same; the different models arrive at conflicting *t*-values, but with identical interpretation; H_1 can thus be accepted.

Case 3:

Table 1.9. Case c

Data	$x_{mean, 1} = 101.26$	$s_{x, 1} = 8.328$	$n_1 = 7$	Interpretation:
	$x_{mean, 2} = 109.73$	$s_{x, 2} = 3.674$	$n_2 = 7$	
F-test		$F = 4.39$	$F_c = 4.29$	H_1
t-tests	Eq. (1.21)	$t = 2.43$		
	Eq. (1.22+), f = 9		$t_c = 2.26$	H_1
	Eq. (1.22-), f = 6		$t_c = 2.45$	H_0
	Eq. (1.23), f = 8		$t_c = 2.31$	H_1

Table 1.10. Various Forms of the Student's t-test. For Details and Examples see Text. The Expression "$V_1 = V_2$" is to be Interpreted as "Indistinguishable Variances"

Test Statistic	Conditions	Variance, Degrees of Freedom	Eq.	Reference, Comments
$t = \dfrac{x_{\text{mean}} - \mu}{\sqrt{V_x}}$	none [case **a**]	$V = V_x/n, f = n - 1$	(1.13)	$V_x = S_{xx}/f$
$t = \dfrac{\Delta x}{\sqrt{V_d}}$	$V_1 = V_2$	$V_d = V_p \cdot k$	(1.14)	$\Delta x = x_{\text{mean}, 1} - x_{\text{mean}, 2}$
	$n_1 \neq n_2$ [case **b3**]	$V_p = ((n_1 - 1)\cdot V_1 + (n_2 - 1)\cdot V_2)/f$ $k = 1/n_1 + 1/n_2$		Refs. {24} {34} {35} V_p: pooled variance
		$V_d = V_1/n_1 + V_2/n_2$	(1.15)	Refs. {32} {36}
		$V_d = V_p' \cdot k$ $V_p' = (n_1 \cdot V_1 + n_2 \cdot V_2)/f$ $k = 1/n_1 + 1/n_2$	(1.16)	Refs. {33} {37}
		$f = n_1 + n_2 - 2$	(1.17)	Refs. {24}, {32–37}
	$V_1 = V_2$ $n_1 = n_2 (= n)$ [case **b1**]	$V_d = V_1/n_1 + V_2/n_2$ $V_d = (V_1 + V_2)/n$ $f = 2n - 2$	(1.18) (1.19) (1.20)	Refs. {33} {36} Refs. {34} Refs. {24} {34} {36}
	$V_1 \neq V_2$	$V_d = V_1/n_1 + V_2/n_2$	(1.21)	Not permitted according to Ref. {32}
	[case **c**]	$f = \dfrac{(Q_1 + Q_2)^2}{\left(\dfrac{Q_1^2}{n_1 \pm 1} + \dfrac{Q_2^2}{n_2 \pm 1}\right)} - 2$	(1.22−)	Ref. {34} $(n - 1)$
		$Q_i = V_i/n_i$		
		$f = \dfrac{1}{k^2/f_1 + (1 - k)^2/f_2}$	(1.22+) (1.22+)	Ref. {35} $(n + 1)$ Ref. {37} $(n + 1)$
		$k = \dfrac{n_2 \cdot V_1}{n_2 \cdot V_1 + n_1 \cdot V_2}$	(1.23)	Ref. {24}
$t = d/\sqrt{V_d}$	$n_1 = n_2$ [case **b2**]	$d_i = x_{1i} - x_{2i}, f = n - 1$	(1.24)	Refs. {32} {35} {36}

Comment: All models yield the same *t*-value, but differ in the number of degrees of freedom to be used. The difference between the means is barely significant in two cases. Suggestion: acquire more data to settle the case. Program TTEST automatically picks the appropriate equation(s) and displays the result(s). Equation (1.21) is used to scan the parameter space (x_{mean}, s_x, n) in the vicinity of the true values to determine whether a small change in experimental protocol (n) or measurement noise could have changed the interpretation from H_0 to H_1 or *vice versa*.

1.5.3   Extension of the *t*-Test to More Than Two Series of Measurements

The situation of having more than two series of measurements to compare is frequently encountered. One possibility resides in doing a *t*-test as discussed above for every pairing of measurement series; this not only is inefficient, but also does not answer the question of whether all series belong to the same population. The technique that needs to be employed is discussed in detail later (Section 1.5.4) and is fully integrated into program MULTI. The same "how-to" format of questions and instructions is used as previously.

Data. Several groups of n_i replicate measurements of a given property on each of m different samples (for a total of $n - \Sigma(n_i)$). The group sizes n_i need not be identical.

Question 1. Do all m groups have the same standard deviation?
To answer this, do the Bartlett test (Section 1.7.3). If there is one group variance different from the rest, stop testing and concentrate on finding and eliminating the reason, if any, for

- Systematic differences in the application of the measurement technique used, or
- Inhomogeneities and improper sampling techniques, or
- Chance result ($s_x \approx 0$) due to digitization effects (cf. Fig. 1.21, case F), if all measurements collapse on one or two levels, etc.

If the variances are indistinguishable, continue by pooling them (V_1, Tables 1.14 and 1.15).

Question 2. Do all m groups have the same mean?
Do the simple ANOVA test (Section 1.5.6) to detect variability between the group means in excess of what is expected due to chance alone.

If no excess between-group variance is found, stop testing and pool all values, because they probably all belong to the same population. If significant excess variance is detected, continue testing.

Question 3. Which of the m groups is significantly different from the rest? Do the multiple-range test (Section 1.5.4) to find sets of means that are indistinguishable among themselves; it may occur that a given mean belongs to more than one such set of similar means.

Note concerning question 2: For m groups of objects that all belong to one population, the means $x_{mean, 1}, \dots x_{mean, m}$ are expected to be distributed as a ND with a SD equal to the average within-group SD divided by the square root of the average group size; if excess variance is detected, then this is interpreted as being due to one or more of the group means differing significantly from the rest.

1.5.4   Multiple-Range Test

A setting that turns up quite often is the following: A series of m batches of a given product have been produced, and a certain parameter, such as the content of a particular compound, was measured n_i times for each batch. The largest and the smallest means, $x_{mean, max}$ resp. $x_{mean, min}$, appear to differ significantly. Which of the two is aberrant? A simple t-test cannot answer this question. The multiple-range test combines several t-tests into one simultaneous test.[50]

Provided that the m variances $V_j = s_j^2$ are roughly equal (Bartlett's test, see Section 1.7.3), the m means are ordered (cf. subroutine SORT, Table 5.17). The smallest mean has index 1, the largest has index m. A triangular matrix (see Tables 4.9, 4.10) is then printed that gives the $m \cdot (m-1)/2$ differences $\Delta x_{mean, uv} = x_{mean, u} - x_{mean, v}$ for all possible pairings. Every element of the matrix is then transformed into a q-value as

$$q = \Delta x_{mean, uv} \cdot \sqrt{2 \cdot n_u \cdot n_v / (D \cdot (n_u + n_v))} \qquad (1.25)$$

with $D = \Sigma\Sigma(x_{iv} - x_{mean, v})^2 / f, \qquad i = 1 \dots n, v = 1 \dots m$

$f = \Sigma(n_v) - m, \qquad D$ is identical to V_1 in Section 1.5.6

The calculated q-value must be compared to a critical q that takes account of the "distance" that separates the two means in the ordered set: if $x_{mean, u}$ and $x_{mean, v}$ are adjacent, the column labeled "2" in Table 1.11 must be used,

Table 1.11. Critical q-Values for Two Means with Index Number Differences of 2, 3, 10, Resp. 20

Degr. of Freedom	Difference $|u - v| + 1$			
	2	3	10	20
1	17.97	17.97	17.97	17.97
2	6.085	6.085	6.085	6.085
10	3.151	3.293	3.522	3.526
∞	2.772	2.918	3.294	3.466

and if $x_{mean, u}$ and $x_{mean, v}$ are separated by eight other averages, the column labeled "10" is used.

Critical q-values for $p = 0.05$ are available.[34,51] In lieu of using these tables, the calculated q-values can be divided by the appropriate Student's $t(f, 0.05)$ and $\sqrt{2}$ and compared to the reduced critical q-values (see Table 1.12), and data file QRED_TBL.dat. A reduced q-value that is smaller than the appropriate critical value signals that the tested means belong to the same population. A fully worked example is found in Chapter 4, Process Validation. Data file MOISTURE.dat used with program MULTI gives a good idea of how this concept is applied. MULTI uses Table 1.12 to interpolate the cut-off point for $p = 0.05$. With little risk of error, this table can also be used for $p = 0.025$ and 0.1 (divide q by $t(f, 0.025) \cdot \sqrt{2}$ respectively $t(f, 0.1) \cdot \sqrt{2}$, as appropriate.

The MRT procedure can be applied to two-dimensional data if there is reason to suspect that length- or cross-wise effects operate. For example, a coating process could periodically change the thickness along the length of the web if the rollers are slightly excentric, and could vary systematically across the width of the web if the rollers deviated from cylindrical shape or the coating gap were different on one side of the web from the other; for an example, see file COAT_WEB.dat. Use program DATA to transpose rows and columns.

1.5.5 Outlier Tests

The rejection of "outliers" is a deeply rooted habit; techniques range from the haughty "I know the truth" attitude, to "looks different from what we are used to, somebody must have made a mistake", to attempts at objective proof of having observed an aberration. The first attitude is not only unscientific, but downright fraudulent if "unacceptable" values are discarded and forgot-

Table 1.12. Reduced Critical q-Values

Degrees of Freedom	Separation $\|u - v\| + 1$ (Difference between index numbers +1)								
	4	6	8	10	12	16	18	20	40
3	1.00	1.00	1.00	1.00	1.00	1.00	1.00	1.00	1.00
5	1.05	1.05	1.05	1.05	1.05	1.05	1.05	1.05	1.05
6	1.05	1.07	1.07	1.07	1.07	1.07	1.07	1.07	1.07
7	1.06	1.08	1.09	1.09	1.09	1.09	1.09	1.09	1.09
8	1.07	1.09	1.10	1.10	1.10	1.10	1.10	1.10	1.10
9	1.07	1.09	1.10	1.11	1.11	1.11	1.11	1.11	1.11
10	1.07	1.10	1.11	1.12	1.12	1.12	1.12	1.12	1.12
20	1.08	1.12	1.14	1.16	1.17	1.17	1.18	1.18	1.18
40	1.08	1.13	1.15	1.17	1.19	1.20	1.21	1.21	1.22
∞	1.09	1.14	1.17	1.19	1.20	1.23	1.24	1.25	1.30

ten, something that is forbidden under GMPs and would land the responsible person in court.

The second perspective is closer to the mark: Measurements that apparently do not fit model or experience should always be investigated in the light of all available information. While there is the distinct possibility of a discovery about to be made, the other outcome of a sober analysis of the circumstances is more probable: a deviation from the experimental protocol. If this is documented, all the better: the probable outlier can, in good conscience, be rejected and replaced by a reliable repeat result.

Over the years an abundance of outlier tests have been proposed that have some theoretical rationale at their roots.[20] Such tests have to be carefully adjusted to the problem at hand because otherwise one would either not detect true outliers (false negatives) in every case, or then throw out up to 50% of the "good" measurements as well (false positives).[5,20] Robust methods have been put forward to overcome this.[52] Three tests will be described:

1. A well-known test is Dixon's: the data are first ordered according to size, and a range $(x_{1+j} - x_n)$ and a subrange $(x_{n-i} - x_n)$ are compared. The ease of the calculations, which probably strongly contributed to the popularity of this test, is also its weakness: since any out of a number of subrange ratio models (combinations of i and j) can be chosen, there is an arbitrary element involved. Obtaining numerically correct tables of critical quotients for convenient values of p is a problem; the use of this test is increasingly being discouraged.[53,54] The Dixon tests build on and are subject to the stochastic nature of range measures; they use only a small portion of the available information and lack ruggedness.

Example 14: For $x() = 1.53$, 1.70, 1.71, 1.73, and 1.78, $D = |1.53 - 1.70| \div |1.53 - 1.78| = 0.680$, $D_{crit} = 0.642$ ($p = 0.05$); as a matter of fact, this result is significant on the $p = 0.035$ level.

2. As in the case of the detection limit (Section 2.2.7), one commonly used algorithm is based on the theory that any point outside $\pm z \cdot s_x$ is to be regarded as an outlier; if recalculation of $\mathbf{x}_{mean}$ and s_x without this questionable point confirms the decision, the "outlier" is to be cast out. The coefficient z is often fixed at 2.0 or 3.0; the t-function (see the two thin curves in Fig. 1.24) and other functions have also been proposed. An obvious disadvantage of these approaches is that extreme values strongly affect s_x, and that more or less symmetrical "outliers" cannot be detected.

Example 15: For the preceding data, $\mathbf{x}_{mean} = 1.69$ and $s_x = 0.0946$ ($n = 5$), which results in $t = 1.69$ (not significant).

3. A wholly different approach is that of Huber,[21] who orders the values according to size, and determines the median (cf. Section 1.1.1); then the absolute deviations $|x_i - x_m|$ are calculated and also ordered, the median absolute deviation (MAD) being found. MAD is then used as is s_x earlier, the coefficient k being chosen to be between 3 and 5. This algorithm is much more robust than the ones described before.

Example 16: because $x_m = 1.71 \longrightarrow \text{MAD} = 0.02 \longrightarrow k = |1.53 - 1.71| \div 0.02 = 9$, which is significant.

Example 17: The $n = 19$ values in Table 1.23 yield a median of 2.37, absolute deviations ranging from 0.00 to 2.99, and a MAD of 0.98. The coefficient k can be as low as 3.05 before a single point is eliminated (-0.614): Use data file HISTO.dat in conjunction with program HUBER.

For the reasons described, no specific test will be advanced here as being superior, but Huber's model and the classical one for $z = 2$ and $z - 3$ are incorporated into program HUBER; the authors are of the opinion that the best recourse is to openly declare all values and do the analysis twice, once with the presumed outliers included, and once excluded from the statistical analysis; in the latter case the excluded points should nonetheless be included in tables (in parentheses) and in graphs (different symbol). "Outliers" should not be labeled as such solely on the basis of a fixed (statistical) rule; the decision should primarily reflect scientific experience.[19] The justification must be explicitly stated in any report; cf. Sections 4.18 and 4.19. If the circumstances demand that a rule be set down, it is best to use a robust model such as Huber's; its sensitivity for the problem at hand, and the typical rate for false positives, should be investigated by, for example, a Monte

Carlo simulation. Program HUBER rejects points on the basis of Huber's rule and shows the results. For completeness, the means and standard deviations before and after the elimination are given, and the equivalent z-values for the classical mean $\pm z \cdot SD$ are calculated. The sensitivity of the elimination rules toward changes in the k- resp. z-values are graphically indicated.

An example, arrived at by numerical simulation, will be given here to illustrate the high probability of rejecting good data. Figure 1.24 shows that the largest residual of every series is around $1 \cdot s_x$ for $n = 3$, around $2 \cdot s_x$ for $n = 13$, and close to $3 \cdot s_x$ for $n = 30$. This makes it virtually certain that

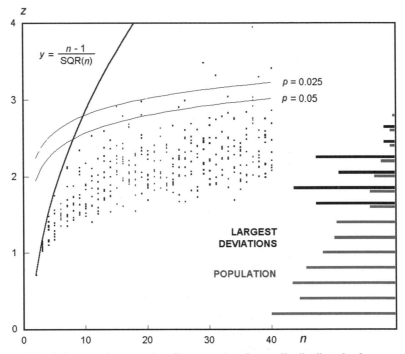

Figure 1.24. Rejection of suspected outliers. A series of normally distributed values was generated by the Monte Carlo technique; the mean and the standard deviation were calculated; the largest normalized absolute deviate (residual) $z = |x_i - \mu|^{\max}/s_x$, is plotted *versus* n (black histogram, all abs. deviates: gray histogram). The simulation was repeated $m = 20$ times for every $n = 2 \ldots 40$. The maximally possible z is given by the curve $y = (n-1)/\sqrt{n}$; however, $y < 2$ for $n < 6$ and $y < 3$ for $n < 11$! A rejection strategy denoted by "$\pm z \cdot s_x$" ($z = 2$ or 3) is implemented by drawing a horizontal line at $y = 2$ or 3. Repetitive application of this strategy can easily lead to the loss of 3–6 values out of $n = 20$, because after each elimination s_x is reduced. Huber's rule with limits set at $x_m \pm k \cdot MAD$, $k = 3.5$, rejects only about 1 out of $n = 20$ values. The range R is not quite twice the largest residual, $R \approx 2 \cdot r^{\max}$; in this connection see Figure 1.3.

for a series of $n > 10$ measurements at least one will be rejected by a "$\pm 2 \cdot s_x$" recipe, which only makes sense if the individual measurement is to be had at marginal cost. Rejection rates become higher still if the observation is taken into account that experimental distributions tend to be broader-tailed than the normal distribution; a t-distribution for $f = 4$ to 9 is said to give the best fit.[20] (See Figs. 1.14 and 1.29.) The distributions for the extreme (black) and all (gray) deviations were simulated for $n = 20$; the frequency scales are different!

Because outlier elimination is something that is not to be taken lightly, the authors have decided to *not* provide on-line outlier deletion options in the programs. Instead, the user must first decide which points he regards as outliers, for example, by use of program HUBER, then start program DATA and use options ⟨Edit⟩ or ⟨Delete Row⟩, and finally create a modified data file with option ⟨Save⟩. This approach was chosen to reinforce GMP-like procedures and documentation.

The situation of the pharmaceutical industry is today governed by the Barr ruling. The foregoing suggestions concerning the use of outlier tests are expressly aimed only at those users and situations not subject to the Barr ruling.

Since the 1993 court decision against Barr Laboratories,[55] the elimination of outliers has taken on a decidedly legal aspect in the U.S. (any non-U.S. company that wishes to export pharmaceuticals or precursor products to the U.S. market must adhere to this decision concerning out-of-specification results, too); the relevant section states that "... An alternative means to invalidate an individual OOS result ... is the (outlier test). The court placed specific restrictions on the use of this test. (1) Firms cannot frequently reject results on this basis, (2) The USP standards govern its use in specific areas, (3) The test cannot be used for chemical testing results ... ." A footnote explicitly refers only to a content uniformity test,[55] but it appears that the rule must be similarly interpreted for all other forms of inherently precise physicochemical methods. For a possible interpretation, see Section 4.24.

1.5.6 Analysis of Variance (ANOVA)

Analysis of variance (ANOVA) tests whether one group of subjects (e.g., batch, method, laboratory, etc.) differs from the population of subjects investigated (several batches of one product; different methods for the same parameter; several laboratories participating in a round robin test to validate a method, for examples see Refs. 5, 9, 21, 30. Multiple measurements are necessary to establish a benchmark variability ("within-group") typical for the type of subject. Whenever a difference significantly exceeds this benchmark, at least two populations of subjects are involved. A graphical analogue is the Youden plot (see Fig. 2.1). An additive model is assumed for ANOVA.

The type of problem to which ANOVA is applicable is exemplified as follows: Several groups of measurements are available pertaining to a certain product, several repeat measurements having been conducted on each batch. The same analytical method was used throughout (if not, between-group variance would be distorted by systematic differences between methods; this problem typically crops up when historical data series are compared with newer ones). Were one to do t-tests, the data would have to be arranged according to the specific question asked, i.e., "do batches 1 and 2 differ among themselves?"; tests would be conducted according to cases $b1$ or $b3$.

The question to be answered here is not "do batches x and y differ?", but "are any individual batches different from the group as a whole?". The hypotheses thus have the form:

$$H_0: \mu_1 = \mu_2 = \mu_3 = \ldots \mu_m,$$
$$H_1: \mu_1 \neq \mu_2 = \mu_3 = \ldots \mu_m, \text{ or }$$
$$H_1: \mu_1 \neq \mu_2 \neq \mu_3 = \ldots \mu_m, \text{ etc.}$$

Since a series of t-tests is cumbersome to carry out, and does not answer all questions, all measurements will be simultaneously evaluated to find differences between means. The total variance (relative to the grand mean x_{GM}) is broken down into a component V_1 "variance within groups," which corresponds to the residual variance, and a component V_2 "variance between groups." If H_0 is true, V_1 and V_2 should be similar, and all values can be pooled because they belong to the same population. When one or more means deviate from the rest, V_2 must be significantly larger than V_1.

Table 1.13. Equations for simple ANOVA

The following variables are used:

m: number of groups, e.g., $m = 5$ in Table 1.15

$j = 1 \ldots m$

n_j: number of measurements in group j (column j)

n: total number of measurements ($\sum n_j$)

x_{ij}: i-th measurement in j-th column

The following sums are calculated:

$$u_j = \sum (x_{ij})$$ sum over all measurements in (1.26)
group j

$$u_n = \sum (u_j)$$ sum over all n measurements (1.27)
(m groups)

Since the total number of degrees of freedom must be $n - 1$, and m groups

Table 1.13. (*Continued*)

$x_{\text{mean},j} = u_j/n_j$	mean over group j	(1.28)
$x_{\text{GM}} = u_n/n$	grand mean	(1.29)
$S_2 = \sum(n_j \cdot (x_{\text{mean},j} - x_{\text{GM}})^2)$	sum of weighted squares	(1.30)
$S_1 = \sum(x_{ij} - x_{\text{GM}})^2$	sum of squares of residuals relative to the appropriate group mean $x_{\text{mean},j}$	(1.31)
$S_1 = \sum(\sum(r_i)^2)$	$r_i = x_{ij} - x_{\text{mean},j}$	(1.32)
$S_T = S_1 + S_2$	total sum of squares	(1.33)
$f_1 = n - m$	degrees of freedom within groups	(1.34)
$f_2 = m - 1$	degrees of freedom between groups	(1.35)

are defined, there remain $n - 1 - (m - 1) = n - m$ degrees of freedom for within the groups (see Tables 1.14 and 1.15).

For the data given in Table 1.15, an ANOVA calculation yields the results shown in Table 1.16.

V_1 and V_2 are subjected to an F-test (see Section 1.6.1) to determine whether H_0 can be retained or not. Since V_2 must be larger than V_1 if H_1 is to hold, $F = V_2/V_1$; should V_2 be smaller or equal to V_1, then H_1 could be rejected without an F-test being performed. V_2 *can* be smaller than V_1 because of calculational artifacts.

If H_0 were to be retained, the individual means $x_{\text{mean},j}$ could not be distinguished from the grand mean x_{GM}; V_T would then be taken as the average variance associated with x_{GM} and $n - 1$ degrees of freedom.

Interpretation of Table 1.16: Since V_2 is significantly larger than V_1, the groups cannot all belong to the same population. Therefore, both the *grand mean* x_{GM}, which is equal to $219.93 \div 35 = 6.28$, and the associated standard deviation $\sqrt{(49.28 \div 34)} = \pm 1.2$ are irrelevant. The question of which

Table 1.14. Presentation of ANOVA Results

Sum of Squares	Degrees of Freedom	Variance	Comment
S_1	$f_1 = n - m$	$V_1 = S_1/f_1$	variance within groups
$+S_2$	$f_2 = m - 1$	$V_2 = S_2/f_2$	variance between groups
$= S_T$	$f_T = n - 1$	$V_T = S_T/f_T$	total variance

Table 1.15. Raw Data and Intermediate Results of an ANOVA Test for Simulated Data. (Eq. 1.30)

	Group j						
	1	2	3	4	5		
index 1	7.87	6.35	4.65	8.74	5.29		
$i = 2$	6.36	7.84	5.06	6.02	6.03		
3	5.73	5.31	6.52	6.69	6.06		
4	4.92	6.99	6.51	7.38	5.64		
5	4.60	8.54	8.28	5.82	4.33		
6	6.19	·	4.45	6.88	5.39		sum
7	5.98	·	·	8.65	5.85		
8	6.95	·	·	6.55	5.51		
							$x_{mean,j}$:
	6.08	7.01	5.91	7.09	5.51	group mean	31.60
	1.05	1.26	1.47	1.10	0.56	group standard deviation	s_j
	8	5	6	8	8	number of measurements	n_j: 35
	48.60	35.03	35.47	56.73	44.10	sum	u_j: 219.9
	7.72	6.35	10.80	8.47	2.20	sum of squared residuals	S_1: 35.52
	0.33	2.64	0.83	5.21	4.78	sum of weighted squares	S_2: 13.76

means are indistinguishable among themselves and different from the rest is answered by the multiple range test.

Other forms of ANOVA: The simple ANOVA set out above tests for the presence of one unknown factor that produces (additive) differences between outwardly similar groups of measurements. Extensions of the concept allow one to simultaneously test for several factors (two-way ANOVA, etc.). The

Table 1.16. Results of an ANOVA Test

Sum of Squares	Degrees of Freedom	Variance	Comment
35.520	30	1.184	variance within groups
13.763	4	3.441	variance between groups
49.283	34	1.450	total variance

F-test: $3.441/1.184 = 2.91$ $F_c(4,30,0.05) = 2.69$
The "null" hypothesis is rejected because $2.91 > 2.69$.

limit of ANOVA tests is quickly reached: they do not provide answers as to the type of functional relationship linking measurements and variables, but only indicate the probability of such factors being present. Thus, ANOVA is fine as long as there are no concrete hypotheses to test, as in explorative data analysis, cf. Section 2.1.

1.6 NUMBER OF DETERMINATIONS

Up to this point it was assumed that the number of determinations n was sufficient for a given statistical test. During the discussion of the t-test (case a), an issue was skirted that demands more attention: Is x_{mean} different from μ?

1. If the question refers to the expectation $E(x_{mean}) = \mu$ for $n \to \infty$, finding a difference could mean a deficiency in a theory.
2. On the other hand, if $E(x_{mean}) \leq L$, where L is an inviolable limit, finding "x_{mean} greater than L" could mean the infraction of a rule or an unsalable product. In other words, L would here have to be significantly larger than x_{mean} for a high probability of acceptance of the product in question.

It is this second interpretation that will occupy us for a moment. Two approaches to specification limits, SL, need to be considered:

2a. The more legalistic reasoning postulates fixed SL with method specifics (instrumentation, n, s_x, etc.) already taken account of. Measurements therefore either conform or do not. All pharmacopeial specifications are set up this way, which means a measurement stands as documented.

2b. The statistical approach is to ask: "what is the risk of failing the quality clause in the contract?" and calculating the p-value; the quality clause was set without a specific method or procedure in mind.

Thus, even under Regime 2a, the manufacturer will set in house limits (IHL) that are sufficiently narrow to reduce the risk of a "recall from market" (and all the attendant publicity) to very low levels. (See Fig. 2.13.) The risk resides in the possibility that a customer or a regulatory body reanalyzes samples and declares "fails" even if the manufacturer's QC lab found "complies." Many things could be at fault in either location, such as equipment, operator training, availability of proper standards, etc., but once a suspicion is raised in public, industrial science takes a back seat.

**Number of necessary
determinations *n***

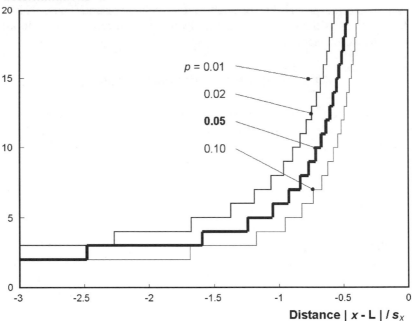

Distance | *x* - L | / s_x

Figure 1.25. The number of measurements n that are necessary to obtain an error probability p of $\mathbf{x}_{\text{mean}}$ exceeding L is given on the ordinate. The abscissa shows the independent variable Q in the range $L - 3 \cdot s_x \ldots L$ in units of s_x. It is quite evident that for a typical p of about 0.05, $\mathbf{x}_{\text{mean}}$ must not be closer than about 0.5 standard deviations s_x from L in order that the necessary number of repeat measurements remains manageable. The enhanced line is for $p = 0.05$; the others are for 0.01 (left), 0.02, and 0.1 (right).

Replacing μ in Eq. 1.13 by L yields

$$t_c < t = \frac{|E(\mathbf{x}_{\text{mean}}) - L|}{s_{x,\text{mean}}} = \frac{|\mathbf{x}_{\text{mean}} - L| \cdot \sqrt{n}}{s_x} \tag{1.36}$$

Most likely, Student's t_c will be fixed by an agreed-on confidence level, such as 95%, and L by a product specification; $E(\mathbf{x}_{\text{mean}})$ is the true value of the parameter. The analyst has the option of reducing s_x or increasing n in order to augment the chances of obtaining a significant t. The standard deviation s_x is given by the choice of test method and instrumentation, and can be influenced to a certain extent by careful optimization and skillful working habits. The only real option left is to increase the number of determinations n. A look at Table 1.3 and Eq. 1.36 reveals that the critical t-value, t_c, is a function of n, and n is a function of t_c; thus a trial-and-error approach must be implemented that begins with a rough estimate based on experience, say, $n = 5$; t is calculated for given $\mathbf{x}_{\text{mean}}$, L, and s_x and compared to the tabulated t_c. Depending on the outcome, n is increased or decreased. The search procedure can be automated by applying the polynomial $t = f(n - 1)$ from Section 5.1.2 and increasing n until $t > t_c$. The idea behind this is that t is proportional to $\sqrt{n}$, and t_c decreases from 12.7 ($n = 2$) to 1.96 ($n = \infty$) for a 97.5% one-sided confidence level. Note that a one-sided test is applied, because it is expected that $\mathbf{x}_{\text{mean}} < L$. Equation (1.36) is rearranged to yield:

$$t_c/\sqrt{n} < |\mathbf{x}_{\text{mean}} - L|/s_x = Q \tag{1.37}$$

The quotient Q is fixed for a given situation. The problem could also be solved graphically by drawing a horizontal at $y = Q$ in Figures 1.17 or 1.19 und taking the intercept as an estimate of n. The necessary number of repeat determinations n is depicted in Figure 1.26.

Example 18: Consider the following situation: The limit is given by $L = 100$, and the experimental average is $\mathbf{x}_{\text{mean}} = 90$; how many measurements n are necessary to find a significant difference between L and $\mathbf{x}_{\text{mean}}$ (Eq. 1.13, Table 1.10)? The error probability can be $p = 0.25 \ldots 0.00005$ (one-sided) and the experimental standard deviation is $\pm 0.1 \ldots \pm 200$. (See Table 1.17.)

Example 19: For $s_x = \pm 5$ and an error probability of falsely accepting H_1 of $p = 0.005$ would require $n \geq 6$ because

for $n = 6, t = 4.0321, \text{CL} = 90 + 5 \cdot 4.0321 \div \sqrt{6} = 98.23$, and

for $n = 5, t = 4.6041, \text{CL} = 90 + 5 \cdot 4.6041 \div \sqrt{5} = 100.30$.

The following statements are equivalent to each other:

- The upper confidence limit $\text{CL}_u(\mathbf{x}_{\text{mean}} = 90)$ is less than 100.
- The true value μ is in the interval $x = 80$ (of no interest here) $\ldots$ 100 with a probability of at least 99% ($100 - 2 \cdot 0.5 = 99$).
- The null hypothesis H_0: $\mu < L$ has a probability of more than 99.5% of being correct.
- The probability of μ being larger than L is less than 0.5% or $p \leq 0.005$.

95% CL(s_x)

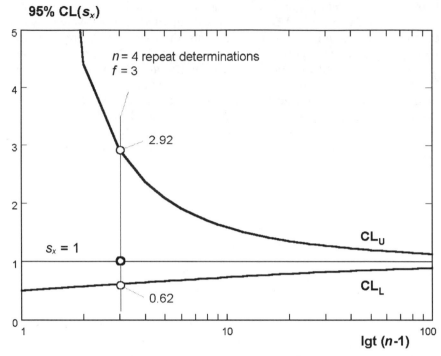

Figure 1.26. Confidence limits of the standard deviation for $p = 0.05$ and $f = 1 \ldots 100$. The f-axis is logarithmically transformed for a better overview. For example, at $n = 4$, the true value σ_x is expected between 0.62 and 2.92 times the experimental s_x. The ordinate is scaled in units of s_x.

Table 1.17. The Number of Determinations n Necessary to Achieve a Given Discrimination

$s_x =$	0.1	0.2	0.5	1	2	5	10	20	50	100	200
$p = 0.25$	2	2	2	2	2	2	2	3	13	47	183
0.1	2	2	2	2	2	3	4	9	43	166	659
0.05	2	2	2	2	2	3	5	13	70	273	1084
0.025	2	2	2	2	3	4	7	18	99	387	1538
0.01	2	2	3	3	3	5	9	25	139	546	2172
0.005	2	2	3	3	4	6	11	31	171	673	2679
0.001	3	3	3	4	5	7	15	44	244	960	3824
0.0005	3	3	3	4	5	8	17	50	277	1087	4330
0.00005	3	4	4	5	6	11	23	69	360	1392	5518

- Because for $n = 6$ ($f = 5$), the critical t-factor is 4.0321, 99% of the individual measurements (two-sided) would be expected in the interval 69.8 ... 110.2.
- with $n = 6$ and $f = 5$, $t = (100 - 90) \div 5 = 2$, and the critical $t(f = 5, p = 0.05,$ one-sided) $= 2.0150$; thus, about $1 - 0.05 = 0.95$ (95%) of all individual measurements would be expected to be below 100.0.

1.7 WIDTH OF A DISTRIBUTION

As was discussed in Section 1.1.2, there are several ways to characterize the width of a distribution:

1. From a purely practical point of view the range or a quantile can serve as indicator. Quantiles are usually selected to encompass the central 60–90% of an ordered set; the influence of extreme values diminishes the smaller this %-value is. No assumptions as to the underlying distribution are made.

2. Theoretical models are a guide in setting up rules and conventions for defining a distribution's width, the standard deviation being a good example. Simply the fact of assuming a certain form of distribution, however, undermines whatever strength resides in the concept, unless theory and fact conform.

Measured distributions containing less than 100 events or so just as likely turn out to appear skewed as symmetrical, c.f. Fig. 1.10.

For this reason alone the tacit assumption of a normal distribution when contemplating analytical results is understandable, and excusable, if only because there is no practical alternative (alternative distribution models require more complex calculations, involve more stringent assumptions, or are more susceptible to violations of these basic assumptions than the relatively robust normal distribution).

1.7.1 The *F*-Test

The *F*-test is based on the normal distribution and serves to compare either

- An experimental result in the form of a standard deviation with a fixed limit on the distribution width, or
- Such an experimental result with a second one in order to detect a difference.

Both cases are amenable to the same test, the distinction being a matter of the number of degrees of freedom f. The F-test is used in connection with the t-test. (See program TTEST.)

The test procedure is as follows:

Table 1.18. Calculation of a F-test Statistic

	Test Distribution	Reference Distribution
Standard deviation	$s_t = \sqrt{V_t}$	$s_r = \sqrt{V_r}$
Degrees of freedom	f_t	f_r
Null hypothesis H_0	the standard deviations s_t and s_r are indistinguishable	
Alternative hypothesis H_1	s_t and s_r are distinguishable on the confidence level given by p	
Test statistic	$F = V_t/V_r$ or $F = V_r/V_t \geq 1.00$ Eq. (1.38)	
Nomenclature	$F = V_1/V_2$ with f_1 resp. f_2 degrees of freedom	

The critical value F_c is taken from an F-table or is approximated (cf. Section 5.1.3); if

$$F > F_c: \text{accept } H_1, \text{reject } H_0$$
$$F \leq F_c: \text{retain } H_0, \text{reject } H_1 \qquad\qquad (1.39)$$

F_c depends on three parameters:

1. The confidence level p must be chosen beforehand; most statistical works tabulate F_c for only a few levels of p, 0.05, 0.02, and 0.01 being typical.
2. The number of degrees of freedom f_t is given by the test procedure, and can be influenced: Since the analyst for technical reasons ordinarily has only a limited set of analytical methods to choose from, each with its own characteristics, the wish to increase the confidence in a decision can normally only be achieved by drastically increasing the number of measurements.
3. The number of degrees of freedom f_r is often an unalterable quantity, namely
3a. f_r is fixed if the reference measurements originate from outside one's laboratory, e.g., the literature,
3b. $f_r = \infty$ if the limit is theoretical or legal in nature.

Example 20: To illustrate the concept ($p = 0.05$ will be used):

1. A reference analytical method, for which much experimental evidence is at hand, shows a standard deviation $s_r = \pm 0.037$ with $n_r = 50$. A few tests with a purportedly better method yield $s_t = \pm 0.029$ for
   - $F = (0.037 \div 0.029)^2 = 1.28$
   - $F_c = 5.7$ for $f_1 = 49, f_2 = 4$
   - On this evidence no improvement is detectable. If—for whatever reason—the reference method is to be proven inferior, for the same F ratio, over 45 measurements with the better method would be necessary (F_c is larger than 1.28 if $f_1 = 49$ and f_2 is less than 45), or, alternatively,
   - s_t would have to be brought down to less than ± 0.0155 (($0.037 \div 0.0155)^2 = 5.7$) in order to achieve significance for the same number of measurements. (*Note:* For a small series of measurements, tight control and a little bit of correlation (mostly the later measurement's dependence on the earlier ones), very low repeatabilities can be achieved (see file CONV.xls and program CONVERGE); the justification for switching methods should be based on reproducibilities, not repeatabilities.

2. A system suitability test prescribes a relative standard deviation of not more than 1% for the procedure to be valid; with $x_{mean} = 173.5$ this translates into $V_r = (1.735)^2$; because the limit is imposed, this is equivalent to having no uncertainty about the numerical value, or in other words, $f_r = \infty$. Since s_t was determined to be ± 2.43 for $n_t = 7$:
   - $F = (2.43 \div 1.735)^2 = 1.96$ and $F_c(0.05, 6, \infty) = 2.1$.
   - Statistically, the two standard deviations cannot be distinguished.
   - Assuming the $\pm 1\%$ limit was set down in full cognizance of the statistics involved, the system suitability test must be regarded as failed because $2.43 > 1.74$. This would be the legalistic interpretation under GMP rules. Statistically it would have made more sense to select the criterion as $s_t \leq 0.01 \cdot x_{mean} \cdot \sqrt{F_c(0.05, f_t, \infty)}$ for acceptance and demanding, say, $n \geq 5$; in this particular case, s_t could have been as large as 3.6.

Unfortunately, few chemists are aware of the large confidence interval a standard deviation carries (see Section 4.34) and thus are prone to setting down unreasonable specifications, such as the one in the previous example. The only thing that saves them from permanent frustration is the fact that if n is kept small enough, the chances for obtaining a few similar results in a row because of tight correlation between them, and hence a small s_x, are relatively good. (See Fig. 1.24.)

For identification, an excerpt from the F-table for $p = 0.05$ is given as follows:

Table 1.19. F-Table for p = 0.05

$f_1 =$	1	2	4	8	16	32	inf
$f_2 = 1$	161.5	199.5	224.6	238.9	246.5	250.4	254.3
2	18.51	19.00	19.25	19.37	19.43	19.46	19.50
4	7.709	6.944	6.388	6.041	5.844	5.739	5.628
8	5.318	4.459	3.838	3.438	3.202	3.070	2.928
16	4.494	3.634	3.007	2.591	2.334	2.183	2.010
32	4.149	3.295	2.668	2.244	1.972	1.805	1.594
inf	3.842	2.996	2.372	1.938	1.644	1.444	1.000

1.7.2 Confidence Limits for a Standard Deviation

In Section 1.3.2, confidence limits are calculated to define a confidence interval within which the true value μ is expected with an error probability of p or less.

For standard deviations, an analogous confidence interval $CI(s_x)$ can be derived via the F-test. In contrast to $CI(x_{mean})$, $CI(s_x)$ is not symmetrical around the most probable value because s_x by definition can only be positive. The concept is as follows: an upper limit, s_u, on s_x is sought that has the quality of a very precise measurement, that is, its uncertainty must be very small and therefore its number of degrees of freedom f must be very large. The same logic applies to the lower limit s_l:

Table 1.20. Calculation of Confidence Limits for a Variance

case	$s_l < s_x$	$s_x < s_u$	
	$F_1 = V_x/V_l$	$F_u = V_u/V_x$	
	$F_l = F(n-1, \infty, p)$	$F_u = F(\infty, n-1, p)$	
thus	$V_l = V_x/F(n-1, \infty, p)$	$V_u = V_x \cdot F(\infty, n-1, p)$	(1.40)
and since	$F(f, \infty, p) = \chi^2(f, p)/f$	$F(\infty, f, p) = f/\chi^2(f, 1-p)$	(1.41)
	$V_l = V_x \cdot f/\chi^2(f, p)$	$V_u = V_x \cdot f/\chi^2(f, 1-p)$	(1.42)
where $f = n-1$.			

The true standard deviation σ_x is expected inside the confidence interval $CI(s_x) = \sqrt{V_l} \ldots \sqrt{V_u}$ with a total error probability $2 \cdot p$ (in connection with F and χ^2, p is taken to be one-sided).

F-values can be calculated according to Section 5.1.3 and χ^2-values according to Section 5.1.4 (see also programs MSD and CALCVAL); both could also be looked up in the appropriate tables.

Example 21: Let $s_x = 1.7$, $n = 8$, $f = 7$, $2 \cdot p = 0.1$, and $p = 0.05$:

Table 1.21. Numerical Example for CL(s_x)

Tabulated value	Tabulated value	Control, see Eq. (1.41)
$F(7, \infty, 0.05) = 2.0096$	$\chi^2(7, 0.05) = 14.067$	$14.067/7 = 2.0096$
$F(\infty, 7, 0.05) = 3.2298$	$\chi^2(7, 0.95) = 2.167$	$7/2.167 = 3.2298$
$s_l = 1.7/\sqrt{2.0096} = 1.2$		
$s_u = 1.7 \cdot \sqrt{3.2298} = 3.0$		

Thus with a probability of 90% the true value $E(s_x) = \sigma_x$ is within the limits: $1.2 \leq \sigma_x \leq 3.0$. For a depiction of the confidence limits on s_x for a given set of measurements and a range of probabilities $0.0002 \leq p \leq 0.2$, see program MSD, option ⟨Display Standard Deviation⟩.

1.7.3 Bartlett Test

Several series of measurements are to be compared as regards the standard deviation. It is of interest to know whether the standard deviations could all be traced to one population characterized by σ (H_0: no deviation observed), and any differences *versus* σ would only reflect stochastic effects, or whether one or more standard deviations belong to a different population (H_1: difference observed):

$$H_0: \sigma_1 = \sigma_2 = \ldots = \sigma_m, \text{and}$$
$$H_1: \sigma_1 \neq \sigma_2 = \sigma_3 = \ldots = \sigma_m, \text{etc.}$$

For example, a certain analytical procedure could have been repeatedly performed by m different technicians. Do they all work at the same level of proficiency, or do they differ in their skills?

The observed standard deviations are $s_1, s_2, s_3, \ldots s_m$ and the corresponding number of degrees of freedom are $f_1 = n_1 - 1, \ldots f_m = n_m - 1$, with f_j larger than 2. The interpretation is formulated in analogy to Section 1.5.6.

The following sums have to be calculated:

Table 1.22. Equations for Bartlett Test

$A = \sum(f_i \cdot V_i)$	sum of all residuals; A/D is the total variance	(1.43)
$B = \sum(f_i \cdot \ln(V_i))$		(1.44)
$C = \sum(1/f_i)$		(1.45)
$D = \sum(f_i)$		(1.46)
Then		
$E = D \cdot \ln(A/D) - B$	χ^2	(1.47)
$F = (C - 1/D)/(3 \cdot m - 3) + 1$	correction term	(1.48)
$G = E/F$	corrected χ^2	(1.49)

If the found G exceeds the tabulated critical value $\chi^2(p, m - 1)$, the null hypothesis H_0 must be rejected in favor of H_1: the standard deviations do not all belong to the same population, that is, there is at least one that is larger than the rest. The correction factor F is necessary because Eq. (1.47) overestimates χ^2.

For a completely worked example, see Section 4.4, Process Validation and data file MOISTURE.dat in connection with program MULTI.

1.8   CHARTING A DISTRIBUTION

In explorative data analysis, an important clue is the form of the population's distribution (cf. Figure 1.9); after accounting for artifacts due to the analysis method (usually an increase in distribution width) and sampling procedure (often a truncation), width and shape can offer insight into the system from which the samples were drawn. Distribution width and shape can be judged using a histogram, see program HISTO. The probability chart (program HISTO, option ⟨NPS⟩ = "Normal Probability Scale") tests for normality, and the χ^2-test can be used for any shape.

1.8.1   Histograms

When the need arises to depict the frequency distribution of a number of observations, a histogram (also known as a bar chart), is plotted. The data are assumed to be available in a vector $x()$; the individual values occur within a certain range given by x_{min} and x_{max}, and these two limits can be identical with the limits of the graph, or different. How are the values grouped? From a theoretical point of view, $c = \sqrt{n}$ classes (or bins) of equal width would be optimal. It is suggested that, starting from this initial estimate c, a combination of an integer c, and lower resp. upper bounds on the graph

($a \leq x_{\min}$) and ($b \geq x_{\max}$) be chosen so that the class boundaries defined by $a + i \cdot w$, with $w = (b - a)/c$ and $i = 0, 1, 2, \ldots c$, assume numerical values appropriate to the problem (not all software packages are capable of satisfying this simple requirement).

Example 22: For $x_{\min} = .327$, $x_{\max} = .952$, $n = 21$

$$a = .20, \quad b = 1.20, \quad c = 5, \quad w = .20, \text{or}$$
$$a = .15, \quad b = 1.35, \quad c = 8, \quad w = .15$$
$$a = .30, \quad b = 1.00, \quad c = 7, \quad w = .10$$

This results in sensible numbers for the class boundaries and allows for comparisons with other series of observations that have different extrema, for example (.375, .892) or (.25, 1.11). Strict adherence to the theory-inspired rule would have yielded class boundaries .327, .4833, .6395, .7958, and .952, with the extreme values being exactly on the edges of the graph. That this is impractical is obvious, mainly because the class boundaries depend on the stochastic element innate in $x_{\min}$ resp. $x_{\max}$. Program HISTO, option $\langle$Scale$\rangle$, allows for an independent setting of a subdivided range R with C bins of width R/C, and lower/upper limits on the graph.

Next, observations $x_1 \ldots x_n$ are classed by first subtracting the x-value of the lower boundary of bin 1, a, and then dividing the difference by the class width w. The integer $\text{INT}((x_i - a)/w + 1)$ gives the class number (index) j. The number of events per class is counted and expressed in form of an absolute, a relative, and a cumulative result (E, $100 \cdot E/n$, resp. $100 \cdot \Sigma(E/n)$).

Before plotting the histogram, the vertical scale (absolute or relative frequency) has to be set; as earlier, practical considerations like comparability among histograms should serve as a guide. The frequencies are then plotted by drawing a horizontal line from one class boundary to the other, and dropping verticals along the boundaries.

Artistic license should be limited by the fact that the eye perceives the area of the bar as the significant piece of information. The ordinate should start at zero. A violation of these two rules—through purpose or ignorance —can skew the issue or even misrepresent the data. Hair-raising examples can be found in the "Economics" sections of even major newspapers.

Example 23: A histogram analysis. (See Table 1.23 and data file HISTO.dat.)

Note that the single event to the left of class 1, equivalent to 5.6%, must be added to the first number in the % Events column to obtain the correct $\Sigma\%$, unless a separate class is established. (Here, it is Class 0.)

Table 1.23. Data and Results of a Histogram Analysis.

Values:	3.351	1.971	2.681	2.309	0.706
	2.973	2.184	−.614	1.848	3.749
	3.431	−.304	4.631	3.010	3.770
	2.996	0.290	1.302	2.371	

Results:	Number of values n:	19	Number of classes:	6
	left boundary a:	−0.5	right boundary b:	5.499
	smallest value:	−.61	largest value:	4.63
	to left of class 1:	1 event	to right of class 6:	0 events
	mean:	2.245	std. deviation s_x:	±1.430

Class	Events	% Events	$\sum\%$	lcb	ucb
0	1	5.26	5.26	−inf	−.501
1	2	10.53	15.79	−.500	0.499
2	2	10.53	26.32	0.500	1.499
3	5	26.32	52.63	1.500	2.499
4	6	31.58	84.21	2.500	3.499
5	2	10.53	94.74	3.500	4.499
6	1	5.26	100.0	4.500	5.499
Total	19	100.00			

Class 0 covers the events below the lower limit of the graph, i.e. from $-\infty$ to -0.5. lcb: lower class boundary; ucb: upper class boundary.

The appropriate theoretical probability distribution can be superimposed on the histogram; ideally, scaling should be chosen to yield an area under the curve (integral from $-\infty$ to $+\infty$ equal to $n \cdot w$. (See program HISTO, option $\langle ND \rangle$.)

In Figure 1.27 the preceding data are plotted using 6, 12, resp. 24 classes within the x-boundaries $a = -0.5$, $b = 5.5$. The left panel gives a common-sense subdivision of this x-range. If, for comparison purposes, more classes are needed, the two other panels depict the consequences; many bins contain only one or two events.

1.8.2 χ^2-Test

This test is used to judge both the similarity of two distributions and the fit of a model to data.

The distribution of a data set in the form of a histogram can always be plotted, without reference to theories and hypotheses. Once enough data have accumulated, there is the natural urge to see whether they fit an expected distribution function. To this end, both the experimental frequencies and the theoretical probabilities must be brought to a common scale: a very convenient

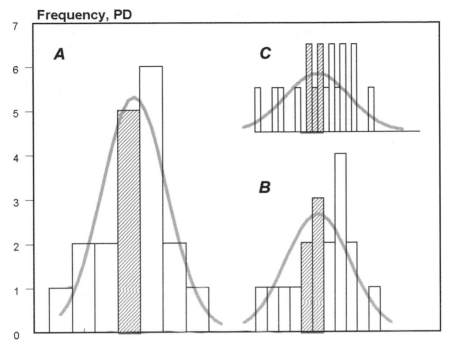

Figure 1.27. Histograms of a data set containing $n = 19$ values, the x-range depicted is in all cases -0.5 to 5.5 inclusive, leaving one event to the left of the lower boundary. The number of classes is 6, 12, resp. 24; in the latter case the observed frequency never exceeds two per class, which is clearly insufficient for a χ^2-test. (See text.) The superimposed normal distribution has the same area as the sum of events, n, times the bin width, namely 19, 8.5, respectively 4.25.

one is that which sets the number of events n equal to certainty (probability $p = 1.00$) = area under the distribution function. The x scale must also be made identical. Next, the probability corresponding to each experimental class must be determined from tables or algorithms. Last, a class-by-class comparison is made, and the sum of all lack-of-fit figures is recorded. In the case of the χ^2-test the weighting model is

$$\chi^2 = \frac{\left[\left(\begin{array}{c}\text{number of}\\\text{observations/class}\end{array}\right) - \left(\begin{array}{c}\text{expected number of}\\\text{observations/class}\end{array}\right)\right]^2}{\left(\begin{array}{c}\text{expected number of}\\\text{observations/class}\end{array}\right)} \qquad (1.50)$$

Example 24: Consider as an example the set of 19 Monte Carlo generated, normally distributed values with a mean = 2.25 and a standard deviation = ±1.43 used in Section 1.8.1: Table 1.24 is constructed in six steps. The experimental (observed) frequencies are compared with the theoretical (expected) number. The critical χ^2-value for $p = 0.05$ and $f = 4$ is 9.49, thus no difference in distribution function is detected. Note that the first and last classes extend to infinity; it might even be advisable to eliminate these poorly defined classes by merging them with the neighboring ones: χ^2 is found as 1.6648 in this case; the critical χ^2 is 5.991. Columns z_l, z_r, ΔCP, expected events, and χ^2 are provided in program HISTO, option ⟨Display⟩:

1. Calculate normalized deviations $z_l = (\text{lcb} - x_{\text{mean}})/s_x$ resp. $z_r = (\text{ucb} - x_{\text{mean}})/s_x$ for each class, e.g., $(1.5 - 2.25)/1.43 = -0.521$. The number 1.5 is from Table 1.23, column lower class boundary (lcb).

2. The cumulative probabilities are either looked up or are calculated according to Section 5.1.1: $z = -0.521$ yields CP = 0.3011.

3. Subtract the CP of the left from that of the right class boundary to obtain the ΔCP for this class (this corresponds to the hatched area in Figure 1.9), e.g., $0.5708 - 0.3011 = 0.2696$.

4. ΔCP is multiplied by the total number of events to obtain the number of events expected in this class, e.g., $0.2696 \cdot 19 = 5.123$.

5. From columns 1 and 7, χ^2 is calculated, $\chi^2 = (\text{Obs} - \text{Exp})^2/\text{Exp}$, e.g., $(5.0 - 5.123)^2 \div 5.123 = 0.0030$.

6. The degrees of freedom f is calculated as $f = m - 3$ (m: number of classes).

Table 1.24. Intermediate and Final Results of a χ^2-Test

Obs. events	z-values		Probability			Expected events	χ^2
	z_l	z_r	CP_l	CP_r	ΔCP		
1	$-\infty$	−1.920	.0000	.0274	.0274	0.521	.4415
2	−1.920	−1.221	.0274	.1111	.0837	1.590	.1056
2	−1.221	−0.521	.1111	.3011	.1900	3.610	.7184
5	−0.521	0.178	.3011	.5708	.2696	5.123	.0030
6	0.178	0.878	.5708	.8100	.2392	4.545	.4658
2	0.878	1.577	.8100	.9426	.1327	2.520	.1074
1	1.577	$+\infty$	.9426	.9999	0.0574	1.091	0.0075
19				.9999		19.000	1.8492

In terms of the weighting model (see Section 3.1) the χ^2-function is intermediate between a least square fit using absolute deviations (measure $= \Sigma(x_i - E(x_i))^2$) and one using relative deviations (measure $= \Sigma((x_i - Ei)/E(x_i))^2$). The effect is twofold: as with the least-squares model, both positive and negative deviations are treated equally, and large deviations contribute the most. Secondly, there is a weighting component with $w_i = 1/E(x_i)$ that prevents a moderate (absolute) error of fit on top of a large expected $E(x_i)$ from skewing the χ^2-measure, and at the same time, deviations from small $E(x_i)$ carry some weight.

The critical χ^2-values are tabulated in most statistical handbooks. An excerpt from the table for various levels of p is given:

Table 1.25. Critical χ^2-Values for p = 0.975, 0.95, 0.9, 0.1, 0.05, and 0.025

p=	0.975	0.95	0.90	0.10	0.05	0.025
f = 1	0.000982	0.00393	0.0158	2.706	3.841	5.024
2	0.0506	0.103	0.211	4.605	5.991	7.378
3	0.216	0.352	0.584	6.251	7.815	9.348
10	3.247	3.940	4.865	15.987	18.307	20.307
20	9.591	10.851	12.443	28.412	31.410	34.170
50	32.357	34.764	37.689	63.167	57.505	71.420
100	74.222	77.929	82.358	118.498	124.342	129.561

This table is used for the two-sided test, that is one simply asks "are the two distributions different?" Approximations to tabulated χ^2-values for different confidence levels can be made using the algorithm and the coefficients given in section 5.1.4.

Because of the convenient mathematical characteristics of the χ^2-value (it is additive), it is also used to monitor the fit of a model to experimental data; in this application the fitted model $Y = ABS(f(x, \ldots))$ replaces the expected probability increment ΔCP (see Eq. 1.7) and the measured value y_i replaces the observed frequency. Comparisons are only carried out between successive iterations of the optimization routine (e.g. a simplex-program), so that critical χ^2-values need not be used. For example, a mixed logarithmic/exponential function $Y - A1 *LOG(A2 + EXP(X - A3))$ is to be fitted to the data tabulated below: do the proposed sets of coefficients improve the fit? The conclusion is that the new coefficients are indeed better. The y-column shows the values actually measured, while the Y-columns give the model estimates for the coefficients $A1, A2$, and $A3$. The χ^2-columns are calculated as $(y - Y)^2 \div Y$. The fact that the sums over these terms, 4.783, 2.616, and 0.307 decrease for successive approximations means that the coefficient set 6.499 ... yields a better approximation than either the initial or the first proposed set. If the χ^2 sum, e.g., 0.307, is smaller than a benchmark figure, the optimization can be stopped; the reason for this is that there is no point in fitting a function to a particular set of measurements and thereby reducing the residuals to values that are much smaller than the repeatability of the measurements. The benchmark is obtained by repeating the measurements and comparing the two series as if one of them was the model Y. For example, for -7.2, -4.4, $+0.3$, $+5.4$, and $+11.1$ and the previous set, a figure of χ^2 of 0.163 is found. It is even better to use the averages of a series of measurements for the model Y, and then compare the individual sets against this.

Table 1.26. Results of a χ^2-Test for Testing the Goodness-of-fit of a Model

Data		Initial		Proposal 1		Proposal 2	
x_i	y_i	$Y1$	χ^2	$Y2$	χ^2	$Y3$	χ^2
1	-7.4	-6.641	0,087	-9.206	0,354	-7.456	0,000
2	-4.6	-2.771	1,207	-6.338	0,477	-4.193	0,040
3	0.5	2.427	1,530	-1.884	1,017	0.545	0,004
4	5.2	8.394	1,215	3.743	0,567	6.254	0,178
5	11.4	14.709	0,744	9.985	0,201	12.436	0.086
Total			4,783		2,616		0,307
Coefficients	A1	6.539		6.672		6,499	
	A2	0.192		0.173		0.197	
	A3	2.771		3.543		3.116	

As noted earlier, the χ^2-test for goodness-of-fit gives a more balanced view of the concept of "fit" than does the pure least-squares model; however, there is no direct comparison between χ^2 and the reproducibility of an analytical method.

Example 25: The first contribution is calculated as $(|7.4|-|6.641|)^2/|6.641|$ = 0.087, with $6.641 = 6.539 \cdot \ln\{0.192 + e^{(1-2.771)}\}$.

A simplex-optimization program that incorporates this scheme is used in the example "nonlinear fitting" (Section 4.2).

1.8.3 Probability Charts

The χ^2-test discussed in the preceding needs a graphical counterpart for a fast, visual check of data. A useful method exists for normally distributed data that could also be adapted to other distributions. The idea is to first order the observations and then to assign each one an index value $1/n$, $2/n$,

... n/n. These index values are then entered into the $z = f(CP)$ algorithm (Section 5.1.1) and the result is plotted versus the observed value. If the experimental distribution is perfectly normal, a straight line will result, see Fig. 1.28. A practical application is shown in Section 4.1. This technique effectively replaces the probability-scaled paper.

Example 26: If 27 measurements are available, the fourth smallest one corresponds to a cumulative probability $CP = 4/27 = 0.148$ and the z-value is -1.045; a symbol would be plotted at the coordinates $(0.148, -1.045)$. The last value is always off scale because the z-value corresponding to $CP = 1.000$ is $+\infty$. (Use program HISTO, option $\langle NPS \rangle$.)

If nonnormal distributions were to be tested for, the $z = f(CP)$ algorithm in Section 5.1.1 would have to be replaced by one that linearizes the cumulative CP for the distribution in question.

1.8.4 Conventional Control Charts (Shewhart Charts)

In a production environment, the quality control department does not ordinarily concern itself with single applications of analytical methods, that

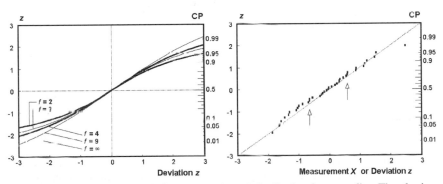

Figure 1.28. Probability-scaled chart for testing of distribution for normality. The abscissa is in the same units as the measurements, the ordinate is nonlinear in terms of cumulative probability. The left figure shows a straight line and four curves corresponding to t-distributions with $f = 2$ and 4 (enhanced), and $f = 3$ and 9 degrees of freedom. The straight line corresponding to a normal distribution evolved from the CP-function familiar from Section 1.2.1 through nonlinear stretching in the vertical direction. The median and the standard deviation can be graphically estimated (regression line, interpolation at 50, respectively 17 and 83%). The right figure depicts 40 assay values for the active principle content in a cream. (Three values are off-scale, namely, 2.5, 97.5, and 100%.) The data by and large conforms to a normal distribution. Note the nearly vertical runs of two to four points (arrow!) at the same x-value; this can be attributed to a quantization raster that is impressed on the measurement because only certain absorbance values are possible due to the three-digit display on the employed instrument.

being a typical issue for the Analytical R&D Department, but concentrates on redoing the same tests over and over in order to release products for sale or reject them. A fine example is the output from a weighing station that determines whether a product has been properly packaged, see Fig. 1.29. Obvious deviations (points A ... L) are used to eject and inspect the packages in question. By filtering the data, the reject/retain decision can be fine-tuned, see Fig. 1.30 and Section 3.6.

When data of a single type accumulate, new forms of statistical analysis become possible. In the following, conventional control and Cusum charts will be presented. In the authors' opinion, newer developments in the form of tight (multiple) specifications and the proliferation of PCs have increased the value of control charts; especially in the case of on-line in-process controlling, monitors depicting several stacked charts allow floor supervi-

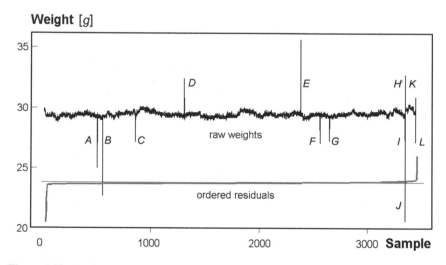

Figure 1.29. Example for a control chart. 3442 boxes passed a control-balance at the end of a packaging line in order to detect missing or extra components. A box should contain a certain number of pouches filled with the medication, and a brochure that explains the proper way of taking it. A brochure is heavier than a pouch. Spike E was due to a second brochure. The other spikes relate to $1 - 4$ missing or extra pouches. If the components had very tightly defined weights, hard reject limits could be set to catch all such occurrences. Unfortunately, such pouch weights vary a bit depending on where on the foil stock they come from and how they were cut. This leads to an irregularly shifting average weight that is compensated for by having the balance's software calculate a box-car average over $n = 10$ boxes (see Section 3.6) that is used as reference level for the next 10 boxes. When these reference values are subtracted, the typical residual becomes much smaller than if a global average had been used, see Fig. 1.30, and the rejection rate due to noise is dramatically cut. Sorting the residuals yields the line at the bottom of the figure; the few large residuals are now concentrated at the extreme left (missing components), respectively at the right end (extra components).

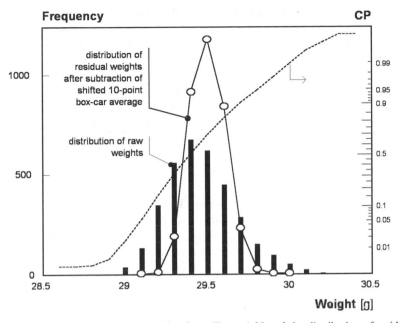

Figure 1.30. A histogram of raw weights from Figure 1.29 and the distribution of residuals that resulted after subtraction of a shifted box-car average are superimposed. The CP-curve, plotted with the ⟨NPS⟩ option in HISTO, is for the raw weights; the corresponding curve for the residuals would be about twice as steep. The asymmetry of the raw-weight distribution is evident both in the histogram and the lack of linearity of the CP-curve; it is due to many subpopulations of product being lumped into one batch. Every time a mechanic makes an adjustment on a knife, a new subpopulation is created. The residuals appear to be normally distributed, however.

sors to recognize the trends before they become dangerous. V-templates (a graphical device to be superimposed on a Cusum chart that consists of a v-shaped confidence limit[32]) can be incorporated for warning purposes, but are not deemed a necessity because the supervisor's experience with the process gives him the advantage of being able to apply several (even poorly defined) rules of thumb simultaneously, while the computer must have sharp alarm limits set. The supervisor can often associate a pattern of deviations with the particular readjustments that are necessary to bring the process back in line. A computer would have to run a fuzzy-logic expert system to achieve the same effect, but these are expensive to install and need retraining every time a new phenomenon is observed and assigned to a cause. Such refinements are rather common during the initial phases of a product's lifecycle; in today's global markets, though, a process barely has a chance of reaching maturity before it is scaled-up or replaced, and so the expert system would most of the time be incompletely validated (QA's point of view) or unavailable because of retraining (according to the production department).

Although a control or a Cusum chart at first glance resembles an ordinary two-dimensional graph, one important aspect is different: the abscissa values are an ordered set graphed according to the rank (index), and not according to the true abscissa values (in this case time) on a continuous scale. Because of this, curve fitting to obtain a "trend" is not permissible. There is one exception, namely when the samples are taken from a continuous process at strictly constant intervals. A good example is a refinery running around the clock for weeks or months at a time with automatic sampling of a process stream at, say, 10-minute intervals. Consider another process run on an 8-or 16-hour shift schedule with a shut-down every other weekend: curve fitting would here only be permissible on data taken on the same day (e.g., at 1-hour intervals), or on data points taken once a day between shut-downs (e.g., always at 10 A.M.). Sampling scheduling is often done differently, though, for good reasons, *viz.* (a) every time a new drum of a critical reagent is fed into the process; (b) every time a receiving or transport container is filled; (c) every time a batch is finished, etc.

The statistical techniques applicable to control charts are thus restricted to those of Section 1.5, that is detecting deviations from the long-term mean respectively crossing of the specified limits.

The conventional control chart is a graph having a "time" axis (abscissa) consisting of a simple raster, such as that provided by graph or ruled stationary paper, and a measurement axis (ordinate) scaled to provide six to eight standard deviations centered on the process mean. Overall standard deviations are used that include the variability of the process and the analytical uncertainty. (See Fig. 1.8.) Two limits are incorporated: the outer set of limits corresponds to the process specifications and the inner one to "warning" or "action" levels for in-house use. Control charts are plotted for two types of data:

1. A standard sample is incorporated into every series, e.g. daily, to detect changes in the analytical method and to keep it under statistical control.

2. The data for the actual unknown samples are plotted to detect changes in the product and/or violation of specification limits.

Keeping track of the standards (Case 1) helps avoid a nasty situation that generally turns up when one is close to the submission deadline or has an inspector asking questions: the actual event happened months ago, the responsible people no longer remember what really happened (who has a lawyer's memory for minutae?), and a specification-conforming point is prominently out-of-trend. Poor product? No, just a piece of evidence pointing to lack of control, for instance sloppy calibration. (See Figures 4.39, 4.9, 4.49, and 4.10.)

The specification limits (Case 2) can either be imposed by regulatory agencies, agreed-upon standards, or market pressure, or can be set at will. In the latter case, given the monetary value associated with a certain risk of rejection, the limits can be calculated by statistical means. More likely, limits are imposed: in pharmaceutical technology ± 2, ± 5, or $\pm 10\%$ about the nominal assay value are commonly encountered; some uses to which chemical intermediates are put dictate asymmetrical limits, reflecting perhaps solubility, purity, or kinetic aspects; a reputation for "high quality" demands that tight limits be offered to customers, which introduces a nontechnical side to the discussion. The classical $\pm 2\sigma$ and $\pm 3\sigma$ SL are dangerous because for purely mathematical reasons, a normalized deviate $z = r_i/s_x \geq 3$ can only be obtained for $n \geq 11$,[53] cf. Fig. 1.24.

Action limits must be derived from both the specification limits and the characteristics of the process: they must provide the operators with ample leeway (time, concentration, temperature, etc.) to react and bring the process back to nominal conditions without danger of the specification limits being exceeded. An important factor, especially in continuous production, is the time constant: a slow-to-react process demands narrow action limits relative to the specification limits, while many a tightly feed-back controlled process can be run with action limits close to the specification limits. If the product conforms to the specifications, but not to the action limits, the technical staff is alerted to look into the potential problem and come up with improvements before a rejection occurs. Generally, action limits will be moved closer to the specification limits as experience accrues.

For an example of a control chart see Fig. 1.31 and Sections 4.1 and 4.8. Control charts have a grave weakness: the number of available data points must be relatively high in order to be able to claim "statistical control". As is often the case in this age of increasingly shorter product life cycles, decisions will have to be made on the basis of a few batch release measurements; the link between them and the more numerous in-process controls is not necessarily straight-forward, especially if IPC uses simple tests (e.g. absorption, conductivity) and release tests are complex (e.g. HPLC, crystal size).

1.8.5 Cusum Charts

A disadvantage of the conventional control charts is that a small or gradual shift in the observed process parameter is only confirmed long after it has occurred, because the shift is swamped in statistical (analytical) noise. A simple way out is the Cusum chart (cumulated sum of residuals, see program CUSUM.exe), because changes in a parameter's average quickly show up, see Fig. 1.32. The

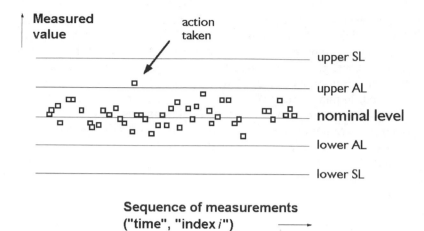

Figure 1.31. Typical control chart showing the specification and action limits. The four limits can be arranged asymmetrically relative to the nominal level, depending on the process characteristics and experience.

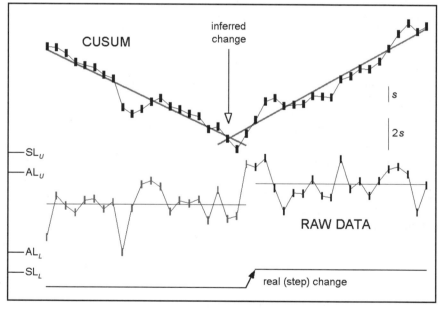

Figure 1.32. Typical Cusum-chart showing a change in process mean. The base-line average *a* is the average over the 41 displayed points. The inferred step (intersection of the two linear regression lines) appears to precede the actual change from level *A* to level *B* because the last point in segment *A* was by chance very high and actually exceeded the AL. The corrective action took hold three points later (gray ellipse), but was not strong enough.

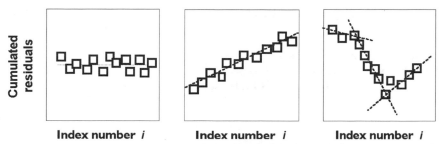

Figure 1.33. Cusum charts (schematic). Type 1, a horizontal line, indicates that the system comprising the observed process and the analytical method is stable and that results stochastically vary about the average. Type 2, a sloping straight line, indicates a nonoptimal choice of the average a, e.g., by using too few decimal places. Type 3, a series of straight-line segments of changing slope, shows when a change in any of the system variables resulted in a new average a' different from the previous average a. Due to the summation process the system noise is damped out and changes in the trend rapidly become obvious.

basic idea is to integrate (sum) the individual values x_i. In practice, many consecutive results are electronically stored; a reference period is chosen, such as the previous month, and the corresponding monthly average a is calculated. This average is then subtracted from all values in the combined reference and observation period; thus $r_i = x_i - a$. On the Cusum chart, the sum $\Sigma(r_i)$, $i = 1 \ldots j$, is plotted at time i. Figure 1.33 examines different types of graphs that can be distinguished[56].

Scales should optimally be chosen so that the same distance (in mm) that separates two points horizontally corresponds to about $2 \cdot s_y/\sqrt{m}$ vertically, that is twice the standard deviation of the mean found for m repeat measurements.[32] A V-formed mask can be used as a local confidence limit.[32] This approach, however, is of statistical nature; the combination of intuition, experience, and insider's knowledge is much more revealing. For examples, see Ref. 57 and Section 4.8.

1.9   ERRORS OF THE FIRST AND SECOND KIND

The two error types mentioned in the title are also designated with the Roman numerals I and II; the associated error probabilities are termed alpha (α) and beta (β).

When one attempts to estimate some parameter, the possibility of error is implicitly assumed. What sort of errors are possible? Why is it necessary to distinguish between two types of error? Reality (as hindsight would show later on, but unknown at the time) could be "red" or "blue," and by the same token, any assumptions or decisions reached at the time were either "red" or "blue".[19] This gives four outcomes that can be depicted as indicated in Table 1.27:

Table 1.27. Null and Alternative Hypotheses. Calculations Involving the Error Probability β are Demonstrated in Section 4.1. The Expression $'X = Y'$ is to be Read as $'X$ is Indistinguishable from Y'

		Real situation	
		$X = Y$	$X \neq Y$
Decision	"null" hypothesis		"false negative"
taken,	H_0	correctly retain H_0	falsely retain H_0
		correctly reject H_1	falsely reject H_1
answer given,	$'X = Y'$	probability $1 - \alpha$	probability β
or	alternative hypothesis	"false positive"	
	H_1	falsely reject H_0	correctly reject H_0
situation		falsely accept H_1	correctly accept H_1
assumed	$'X \neq Y'$		
			"power" of a test:
		probability α	probability $1 - \beta$

The different statistical tests discussed in this book are all defined by the left column, that is, the initial situation H_0 is known and circumscribed, whereas H_1 is not (accordingly one should use the error probability α).

In this light, the type II error is a hypothetical entity, but very useful. A graphical presentation of the situation will help (see Figure 1.34).

Assume the ensemble of all results A (e.g., reference Method A) to be fixed and the ensemble B (e.g., test Method B) to be movable in the horizontal direction; this corresponds to the assumption of a variable bias $\Delta x = x_{mean, B} - x_{mean, A}$, to be determined, and usually, to be proven noncritical. The critical x-value, x_c, is normally defined by choosing an error probability α in connection with H_0 (that is, B is assumed to be equal to A). In the case presented here, a one-sided test is constructed under the hypothesis "H_1: B larger than A". The shaded area in Figure 1.34(left) gives the probability of erroneously rejecting H_0 if a result $x_{mean} > x_c$ is found ("false positive"). In Figure 1.34(right) the area corresponding to an erroneous retention of H_0, β ("false negative" under assumption "B different from A") is hatched. Obviously, for different exact hypotheses H_1, i.e., clearly defined position of B, β varies while α stays the same because the limit is fixed. Thus, the power of the test, 1-β, to discriminate between two given exact hypotheses H_0 and H_1 can for all intents and purposes only be influenced by the number of samples, n, that is by narrowing the distribution function for the observed mean.

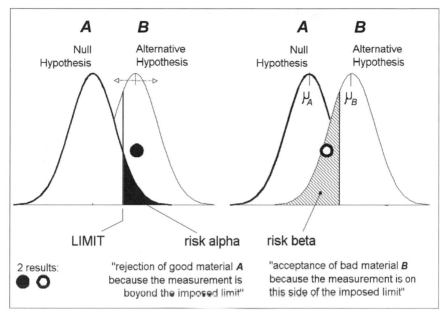

Figure 1.34. Alternative hypothesis and the power of a t-test. Alpha (α) is the probability of rejecting an event that belongs to the population associated with H_0, it is normally in the range 0.05 ... 0.01. Beta (β) is the probability that an event that is effectively to be associated with H_1 is accepted as belonging to the population associated with H_0. Note that the power of the test to discriminate between hypotheses increases with the distance between μ_A and μ_B. μ_A is fixed either by theory or by previous measurements, while μ_B can be adjusted (shifted along the x-axis), for examples see $H_1 - H_4$, Section 4.1. Compare with program HYPOTHESIS.

(See Fig. 1.16.) If there is any reason to believe that an outcome of a decision would either carry a great risk (e.g., in the sense of safety), or would have immense impact (medication X better than Y), H_0 and H_1 will have to be appropriately assigned (H_0: $X = Y$ and H_1: $X \neq Y$) or (H_0: $X \neq Y$ and H_1: $X = Y$), and α will have to be set to reflect the concern associated with not obtaining decision H_0 when H_0 is deemed to be the right decision. For example, if a chemical's properties are to be used for promotional purposes, there had better be hard evidence for its superiority over the competitor's product, otherwise the company's standing as a serious partner will soon be tarnished. Thus, instead of postulating "H_0: our product is better" and "proving" this with a few measurements, one should assign "H_0: our product is the same". The reason is that in the former case one would, by choosing a small α, make it hard to reject H_0 and herewith throw a pet notion overboard, and at the same time, because of an unknown and possibly large β, provoke the retention of H_0 despite the lack of substantiation. In the case "H_0: ours is

Table 1.28. Interpretation of a Null/Alternative Hypothesis Situation

		Real situation	
		Our product is the same	Our product is better
Decision taken,	"null" hypothsis H_0: our product is the same	have R&D come up with new ideas	"false negative" loss of a good marketing argument; hopefully the customer will appreciate the difference in quality Risk is hard to estimate
answer given, or	Alternative hypothesis H_1:	"false positive" the Marketing Department launches an expensive advertising campaign; the the company is perceived to be insincere This risk can be defined	capture the market, you have a good argument
situation assumed	our product is better		

the same," the tables are turned: "H_1: superiority" must be proven by hard facts, and if found, is essentially "true" (α small), while the worst that could happen due to a large β would be the retention of an unearned "the same," that is, the loss of a convenient marketing argument.

CHAPTER

2

BI- AND MULTIVARIATE DATA

Data sets become two dimensional in any of the following situations (in order of decreasingly rigorous theoretical underpinnings) and are analyzed by the indicated methods.

Regression
- One-dimensional data are plotted *versus* an experimental variable; a prime example is the Lambert-Beer plot of absorbance vs. concentration, as in a calibration run. The graph is expected to be a straight line over an appreciable range of the experimental variable. This is the classical domain of linear regression analysis.
- The same sample is repeatedly tested during a period of time, as in stability analysis; here, we hope, only a slight change of signal is observed. If there is a significant change, the form of the function (straight line, exponential curve, etc) gives information about the underlying mechanism.[58]

Correlation
- More than one dimension, i.e., parameter, of the experimental system is measured, say absorbance and pH of an indicator solution; the correlation behavior is tested to find connections between parameters; if a strong one is found, one measurement could in the future serve as a surrogate for the other, less accessible one.

Youden plot, ANOVA
- The same parameter is tested on at least two samples by each of several laboratories using the same method (round-robin test[9,13,26]), or
- At least two parameters are tested by the same laboratory on many nominally similar samples. In both cases, the simplest outcome is a round patch in the Youden plot,[32] see Fig. 2.1, of points that signifies "just noise, no correlation ... no participating laboratory (or sample or point in time) is exceptional." On the other hand, an elliptical patch, especially if the slope deviates from what could be expected, shows that some effects are at work that need further investigation. After "just noise," the

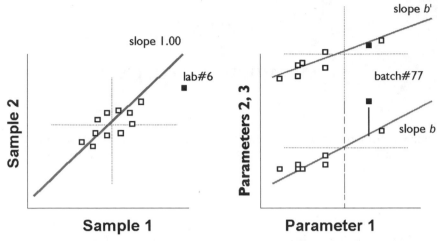

Figure 2.1. Youden's plot. Dotted horizontals and verticals: expected averages; full line: diagonal with slope 1.00 if the same parameter is tested on two samples, or slope b if two, three, or more parameters are tested on the same sample. A dispersion of the points along the diagonal in the left panel indicates a factor common to both samples. As an example, laboratory no. 6 could have trouble calibrating its methods or does not employ an internal standard. (See Section 4.14.) In a modified form of Youden's plot, right panel, two or more parameters are tested on each of a large number of similar samples, e.g., a mix of three colorants, where batch #77 contains too much of one colorant relative to the two others, or an analytical bias crept in. (See also Figure 4.49.) Dispersion along the lines signals sampling, weighing, or homogeneity problems.

next most complicated assumption is that some of the laboratories/some of the samples are afflicted by a factor that has a simple additive effect on the result. If true, this would be confirmed by ANOVA.

2.1 CORRELATION

Two statistical measures found in most software packages are the *correlation coefficient*, r, and the *coefficient of determination*, r^2. The range of r is bounded by $-1 \leq r \leq +1$; $|r| = 1$ is interpreted as "perfect correlation," and $r = 0$ as "no correlation whatsoever."

$$r = \frac{S_{xy}}{\sqrt{S_{xx} \cdot S_{yy}}} \tag{2.1}$$

For the definition of S_{xx}, S_{xy}, and S_{yy}, see Eqs. 2.4 through 2.6. Under what circumstances is r to be used? When hidden relations between different parameters are sought, as in explorative data analysis, r is a useful measure. Any pairing of parameters that results in a statistically significant r is best reviewed in graphical form, so that a decision about further data treatment can be reached. Even here caution must be exercised, because r is a summary measure that can suggest effects when none are present, and *vice versa*, i.e., because of an outlier or inhomogeneous data. Program CORREL provides the possibility of finding and inspecting the correlations in a large data set.

Example 27: Parameters A and B (see data file JUNGLE2.dat) appear to be strongly correlated ($r^2 = 0.877$, $p < 0.0002$, $n = 48$) whereas parameters HPLC and TITR achieve $r^2 = 0.646$, $p = 0.0002$, $n = 48$ (JUNGLE2.dat) and $r^2 = 0.762$, $p = 0.0181$, $n = 5$ (JUNGLE1.dat). For most other applications, and calibration curves in particular, the correlation coefficient must be viewed as a relic of the past[59]: many important statistical theories were developed in sociological or psychological settings, along with appropriate measures such as r^2. There the correlation coefficient indicates the degree of parallelism between one parameter and another, say, reading skills and mathematical aptitude in IQ tests. The question of cause and effect, so important in the physical sciences, is not touched upon: one parameter might be the (partial) cause and the other the effect, or, more likely, both are effects of several interacting and unquantifiable factors, say intelligence, upbringing, and heredity.

The situation in analytical chemistry is wholly different: cause and effect relationships are well characterized; no one is obliged to rediscover the well-documented hypothesis that runs under the name Lambert–Beer "law." What is in demand, though, is a proof of adherence. With today's high-precision instrumentation, coefficients of determination larger than 0.999 are commonplace. The absurdity is that the numerical precision of the algorithms used in some soft-/firmware packages limits the number of reliable digits in r^2 to two or three, but show four or more digits on the read-out; however, the authors are not aware of anyone having qualms about which type of calculator to use. Furthermore, it is hard to develop a "feeling" for numbers when all the difference resides in the fourth or fifth digit see Fig. 2.2. As an alternative goodness-of-fit measure, the residual standard deviation is proposed because it has a dimension the chemist is used to (the same as the ordinate) and can be directly compared to instrument performance [cf. Eq. (2.13)].

Example 28: Using file VALID3.dat ($r^2 = 0.99991038 \ldots$) and a suitably modified program LINREG, depending on whether the means are subtracted as in Eqs. (2.4)–(2.6), or not, as in Eqs. (2.7)–(2.9), whether single- or double-precision is used, and the sequence of the mathematical operations,

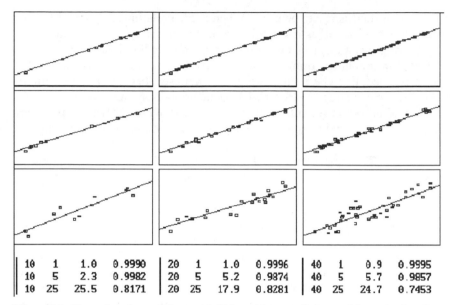

10	1	1.0	0.9990	20	1	1.0	0.9996	40	1	0.9	0.9995
10	5	2.3	0.9982	20	5	5.2	0.9874	40	5	5.7	0.9857
10	25	25.5	0.8171	20	25	17.9	0.8281	40	25	24.7	0.7453

Figure 2.2. Examples of correlations with high and low coefficients of determination. Data were simulated for combinations of various levels of noise ($\sigma = 1, 5, 25$, top to bottom) and sample size ($n = 10, 20, 40$, left to right). The residual standard deviation follows the noise level (for example, 0.9, 5.7, 24.7, from top to bottom). Note that the coefficient 0.9990 in the top left panel is on the low side for many analytical calibrations where the points so exactly fit the theoretical line that $r^2 > 0.999$ even for low n and small calibration ranges.

the last four digits of r^2 can assume the values 1038, 1041, 0792, 0802, and 0797. Plainly, there is no point in quoting r^2 to more than three, or at most four, decimal places, unless one is absolutely sure that the algorithm is not the limiting factor; even then, r^2 at levels above 0.999 does not so much indicate differences in the quality of the experiment as the presence of chance events outside the control of the investigator.

2.2   LINEAR REGRESSION

Whenever one property is measured as a function of another, the question arises of which model should be chosen to relate the two. By far the most common model function is the linear one; that is, the dependent variable y is defined as a linear combination containing two adjustable coefficients and x, the independent variable, namely,

$$Y = a + b \cdot x \tag{2.2}$$

A good example is the absorption of a dyestuff at a given wavelength λ (lambda) for different concentrations, as expressed by the well-known Lambert–Beer's law:

Absorbance = A_{blank} + pathlength × absorptivity × concentration

with the identifications Y = Absorbance, $a = A_{blank}$, x = concentration, and b = pathlength × absorptivity.

If the measurements do not support the assumption of a linear relationship,[18] one often tries transformations to "linearize" it. One does not have to go far for good reasons:

- Because only two parameters need to be estimated, the equation of the straight line is far easier to calculate than that of most curves.
- The function is transparent and robust, and lends itself to manipulations like inversion ($X = f(y)$).
- Relatively few measurements suffice to establish a regression line.
- A simple ruler is all one needs for making or checking graphs. A linear relationship inherently appeals to the mind and is simple to explain.
- Before the advent of the digital computer high-order and nonlinear functions were impractical at best, and without a graphics plotter much time is needed to draw a curve. Interpolation, particularly in the form $X = f(y)$, is neither transparent nor straightforward if confidence limits are requested.

Thus, the linear model is undoubtedly the most important one in the treatment of two-dimensional data and will therefore be discussed in detail.

Overdetermination of the system of equations is at the heart of regression analysis, that is one determines more than the absolute minimum of two coordinate pairs (x_1/y_1) and (x_2/y_2) necessary to calculate a and b by classical algebra. The unknown coefficients are then estimated by invoking a further model. Just as with the univariate data treated in Chapter 1, the least-squares model is chosen, which yields an unbiased "best-fit" line subject to the restriction:

$$\Sigma(r_i)^2 = \text{minimum} \tag{2.3}$$

Here r_i is the residual associated with the ith measurement. The question is

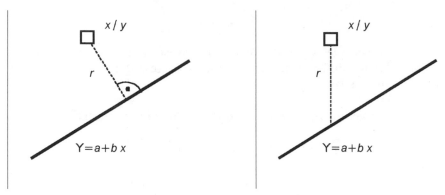

Figure 2.3. The definition of the residual. The sum of the squared residuals is to be minimized.

now how this residual is to be geometrically defined. The seemingly logical thing to do is to drop a perpendicular from the coordinate (x_i/y_i) onto the regression line, as shown by Fig. 2.3 (left):

While it is perfectly permissible to estimate a and b on this basis, the calculation can only be done in an iterative fashion, that is, both a and b are varied in increasingly smaller steps (see Optimization Techniques, Section 3.5) and each time the squared residuals are calculated and summed. The combination of a and b that yields the smallest of such sums represents the solution. Despite digital computers, Adcock's solution, a special case of the maximum likelihood method,[60] is not widely used; the additional computational effort and the more complicated software are not justified by the "improved" (a debatable notion) results, and the process is not at all transparent, i.e., not amenable to manual verification.

2.2.1   The Standard Approach

The current standard approach[34,46,61] is shown in Figure 2.3 (right): The vertical residuals are minimized according to $r_i = y_i - Y = y_i - (a + b \cdot x_i)$. A closed (noniterative) solution is obtained that is easily verifiable by manual calculations. There are three assumptions that must be kept in mind:

1. The uncertainty inherent in the individual measurements of the property Y must be much larger than that for property X, or, in other words, the repeatability s_y relative to the range of the y-values must be much larger than the repeatability s_x relative to the range of the x-values. Thus, if the graph $(x_{min}, x_{max}, y_{min}, y_{max})$ is scaled so as to be approximately square (the regression line is nearly identical with a 45 degree

diagonal), the confidence intervals are related as $CI(y) \gg CI(x)$ if measured in millimeters. Maximum likelihood is the technique to use if this assumption is grossly violated.[60]

2. Homoscedacity must prevail, that is the reproducibilities of y measured for small, medium, and large x-values must be approximately the same, so the uniform weighting scheme can be implemented.

3. The distribution of errors must be Gaussian; regression under conditions of *Poisson*-distributed noise is dealt with in Ref. 62.

Restrictions (1) and (2) are of no practical relevance, at least as far as the slope and the intercept are concerned, when all data points closely fit a straight line; the other statistical indicators are influenced, however.

In the following the standard unweighted linear regression model is introduced. All necessary equations are found in Table 2.1 and are used in program LINREG. In a later section (2.2.10) nonuniform weighting will be dealt with.

The equations were derived by combining Eqs. (2.2) and (2.3), forcing the regression line through the center of gravity (x_{mean}/y_{mean}), and setting the partial derivative $\delta(\Sigma r^2)/\delta b = 0$ to find the minimum.

2.2.2 Slope and Intercept

Estimating the slope b is formally equivalent to the calculation of a mean from a distribution (Section 1.1.1) in that the slope also represents a statistical mean (best estimate). The slope is calculated as the quotient of the sums S_{xy} and S_{xx}. Since the regression line must pass through the coordinate (x_{mean}/y_{mean}), the intercept is given by extrapolating from this point to $x = 0$. The question of whether to force the regression line through the origin ($a \equiv 0$) has been discussed at length.[63–66] In most analytical situations $a \equiv 0$ could be justified by theory. However, reality is rarely as simple as theory, e.g., the lack of selectivity or unexpected interactions between chemical species. Assuming reality is simple, then the trade-off between a lower number of calibration samples and increased variance V_x of the interpolated result has to be discussed on a case-by-case basis.

The confidence interval $CI(b)$ serves the same purpose as $CI(x_{mean})$ in Section 1.3.2; the quality of these average values is described in a manner that is graphic and allows meaningful comparisons to be made. An example from photometry, see Table 2.2, is used to illustrate the calculations (see also data file UV.dat); further calculations, comments, and interpretations are found in the appropriate Sections. Results in Table 2.3 are tabulated with more significant digits than is warranted, but this allows the reader to check

Table 2.1. Linear Regression: Equations

Linear Regression: Equations

Correct formulation (sums Σ are taken over all measurements, $i = 1 \ldots n$):

$$x_{mean} = (\textstyle\sum x_i)/n \qquad y_{mean} = (\textstyle\sum y_i)/n \qquad \text{means} \qquad (1.1)$$

$$S_{xx} = \textstyle\sum (x_i - x_{mean})^2 \qquad\qquad\qquad\qquad\qquad (1.3)$$
$$(2.4)$$

$$S_{yy} = \textstyle\sum (y_i - y_{mean})^2 \qquad\qquad \text{Sums of Squares} \qquad (2.5)$$

$$S_{xy} = \textstyle\sum (x_i - x_{mean}) \cdot (y_i - y_{mean}) \qquad\qquad\qquad (2.6)$$

Algebraically equivalent formulation as used in pocket calculators (beware of numerical artifacts when using Eqs. (2.7–2.9); cf. Table 1.1):

$$S_{xx} = \textstyle\sum (x_i^2) - (\textstyle\sum x_i)^2/n \qquad\qquad\qquad\qquad (2.7)$$

$$S_{yy} = \textstyle\sum (y_i^2) - (\textstyle\sum y_i)^2/n \qquad\qquad \text{Sums of Squares} \qquad (2.8)$$

$$S_{xy} = \textstyle\sum (x_i \cdot y_i) - (\textstyle\sum x_i) \cdot (\textstyle\sum y_i)/n \qquad\qquad\qquad (2.9)$$

$$Y = a + b \cdot x \qquad\qquad\qquad \text{regression model} \qquad (2.10)$$

$$b = S_{xy}/S_{xx} \qquad a = y_{mean} - b \cdot x_{mean} \qquad \text{estimates } a, b \qquad (2.11)$$
$$(2.12)$$

$$V_{res} = s_{res}^2 = \frac{\textstyle\sum (y_i - Y)^2}{n-2} = \frac{S_{yy} - b \cdot S_{xy}}{n-2} \qquad \text{residual variance} \qquad (2.13)$$

$$V_b = V_{res}/S_{xx} \qquad\qquad \text{variance of slope } b \qquad (2.14)$$

$$V_a = V_{res} \cdot \left(\frac{1}{n} + \frac{x_{mean}^2}{S_{xx}} \right) \qquad \text{variance of intercept } a \quad (2.15)$$

$$V_Y = V_{res} \cdot \left(\frac{1}{n} + \frac{(x - x_{mean})^2}{S_{xx}} \right) \qquad \text{variance of estimate } Y \quad (2.16)$$

$$CL(Y) = a + b \cdot x \pm t(f, p) \cdot \sqrt{V_Y} \qquad \text{CL of estimate } Y \qquad (2.17)$$

$$V_X = \frac{V_{res}}{b^2} \cdot \left(\frac{1}{n} + \frac{1}{k} + \frac{(y^* - y_{mean})^2}{b^2 \cdot S_{xx}} \right) \qquad \text{variance of estimate } X \quad (2.18)$$

$$CL(X) = (y^* - a)/b \pm t(f, p) \cdot \sqrt{V_x} \qquad \text{CL of estimate } X \qquad (2.19)$$

---

n	number of calibration points
x_i	known concentration (or other independent variable)
y_i	measured signal at x_i
x_{mean}	mean of all x_i
y_{mean}	mean of all y_i
S_{xx}	sum of squares over Δx
S_{yy}	sum of squares over Δy

Table 2.1. (Continued)

S_{xy}	sum of cross-product $\Delta x \cdot \Delta y$
a	intercept
b	slope
r_i	i-th residual $r_i = y_i - (a + b \cdot x_i)$
s_{res}	residual standard deviation
V_{res}	residual variance
$t(f, p)$	Student's t-factor for $f = n - 2$ degrees of freedom
k	number of replicates y_k^* on unknown
y^*	mean of several y_k^*, $k = 1 \ldots m$, $y^* = (\sum y_k^*)/m$
Y	expected signal value (at given x)
X	estimated concentration for mean signal y^*

recalculations and programs. Figure 2.4 gives the corresponding graphical output.

If the calibration is repeated and a number of linear regression slopes b are available, these can be compared as are means. (See Section 1.5.1, but also Section 2.2.4.)

2.2.3 Residual Variance

The residual variance V_{res} summarizes the vertical residuals from Figure 2.4; it is composed of

$$V_{res} = V_{reprod} + V_{nonlin} + V_{misc} \qquad (2.20)$$

where V_{reprod} is that variance due to repetitive sampling and measuring of an average sample (compare Eq. 1.6); V_{nonlin} stands for the apparent increase if the linear model is applied to an inherently curved set of data (cf. Table

Table 2.2. Data for Linear Regression

Concentration x	Signal y	
50%	0.220 AU	x: concentration in % of
75	0.325	the nominal concentration
100	0.428	
125	0.537	y: measured absorbance
150	0.632	AU: absorbance units

Degrees of freedom: $n = 5$, thus $f = n - 2 = 3$
Critical Student's t: $t(f = 3, p = 0.05) = 3.1824$

Table 2.3. Intermediate and Final Results

Intermediate Results		
Item	Value	Equation
x_{mean}	100	(1.1)
y_{mean}	0.4284	(1.1)
S_{xx}	6250	(2.4)
S_{xy}	25.90	(2.5)
S_{yy}	0.10737	(2.6)

Final Results					
Item	Value	Equation	Item	Value	Equation
b	0.004144	(2.11)	r^2	0.99963	(2.1)
a	0.01400	(2.12)	$t(3, 0.05)$	3.18	
V_{res}	0.00001320	(2.13)	s_{res}	0.00363	(2.13)
s_b	0.0000459	(2.14)	$t \cdot s_b$	0.000146	
s_a	0.00487	(2.15)	$t \cdot s_a$	0.0155	

$t \cdot s_b$ amounts to 3.5% of b, CL(b): 0.0040, and 0.0043, CI(b): 0.0003
$t \cdot s_a$ amounts to 111% of a, CL(a): −0.0015, and 0.029, CI(a): 0.03

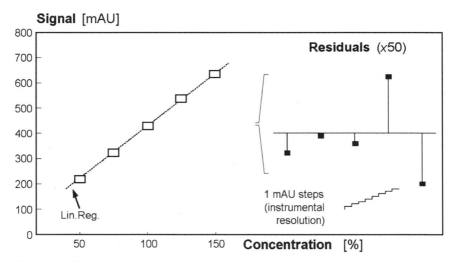

Figure 2.4. Graph of the linear regression line and data points (left), and the residuals (right). The fifty-fold magnification of the right panel is indicated; the digital resolution ±1 mAU of a typical UV-spectrophotometer is illustrated by the steps.

Residuals

REPRODUCIBILITY   NONLINEARITY   HETEROSCEDACITY

Figure 2.5. Three important components of residual variance. The residuals are graphed *versus* the independent variable x.

4.18); V_{misc} contains all other variance components, e.g., that stemming from the x-dependent s_y (heteroscedacity). (See Fig. 2.5.)

Under controlled experimental conditions the first term will dominate. The second term can be assessed by using a plot of the residuals (see the next section and Figure 4.21); for a correctly validated GMP-compatible method, curvature must be avoided and checked for during the R&D phase (justification for the linear model!) by testing a concentration range that goes far beyond the 80 to 120% of nominal that is often used under routine QC conditions. Thus the residual variance should not be much larger than the first term. Depending on the circumstances, a factor of 2–4 could already be an indication of noncontrol. Under GMP conditions heteroscedacity must be avoided or the use of the unweighted model needs to be justified.

Taking the square root of V_{res}, one obtains the residual standard deviation, s_{res}, a most useful measure:

- s_{res} has the same dimension as the reproducibility and the repeatability, namely the dimension of the measurement; those are the units the analyst is most familiar with, such as absorbance units, milligrams, etc.

- s_{res} is nearly independent of the number of calibration points and their concentration values x, cf. Figure 2.8.

- s_{res} is easy to calculate and, since the relevant information resides in the first significant digits, its calculation places no particular demands on the soft- or hardware (cf. Section 3.3) if the definition of r_i in Table 2.1 and Eqs. (1.3a)–(1.3d) is used.

- s_{res} is necessary to obtain other error-propagation information and can be used in optimization schemes (cf. Section 3.3).

Example 29 (see Table 2.1, Section 2.2.1): a residual standard deviation of less than 0.004 relative to $y = 0.428$ indicates that the experiment is relatively well controlled: on the typical UV/VIS spectrometer in use today, three decimal places are displayed, the least significant giving milliabsorbance units; noise is usually $\pm 1 - 2$ mAU. If the residual standard deviation is felt to be too large, one ought to look at the individual residuals to find the contributions and any trend: the residuals are -0.0012, 0.0002, -0.0004, 0.005, and -0.0036. No trend is observable. The relative contributions are obtained by expressing the square of each residual as a percentage of $V_{\text{res}} \cdot (n - 2)$, i.e. $100 \cdot (0.0012)^2/0.0000132/3$, etc., which yields 3.6, 0.1, 0.4, 63, resp. 33%. Since the last two residuals make up over 96% of the total variance, bringing these two down to about 0.002 by more careful experimentation would result in a residual standard deviation of about 0.003, an improvement of 25%.

2.2.4   Testing Linearity and Slope

The test for the significance of a slope b is formally the same as a t-test (Section 1.5.2): if the confidence interval CI(b) includes zero, b cannot significantly differ from zero, thus $b = 0$. If a horizontal line can be fitted between the plotted CL, the same interpretation applies, cf. Figures 2.6a–c. Note that s_b corresponds to $s(\mathbf{x}_{\text{mean}})$, that is, the standard deviation of a mean. In the above example the confidence interval clearly does not include zero; this remains so even if a higher confidence level with $t(f = 3, p = 0.001) = 12.92$ is used.

Two slopes are compared in a similar manner as are two means: the simplest case is obtained when both calibrations are carried out using identical calibration concentrations (as is usual when SOPs are followed); the average variance V_b' is used in a t-test:

$$V_b' = (V_{b,1} + V_{b,2})/2 \tag{2.21}$$

$$t = |b_1 - b_2|/\sqrt{V_b'} \qquad \text{with } f = 2 \cdot n - 4 \text{ degrees of freedom} \tag{2.22}$$

Example 30: Two calibrations are carried out with the results $b = 0.004144 \pm 0.000046$ and $b' = 0.003986 \pm 0.000073$; V' is thus ± 0.000061 and $t = 0.0000158/0.000061 = 2.6$; since $n = n' = 5$, $f = 6$ and $t(6, 0.05) = 2.45$ so that a significant difference in the slopes is found. The reader is reminded to round only final results; if already rounded results had been used here, different conclusions would have been reached: the use of five or four

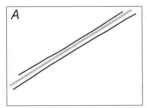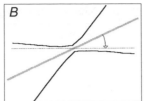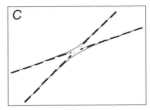

Figure 2.6. A graphical depiction of a significant and a nonsignificant slope (slopes $\pm s_b$ = 4.3 ± 0.5 (A) resp. −0.75 ± 1.3 (B)). If a horizontal line can be fitted between the confidence limits an interpretation $X = f(y^*)$ is impossible. It suffices in many cases to approximate the curves by straight lines (C).

decimal places results in $t = 2.47$ respectively $t = 1.4$, the latter of which is clearly insignificant relative to the tabulated $t_c = 2.45$.

A graphical test can be applied in all cases: the regression lines and their confidence limits for the chosen confidence level are plotted as given by Eq. (2.24) (next Section); if a vertical shift suffices to bring one line within the area defined by the CL of the other, then the slopes cannot be distinguished, but the intercept a might be different. If the intercepts a and a' are indistinguishable, too, the two lines cannot be distinguished. If, as an alternative to plotting the CL(Y) point by point over the whole x interval of interest, an approximation by straight-line segments as shown in Fig. 2.6c will suffice in most cases: the CL(Y) are plotted for x_{min}, x_{mean}, and x_{max}.

Eight combinations are possible with the true/false answers to the following three questions: (1) is $s_{res, 1} = s_{res, 2}$?, (2) is $b_1 = b_2$?, (3) is $y_{mean, 1} = y_{mean, 2}$? A rigorous treatment is given in Ref. 34. First, question 1 must be answered: if H_0 is retained, question 2 must be answered. Only if this also leads to a positive result can question 3 be posed.

There are several ways to test the linearity of a calibration line; one can devise theory-based tests, or use common sense. The latter approach is suggested here because if only a few calibration points are available on which to rest one's judgement, a graph of the residuals will reveal a trend, if any is present,[67] while numerical tests need to be adjusted to have the proper sensitivity. It is advisable to add two horizontal lines offset by the measure of repeatability $\pm s_x$ accepted for the method; unless the apparent curvature is such that points near the middle, respectively the end of the x-range are clearly outside this reproducibility band, no action need to be taken.

Regarding the residuals, many an investigator would be tempted to cast out "outliers"; the reader is advised to consult Section 1.5.5. If values are grouped (i.e. several values y_i are measured at the same x), outlier tests can be applied to the individual group, however, blind reliance on a "rule," such as $y_{mean} \pm 2 \cdot s_y$, is strongly discouraged.

Table 2.4. Comparison of Averages and Slopes

	Univariate Data (Sections 1.1.1 and 1.1.2)	Linear Regression (Section 2.2.1)	Eq.
model	$x_{mean} = \mu$	$Y = a + b \cdot x_{mean}$	(2.2)
Estimated variance	$V_{x, mean} = \dfrac{V_x}{n}$	$V_Y = \dfrac{V_{res}}{n}$	(2.23)
Estimated confidence limits	$CL(x_{mean}) = x_{mean} \pm t \cdot \sqrt{V_{x, mean}}$	$CL(Y) = Y(x_{mean}) \pm t \cdot \sqrt{V_Y}$	(2.24)

A test of linearity as applied to instrument validation is given in Ref. 68.

2.2.5   Interpolating $Y(x)$

The estimate of Y for a given x is an operation formally equivalent to the calculation of a mean, see Table 2.4:

The expression for $V_Y = s_y^2 = V_{res}/n$ is true if $x = \mathbf{x}_{mean}$; however, if x is different from $\mathbf{x}_{mean}$, an extrapolation penalty is paid that is proportional to the square of the deviation. [See Eq. (2.16).] This results in the characteristic "trumpet" shape observed in Figures 2.6 and 2.8. The influence of the calibration design is shown in Figure 2.8, where the corresponding individ-

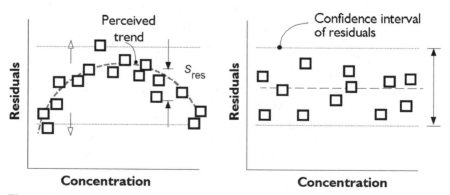

Figure 2.7. Using residuals to judge linearity. Horizontal lines: the accepted variation of a single point, e.g., $\pm 2 \cdot s_{res}$; thick dashed line: perceived trend; note that in the middle and near the ends there is a tendency for the residuals to be near or beyond the accepted limits, that is, the model does not fit the data (arrows). For a numerical example, see Section 4.13. The right panel shows the situation when the model was correctly chosen.

Signal [mAU]

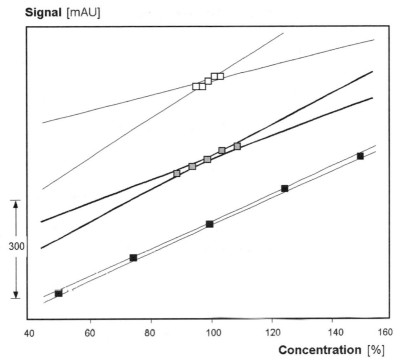

Concentration [%]

Figure 2.8. The slopes and residuals are the same as in Figure 2.4 (50, 75, 100, 125, and 150% of nominal; black squares), but the x-values are more densely clustered: 90, 95, 100, 105, and 110% of nominal (gray squares), respectively 96, 98, 100, 102, and 104% of nominal (white squares). The following figures of merit are found for the sequence bottom, middle, top: the residual standard deviations: ±0.00363 in all cases; the coefficients of determination: 0.9996, 0.9909, 0.9455; the relative confidence intervals of b: ±3.5%, ±17.6%, ±44.1%. Obviously the extrapolation penalty increases with decreasing S_{xx}, and can be readily influenced by the choice of the calibration concentrations. The difference in S_{xx} (6250, 250 resp. 40) exerts a very large influence on the estimated confidence limits associated with a, b, $Y(x)$, and $X(y^*)$.

ual points have the same residuals because they were just shifted along the regression line to increase or decrease S_{xx}.

The confidence limits thus established indicate the y-interval within which $Y(x)$ is expected to fall; the probability that this is an erroneous assumption is $100 \cdot p\%$; in other words, if the measurements were to be repeated and slightly differing values for a and b were obtained, the chances would only be $100 \cdot p\%$ that a Y is found outside the confidence limits $CL(Y)$. Use option $\langle Y(x)\rangle$ in program LINREG. The details of the calculation are found in Table 2.5.

The $CL(Y)$ obviously refer to the expected mean ordinate Y at the given

Table 2.5. Interpolations

Example			
a	0.0140	n	5
b	0.004144	S_{xx}	6250
r_{res}	0.00363	t	3.18
x	120.0	x_{mean}	100.0
Estimate average Y	$0.0140 + 120 \cdot 0.004144 = 0.511$		
Estimate Std. Dev. s_Y	$0.00363 \cdot \sqrt{1/5 + (120 - 100)^2/6250}$		
	$= 0.00187$		
Result Y	$0.511 \pm 0.006 (\pm 1.2\%)$		(2.17)
CL(y_{mean})	$0.505 \le Y(x) \le 0.517$		(2.17)
CL (population)	$0.498 \le y(x) \le 0.524$		(2.25)

abcissa x; if one were interested in knowing within which interval one would have to expect individual measurements y, the CL(y) apply ("Y" refers to an estimate, "y" to a measurement!): Equation (2.16) for V_Y is expanded to read

$$V_y = V_{res} \cdot \left(\frac{1}{n} + 1 + \frac{(x - x_{mean})^2}{S_{xx}} \right) \qquad (2.25)$$

The additional term "+1" is explained in Figure 2.9 and in the following:

If it is assumed that a given individual measurement y_i at x_i is part of the same population from which the calibration points were drawn (same chemical and statistical properties), the reproducibility s_y associated with this measurement should be well represented by s_{res}. Thus, the highest y-value still tolerated for y_i could be modeled by superimposing CI(y_i) on CI($Y(x_i)$) as shown by Figure 2.9 (left). A much easier approach is to integrate the uncertainty in y_i into the calculation of CL(Y); because variances (not standard deviations) are additive, this is done by adding "V_{res}" outside, respectively "+1" inside the parentheses of Eq. (2.16) to obtain Eq. (2.25).

Example 31: In Table 2.5, the term under the root would increase from 0.264 to 1.264; this increase by a factor of 4.8 translates into CI(y) being 2.2 times larger than CI(Y). The corresponding test at $x = 125$ ($0.517 \le y(x) \le 0.547$) shows the measured value in Table 2.2 (0.537) to be well within the tolerated limits. Only if the residual standard deviation (0.00363) was much larger than expected for the analytical method would there be reason to reassess this calculation.

A test for outliers can be based on this concept, for instance by using

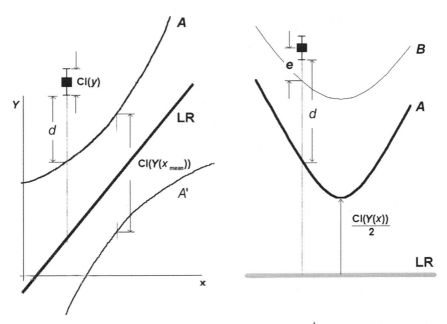

Figure 2.9. The confidence interval for an individual result CI(y^*) and that of the regression line's CL_U A are compared (schematic, left). The information can be combined as per Eq. (2.25), which yields curves B (and B', not shown). In the right panel curves A and B are depicted relative to the linear regression line. If $e > 0$ or $d > 0$, the probability of the point belonging to the population of the calibration measurements is smaller than alpha; cf. Section 1.5.5. The distance e is the difference between a measurement y (error bars indicate 95% CL) and the appropriate tolerance limit B; this is easy to calculate because the error is calculated using the calibration data set. The distance d is used for the same purpose, but the calculation is more difficult because both a CL(regression line) A and an estimate for the CL(y) have to be provided.

an appropriate t-value or by making use of a special table[69] (see Appendix 5.1.2), but as with all outlier tests, restraint is advised: data points should never be suppressed on statistical reasoning alone. A good practice is to run through all calculations twice, once with, and once without the suspected outlier, and to present a conservative interpretation, e.g., "the concentration of the unknown is estimated at 16.3 ± 0.8 mM using all seven calibration points. If the suspected outlier (marked "x" in the graph) were left out, a concentration of 16.7 ± 0.6 mM with $n = 6$ would be found". The reader can then draw his own conclusions. If working under GMPs, read Section 1.5.5 before even considering to touch an outlier.

The precision associated with $Y(x)$ is symmetrical around x_{mean}, see left panel of Fig. 2.10. In practice, the relative precision is more interesting, see the right panel.

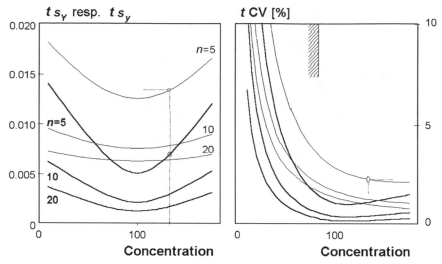

Figure 2.10. For $n = 5$, 10, resp. 20 the estimated CI(y_i) and CI(Y) (bold) are plotted *versus* x. The left figure shows the absolute values $|t \cdot s_Y|$, while the right one depicts the relative ones, namely $100 \cdot t \cdot s_Y/Y$ in %. At $x = 130$ one finds $Y = 0.553$ with a CI of ± 0.013 ($\pm 2.4\%$, circles). It is obvious that it would be inopportune to operate in the region below about 90% of nominal if relative precision were an issue (hatched bar). There are two remedies in such a case: increase n (and costs) or reduce all calibration concentrations by an appropriate factor, say 10%. The bold lines give the estimates for the regression line (Eq. 2.16), while the thin ones are for individual points (Eq. 2.25).

2.2.6   Interpolating $X(y)$

The quintessential statistical operation in analytical chemistry consists in estimating, from a calibration curve, the concentration of an analyte in an unknown sample. If the regression parameters a and b, and the unknown's analytical response y^* are known, the most likely concentration is given by Eq. (2.19), y^* being the average of all repeat determinations on the unknown.

The interpolation can lead to asymmetric distributions, even if all measurements that go into the calculation are normally distributed.[70]

While it is useful to know $X(y^*)$, knowing the CL(X) or, alternatively, whether X is within the preordained limits, given a certain confidence level, is a prerequisite to interpretation, see Figure 2.11. The variance and confidence intervals are calculated according to Eq. (2.18).

Example 32 (see Section 2.2.1): assume that the measurement of a test article yields an absorbance of 0.445; what is the probable assay value? Even for $m = 10$ repeat determinations, the true value of $X(y^*)$ is only loosely

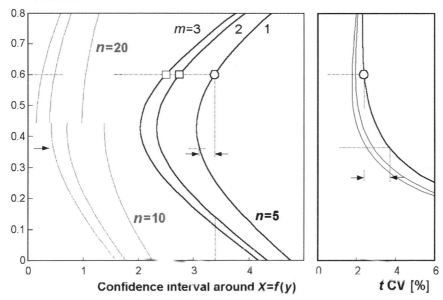

Figure 2.11. For various combinations of n (5; 10; resp. 20) and m (1, 2, resp. 3) the estimated CI(X) is plotted *versus* absorbance y^*. The left figure shows the absolute values $|t \cdot s_X|$, while the right figure depicts the relative ones, namely $100 \cdot t \cdot s_X/X$ in %. It is obvious that it would be inopportune to operate in the region below about 90% of nominal (in this particular case below $y = 0.36$; the absolute error for $y^* = 0.36$ is smaller than that for $y^* = 0.6$, but the inverse is true for the relative error, see arrows). There are three remedies: increase n or m (and costs), or reduce the calibration concentrations to shift the center of mass (x_{mean}, y_{mean}) below 100/0.42. At $y^* = 0.6$ and $m = 1$ (no replicates!) one finds $X = 141.4$ with a CI of ±3.39 (±2.4%, circle).

defined: $102.5 \leq X \leq 105.5$. This large confidence interval implies that the result cannot be quoted other than "104%." (See Table 2.6.) The situation can be improved as follows: in the above equation the three terms in the parentheses come to 1/5, 1/m, resp. 0.0026, that is 16.6, 83.2, and 0.2% of the total for $m = 1$. The third term is insignificant owing to the small difference between y^* and y_{mean}, and the large S_{xx}, so that an optimization strategy would have to aim at reducing $(1/n + 1/m)$, which is best achieved by increasing m to, say, 5. Thus the contributions would come to 49.7, 49.7, resp. 0.65% of the new total. Assuming $n = m$ is the chosen strategy, about $n = 16$ would be necessary to define $X(y^*)$ to ±1 ($n = 60$: ±0.5). Clearly, a practical limit becomes visible here that can only be overcome by improving s_{res} (i.e., better instrumentation and/or more skillful work). Evidently,

Table 2.6. Interpolation

Regression Parameters			
$y^* = 0.445$	$y_{\text{mean}} = 0.4284$	$S_{xx} = 6250$	$s_{\text{res}} = 0.00363$
$a = 0.0140$	$b = 0.004144$	$n = 5$	$m = 1$

Final Results

$X(y^*) = (0.445 - 0.0140)/0.004144 = 104.0\%$ of nominal   (2.1a)

$$Y_X = \left(\frac{.003630}{0.004144} \right)^2 \cdot \left(\frac{1}{5} + \frac{1}{1} + \frac{(0.445 - 0.4284)^2}{(0.004144)^2 \cdot 6250} \right) = 0.924$$

$t \cdot s_x = 3.182 \cdot \sqrt{0.924} = \pm 3.06$ (2.18)

Confidence limits $CL(X(y^*))$:

$t \cdot s_x = \pm 3.06$ for $m = 1$ independent measurement	100.9 ... 107.1
$t \cdot s_x = \pm 2.04$ for $m = 3$ independent measurements	102.0 ... 106.0
$t \cdot s_x = \pm 1.53$ for $m = 10$ independent measurements	102.5 ... 105.5

knowing $X(y^*)$ but ignoring $CL(X)$ creates an illusion of precision that is simply not there.

Program SIMCAL was expressly written to allow these sorts of what-if questions to be explored, with realistic intercepts, slopes, signal noise, digitizer characteristics, and economical factors specified, so one can get a feeling for the achieved precision and the costs this implies.

The $CI(X)$ yields information as to which digit the result should be rounded to. As discussed in Sections 1.1.5 and 1.6, there is little point in quoting $X(y^*)$ to four significant digits and drawing the corresponding conclusions, unless the $CI(X)$ is very small indeed; in the preceding example, one barely manages to demonstrate a difference between $X(y^*) = 104$ and the nominal value $X_n = 100$, so that it would be irresponsible to quote a single decimal place, and it would be out of the question to write down whatever the calculator display indicates, say "104.005792."

The $CL(X)$ are calculated as given in Eqs. (2.18) and (2.19); a comparison with Eq. (2.16) reveals the formal equivalence: the expression $(y^* - y_{\text{mean}})/b$ corresponds to $(x - x_{\text{mean}})$ and dividing s_{res} by b converts a measure of uncertainty in y to one in x.

The estimation of the intersection of two regression lines, as used in titrimetry, is discussed in Refs. 71–73; see program INTERSECT and Section 2.2.11.

A sin that is casually committed under routine conditions is to once and for all validate an analytical method at its introduction, and then to assume $a \equiv 0$ thus, $X(y^*)$ would be calculated from the measurement of a reference,

y_R, and that of the sample, y_S, by means of a simple proportionality. The possible consequences of this are discussed in Ref. 74. The only excuse for this shortcut is the nonavailability of a PC; this approach will have to be abandoned with the increasing emphasis regulatory agencies are putting on statements of precision.

Formalizing a strategy: What are the options the analyst has to increase the probability of a correct decision? V_{res} will be more or less given by the available instrumentation and the analytical method; an improvement would in most cases entail investments, a careful study to reduce sample work-up related errors,[13] and operator training. Also, specification limits are often fixed.

A general course of action could look like this:

1. Assuming the specification limits *SL* are given (regulations, market, etc.): postulate a tentative confidence interval CI(X) no larger than about $\frac{1}{4}$ SI; SI = specification interval. (See Fig. 2.12.)

2. Draw up a list of all analytical methodologies that are in principle capable of satisfying this condition.

3. Eliminate all methodologies that do not stand up in terms of selectivity, accuracy, linearity, and so on.

4. For all methodologies that survive step (3), assemble typical data, such as V_{res}, costs per test, and so on. For examples see Refs. 19, 75.

5. For every methodology set up reasonable "scenarios," that is, tentative analytical protocols with realistic n, m, S_{xx}, estimated time, and costs. Make realistic assumptions about the quality of data that will be delivered under routine conditions, cf. Figure 1.7.

6. Play with the numbers to improve CI(X) and/or cut costs (program SIMCAL).

7. Drop all methodologies that impose impractical demands on human and capital resources: many analytical techniques, while perfectly sound, will be eliminated at this stage because manpower, instrumentation, and/or scheduling requirements make them noncompetitive.

8. If necessary, repeat steps (5)–(7) to find an acceptable compromise.

Note concerning point (7): In the medium to long run it is counterproductive to install methodologies that work reliably only if the laboratory environment is controlled to unreasonable tolerances and operators have to aquire work habits that go against the grain. While work habits can be improved up to a certain point by good training (there is a cultural component in this), automation might be the answer if one does not want to run into GMP com-

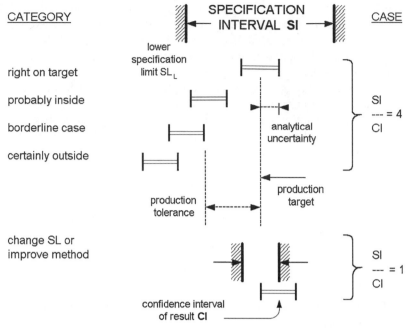

Figure 2.12. The relationship between specification interval, SI, and confidence intervals of the test result, CI. The hatched bars denote the product specifications while the horizontal bars give the test results with confidence limits. A ratio SI/CI ≥ 4 is required if differentiation between result categories is needed.

pliance problems. In the pharmaceutical industry, if compliance cannot be demonstrated, a product license might be revoked or a factory closed, at enormous cost.

Step (6) can be broken down as given in Table 2.7. If the hardware and its operation is under control, and some experience with similar problems is available, experiments need only be carried out late in the selection process to prove/disprove the viability of a tentative protocol. Laboratory work will earnestly begin with the optimization of instrumental parameters, and will continue with validation. In following such a simulation procedure, days and weeks of costly lab work can be replaced by hours or days of desk work.

As shown in Figure 2.12, the specification/confidence interval ratio SI/CI is crucial to interpretation: While SI/CI ≥ 4 allows for distinctions, with SI/CI = 1 doubts will always remain. SI/CI = 4 is the minimum one should strive for when setting up specifications (provided one is free to choose) or when selecting the analytical method, because otherwise the production department will have to work under close to zero tolerance conditions as

Table 2.7. Tactics for Improving a Calibration

Tactic	Target	Costs	Example
shift calibration points to reduce $(y^* - \mathbf{y}_{mean})$ or increase S_{xx}	reduce third term under parenthesis in Eq. (2.18)	organizational	Figs. 2.8 and 2.11
increase n or m	reduce terms 1 or 2 in Eq. (2.18)	time, material	Figs. 2.10 and 2.11
improve skills	decrease one component of V_{res}	training, organizational, time	reduce weighing errors, Fig. 4.10
buy better hardware	decrease other component of V_{res}	capital investment	better balances, mechanical dispensers, better detector
shift calibration points so the y-range within which the interpolation will take place does not include values below about 0.9 $\mathbf{y}_{mean}$	reduce interpolation error for low y^*	organizational	Figs. 2.10 and 2.11
do each test analysis a first time to obtain a rough estimate and repeat it with the optimal sample dilution	$y^* = \mathbf{y}_{mean}$	run a repeat analysis using a non-standard dilution scheme	Section 4.13
control the laboratory environment and/or modify the experimental plan	optimize conditions, reduce signal drift	infrastructure and/or organizational	Section 4.32, Eq. (1.6), Ref. 188

regards composition and homogeneity; cf. Section 4.24 (Fig. 4.35). Once a target CI(X) is given, optimization of experimental parameters can be effected as shown in Section 2.2.8.

Depending on the circumstances, the risk and the associated financial cost of being outside specifications might well be very high, so inhouse limits

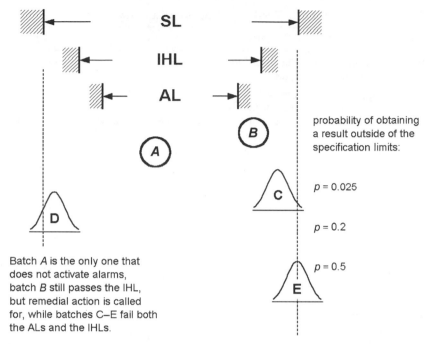

Figure 2.13. The distance between the upper inhouse limit (IHL) and the upper specification (SL) limits is defined by the CI(X). The risk of the real value being outside the SL grows from negligible (measured value A far inside IHL) to 50% (measured value E on SL). Note that the definition of the upper inhouse limit takes into account the details of the analytical method (n, x_i, m, V_{res}) to set the minimal separation between IHL and SL for an error probability for ($x_{mean} >$ SL) of less than $p = 0.025$. The alarm limits (AL), as drawn here, are very conservative; when there is more confidence in man and machine, the AL will be placed closer to the IHL. IHL and AL need not be symmetrical relative to the SL.

(IHL) for X could be set that would guarantee that the risk of a deviation would be less than a given level.

It is evident that the distance between the inhouse and the specification limits is influenced by the quality of the calibration/measurement procedure; a fixed relation, such as "2σ, 3σ," as has been proposed for control charts, might well be too optimistic or too pessimistic (for a single test result exactly on the 2σ inhouse limit, the true value μ would have a $\approx$ 16% chance of being outside the 3σ SL). Note that it takes at least $n = 6$ (resp. $n = 11$) values to make a $z = 2$ ($z = 3$) scheme (see Figure 1.24) even theoretically possible. For $n = 4$, for instance, $|x_{mean}|$ would have to be $\geq 1.5\sigma$ in order that the largest $|x|$ could be beyond 3σ; run a series of simulations on program CONVERGE and concentrate on the first four data points to see that an OOS result ($|x_i| > 3\sigma$) is much more frequent than $|x_{mean}| > 2\sigma$. Action limits (AL) can be identical with the inhouse limits IHL, or even tighter. IHL are a quality assurance concept that reflects a mandated policy "we will play on the safe side," while AL are a production/engineering concept traceable to process validation and a concern to prevent down-time and failed batches. The three sets of limits need not by symmetrically placed with respect to the nominal value, or each other.

2.2.7 Limit of Detection

Analytical measurements are commonly performed in one of two ways:

- When sufficient amounts of sample are available one tries to exploit the central part of the dynamical range because the signal-to-noise ratio is high and saturation effects need not be feared. (Cf. Figures 2.11 and 3.1.) Assays of a major component are mostly done in this manner.
- Smaller concentrations (or amounts) are chosen for various reasons, for example to get by with less of an expensive sample, or to reduce overloading an analytical system in order to improve resolution.

In the second case the limit of detection sets a lower boundary on the accessible concentration range; Ref. 76 discusses some current achievements.

Different concepts of "limit of detection" (LOD) have been advanced over the years[77]:

1. The most well known one defines the LOD as that concentration (or sample amount) for which the signal-to-noise ratio (SNR) is given by $S/N = z$, with $z = 3 \ldots 6$.[78-83] Evidently, this LOD is (a) dependent only on baseline noise, N, and the signal, S, (b) independent of any calibration schemes, and (c) independent of heteroscedacity. While the concept as such has profited the analytical community, the proposal of one or the other z value as being the most appropriate for some situation has raised the question of arbitrariness. This concept affords protection against Type I errors (concluding that an analyte is present when it is not), providing z is set large enough, but not against Type II errors (false negatives), cf. Sections 1.9 and 1.5.5. There are a number of misconceptions about this popular index of quality of measurement; the correct use of the SNR is discussed in Ref. 84.

2. The linear regression line is established and the quantity $q = 100 \cdot s_y/Y$ is determined for $x = x_{max} \rightarrow x_{min}$ until q equals 15%; the corresponding x is the LOD.

3. The FDA mandates that of all the calibration concentrations included in the validation plan, the lowest x for which CV $\leq$ 15% is the LOD (extrapolation or interpolation is forbidden). This bureaucratic rule results in a waste of effort by making analysts run unnecessary repeat measurements at each of a series of concentrations in the vicinity of the expected LOD in order to not end up having to repeat the whole validation because the initial estimate was off by + or − 20%; extrapolation followed by a confirmatory series of determinations would do. The consequences are particularly severe if validation means repeating calibration runs on several days in sequence, at a cost of, say, (6 concentrations) × (8 repeats) × (6 days) = 288 sample work-ups and determinations.

4. Calibration-design-dependent estimation of the LOD. (See next.)

The determination of the LOD in connection with transient signals is discussed in Ref. 85.

The *calibration-design-dependent LOD* approach,[86,87] namely the use of the confidence limit function, is endorsed here for reasons of logical consistency, response to optimization endeavors, and easy implementation. Fig. 2.14 gives a (highly schematic) overview:

The procedure is as follows:

1. The interception point is normally $x = 0$ (arrow A, circle); when a is negative, an instruction guards against an unrealistically low LOD by specifying x to be the interception point with the abscissa so that $Y_{\text{intercept}} \geq 0$ (arrow B, square):
   IF Intercept_A > = 0 THEN $x = 0$ ELSE $x = -\text{Intercept\_A}/\text{Slope\_B}$

2. CL_u is obtained from Eqs. (2.16) and (2.17) using the add sign ("+")

3. A horizontal is drawn through the upper confidence limit marked with a circle or a square, as appropriate, in Figure 2.14

4. CL_u is inserted in Eq. (2.10) ($\text{CL}_u = a + b \cdot X_{\text{LOD}}$). The intercept of the horizontal with the regression line defines the limit of detection, X_{LOD}, any value below which would be reported as "less than X_{LOD}."

5. CL_u is inserted in Eqs. (2.18) and (2.19), with $k = \infty$, and using the "+" sign. The intercept of the horizontal with the lower confidence limit function of the regression line defines the limit of quantitation, x_{LOQ}, any value above which would be quoted as "$X(y^*) \pm t \cdot s_x$"

6. X-values between LOD and LOQ should be reported as "LOD < $X(y^*)$ < LOQ"

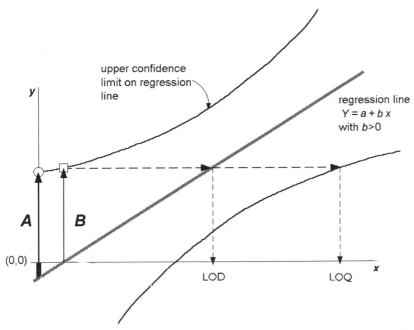

Figure 2.14. The definition of the limits of detection, LOD, respectively quantitation, LOQ (schematic).

This reporting procedure is implemented in program LINREG and its derivatives.

Pathological situations arise only when

- s_{res} is so large that the circle is higher than the square, or
- Slope b is close to zero and/or s_{res} is large, which in effect means the horizontal will not intercept the lower confidence limit function, and
- The horizontal intercepts the lower confidence limit function twice, i.e., if n is small, s_{res} is large, and all calibration points are close together; this can be guarded against by accepting X_{LOQ} only if it is smaller than x_{mean}.

How stringent is this model in terms of the necessary signal height relative to the baseline noise? First, some redefinition of terms is necessary:

- "Signal" is replaced by the calculated analyte concentration X_{LOD} at LOD resp. LOQ

Table 2.8. Reporting Interpolation Results Near the Detection Limit

Example (see Section 2.2.2):
Regression Parameters

$X_{LOD} = (0.0155)/0.004144 = 3.7$
$CL_u(a) = 0.0140 + 0.0155 = 0.0295$
X_{LOQ} is estimated as $3.7 + 3.6 = 7.3$
(Eq. (2.19) with $1/k = 0$ and $y^* = 0.0295$)

Final Results

The results for three unknown samples would be reported as follows:

	LOD 3.7	LOQ 7.3	
ESTIMATED RESULT	2.5	5.2	15.6
REPORTED AS	'<3.7'	'5.2'	'15.6 ± 4.3 ($k = 1$)'

- "Noise" is understood to mean the residual standard deviation expressed in abscissa units, $N_X = s_{res}/b$

Since, for this model, a calibration scheme is part of the definition, the following practical case will be evaluated: for $n = 3 \ldots 50$ the lowest concentration value will be at 50, the highest at 150% of nominal (the expected concentration of the sample); all other points will be evenly spaced inbetween. In Figure 2.15 the quotient "signal/noise" X_{LOD}/N_X is plotted *versus* the logarithm of n. It is quite evident that the limit of detection defined by the above quotient becomes smaller as n grows for constant repeatability s_{res}. The quotient for $p = 0.1$ is around 2.0 for the largest n,[78] and rises to over 7 for $n = 3$. These results are nearly the same if the x-range is shifted or compressed. The obvious value of the model is to demonstrate the necessity of a thoughtful calibration scheme (cf. Section 4.13) and careful measurements when it comes to defining the LOQ/LOD pair. Throughout, this model is more demanding than a very involved, correct statistical theory.[88] (See Figure 2.15.) This stringency is alleviated by redistributing the calibration points closer to the LOD. A comparison of various definitions of LOD/LOQ is given in Figure 4.31 and Section 4.23 (Table 4.28).

2.2.8 Minimizing the Costs of a Calibration

The traditional analyst depended on a few general rules of thumb for guidance while he coped with technical intricacies; his modern counterpart has a multitude of easy-to-use high-precision instruments at his disposal[89] and is

S / N ratio

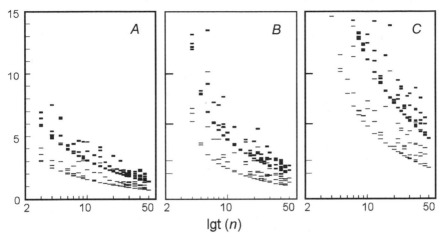

lgt (n)

Figure 2.15. The limit of detection LOD: the minimum signal/noise-ratio necessary according to two models (ordinate) is plotted against $\log_{10}(n)$ under the assumption of evenly spaced calibration points. The three sets of curves are for $p = 0.1$ (A), 0.05 (B), and 0.02 (C). The correct statistical theory is given by the fine points,[88] while the model presented here is depicted with coarser dots.[86] The widely used $S/N = 3 \ldots 6$ models would be represented by horizontals at $y = 3 \ldots 6$.

under constant pressure to justify the high costs of the laboratory. With a few program lines it is possible to juggle the variables to often obtain an unexpected improvement in precision,[90–94] organization, or instrument utilization, cf. Table 2.7. A simple example will be provided here; a more extensive one is found in Section 4.3.

Example 33: Assume that a simple measurement costs 20 currency units; n measurements are performed for calibration and m for replicates of each of five unknown samples. Furthermore, the calibration series of n measurements must be paid for by the unknowns to be analyzed. The slope of the calibration line is $b = 1.00$ and the residual standard deviation is $s_{res} = 3$, cf. Refs. 75, 95. The n calibration concentrations will be evenly spaced between 50 and 150% of nominal, that is for $n = 4$: x_i: 50, 83, 117, 150. For an unknown corresponding to 130% of nominal, s_x should be below ±3.3 units, respectively $V_x < 3.3^2 = 10.89$. What combination of n and m will provide the most economical solution? Use Eq. (2.4) for S_{xx} and Eq. (2.18) for V_x. Solution: since S_{xx} is a function of the x-values, and thus a function of n (e.g. $n = 4$: $S_{xx} = 5578$), solve the three equations in the given order for various combinations of n and m and tabulate the costs per result, $c/5$; then select the

Table 2.9. Costs and Quality of an Interpolation Result

		Currency				Variances V_x			
n	S_{xx}	$m = 1$	2	3	4	$m = 1$	2	3	4
3	5000	<u>32</u>	**52**	72	92	<u>13.6</u>	**9.1**	7.6	6.9
4	5578	36	56	76	96	12.7	8.2	6.7	6.0
5	6250	40	60	80	100	12.1	7.6	6.1	5.3
6	7000	44	64	84	104	11.7	7.2	5.7	4.9
7	7756	48	68	88	108	11.3	6.8	5.3	4.6
8	8572	52	72	92	112	11.1	6.6	5.1	4.3
9	9276	56	76	96	116	10.9	6.4	4.9	4.1
10	10260	60	80	100	120	10.7	6.2	4.7	3.9

combination that offers the lowest cost at an acceptable variance. (See Table 2.9.) Similar simulations were run for more or less than five unknown samples per calibration run. The conclusions are fairly simple: costs per analysis range from 32 to 120 currency units; if no constraints as to precision were imposed, ($n = 3/m = 1$, underlined) would be the most favorable combination. In terms of cost the combination ($n = 3/m = 2$) is better than ($n = 10/m = 1$) for up to six unknowns per calibration run.

This sort of calculation should serve as a rough guide only; nonfinancial reasoning must be taken into account, such as an additional safety margin in terms of achievable precision, or double determinations as a principle of GMP. Obviously, using the wrong combination of calibration points n and replicates m can enormously drive up costs. An alternate method is to incorporate previous calibrations with the most recent one to draw upon a broader data base and thus reduce estimation errors CI(X); one way of weighting old and new data is given in Ref. 96 (Bayesian calibration); whether this would be accepted under GMP rules is open to debate. Program SIMCAL allows more complex cost calculations to be made, including salaries, amortization, and instrument warm-up time.

2.2.9   Standard Addition

A frequently encountered situation is that of no blank matrix being available for spiking at levels below the expected ("nominal") level.

The only recourse is to modify the recovery experiments above in the sense that the sample to be tested itself is used as a kind of "blank," to which further analyte is spiked. This results in at least two measurements, namely "untreated sample" and "spiked sample," which can then be used to establish a calibration line from which the amount of analyte in the untreated sample

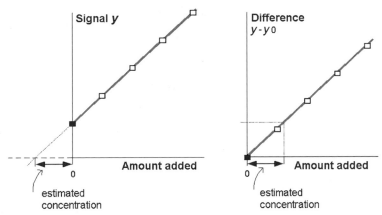

Figure 2.16. Depiction of the standard addition method: extrapolation (left), interpolation (right). The data and the numerical results are given in the following example.

is estimated. It is unnecessary to emphasize that linearity is a prerequisite for accurate results. This point has to be validated by repeatedly spiking the sample. The results can be summarized in two different ways:

The traditional manner of graphing standard addition results is shown in Figure 2.16(left): the raw (observed) signal is plotted *vs.* the amount of analyte spiked into the test sample; a straight line is drawn through the two measurements (if the sample was repeatedly spiked, more points will be available, so that a linear regression can be applied); the line is extrapolated to "zero" signal. A glance at Section 2.2.6 makes it apparent that extrapolation, while perfectly legitimate, widens the confidence interval around conc. $= -X(0)$, the sought result. This effect can be countered somewhat by higher spiking levels and thereby increasing S_{xx}: instead of roughly estimating x and spiking the sample to $2 \cdot X(0)$, the sample is spiked to a multiple of concentration level $X(0)$. This strategy is successful only if spiking to such large levels does not increase the total analyte concentration beyond the linear range. (See Fig. 3.1.)

Furthermore, there is the problem that the signal level to which one extrapolates need not necessarily be $y = 0$; if there is any interference by a matrix component, one would have to extrapolate to a level $y > 0$. This uncertainty can only be cleared if the standard addition line perfectly coincides with the calibration line obtained for the pure analyte in absence of the matrix, i.e. same slope and 100% recovery, see also Figure 3.2. This problem is extensively treated in Refs. 97–101. A modification is presented in Ref. 102.

Another approach to graphing standard addition results is shown in Figure 2.16 (right): the signal for the unspiked test sample is marked off on

Table 2.10. Standards Addition Results

Amount Spiked	Signal	Levels Used	Conventional Estimate (Extrapolation)	Alternative Estimate (Interpolation)
0 mg	58376			
1	132600	0–1*	(–) 0.787*	0.787*
2	203670	0–2	(–) 0.811 ± 48%	0.796 ± 33%
3	276410	0–3	(–) 0.813 ± 10%	0.797 ± 8%

*) If only data points 1 + 2 are used, confidence limits cannot be calculated.

the ordinate as before; at the same time, this value is subtracted from all spiked-sample measurements. The same standard addition line is obtained as in the left figure, with the difference that it passes through the origin $x = 0/y = 0$. This trick enables one to carry out an interpolation instead of an extrapolation, which improves precision without demanding a single additional measurement.

The trade-offs between direct calibration and standard addition are treated in Ref. 103. The same recovery as is found for the "native" analyte has to be obtained for the spiked analyte (see Section 3.2). The application of spiking to potentiometry is reviewed in Refs. 104 and 105. A worked example for the application of standard addition methodology to FIA/AAS is found in Ref. 106. Reference 70 discusses the optimization of the standard addition method.

Example 34: The test sample is estimated, from a conventional calibration, to contain the analyte in question at a level of about 0.8 mg/ml; the measured GC signal is 58 376 area units. (See Table 2.10):

2.2.10   Weighted Regression

In the previous sections of Chapter 2 it was assumed that the standard deviation s_y obtained for a series of repeat measurements at a concentration x would be the same no matter which x was chosen; this concept is termed "homoscedacity" (homogeneous scatter across the observed range).

Under certain combinations of instrument type and operating conditions the preceeding assumption is untenable: signal noise depends on the analyte concentration. A very common form of "heteroscedacity" is presented in Fig. 2.17.

The reasons for heteroscedacity can be manifold, for example:

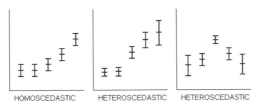

HOMOSCEDASTIC   HETEROSCEDASTIC   HETEROSCEDASTIC

Figure 2.17. Schematic depiction of homo- (left) and heteroscedacity (right).

- The relative standard deviation RSD (or "c.o.v. = coefficient of variation") is constant over the whole range, such as in many GC methods, that is, the standard deviation s_y is proportional to y.
- The RSD is a simple function of y, as in isotope labeling work (RSD proportional to $1/\sqrt{y}$.
- The RSD is small in the middle and large near the ends of the linear range, as in photometry.

Curve-fitting need not be abandoned in this case, but some modifications are necessary so that precisely measured points influence to a greater degree the form of the curve, more so than a similar number of less precisely measured ones. Thus, a "weighting" scheme is introduced. There are different ways of doing this; the most accepted model makes use of the experimental standard deviation,[107,108] namely:

$$w_i = k \cdot \left(\frac{1}{s_y(x_i)} \right)^2 , \text{ with } \Sigma\, w_i = n \qquad (2.26)$$

How does one obtain the necessary s_y-values? There are two ways:

1. One performs so many repeat measurements at each concentration point that standard deviations can be reasonably calculated, e.g., as in validation work; the statistical weights w_i are then taken to be inversely proportional to the local variance. The proportionality constant k is estimated from the data.

2. One roughly models the variance as a function of x using the data that are available[108–110]: the standard deviations are plotted *versus* the concentrations, and if any trend is apparent, a simple curve is fitted, e.g., Eq. (2.27) in Table 2.11. As more experience is accumulated, this relation can be modified.[111,112]

Table 2.11. Calculation of Weighted Sums of Squares

1	$s(x_i) = a_s + b_s \cdot x_i$	model of std. dev.	(2.27)
2	$k = n/\sum V(x_i)^{-1}$	normalization factor	(2.28)
3	$w_i = k/V(x_i)$	statistical weight	(2.29)
4	$\mathbf{x}_{\text{mean},w} = \sum(w_i \cdot x_i)/n$	weighted summation	(2.30)
5	$\mathbf{y}_{\text{mean},w} = \sum(w_i \cdot y_i)/n$	weighted summation	(2.31)
6	$S_{xx,w} = \sum(w_i \cdot (x_i - \mathbf{x}_{\text{mean},w})^2)$	weighted S_{xx}	(2.32)
7	$S_{yy,w} = \sum(w_i \cdot (y_i - \mathbf{y}_{\text{mean},w})^2)$	weighted S_{yy}	(2.33)
8	$S_{xy,w} = \sum(w_i \cdot (x_i - \mathbf{x}_{\text{mean},w}) \cdot (y_i - \mathbf{y}_{\text{mean},w}))$	weighted S_{xy}	(2.34)

It is important to realize that for the typical analytical application (with relatively few measurements well characterized by a straight line) a weighting scheme has little influence on slope and intercept, but appreciable influence on the confidence limits of interpolated $X(y)$ resp. $Y(x)$.

The step-by-step procedure for option (2) is nearly the same as for the standard approach for option (1); only Eqs. (2.27) and (2.29) have to be appropriately modified to include the experimental values. All further calculations proceed as under Section 2.2.1, standard approach.

Example 35: The following demonstrates the difference between a weighted and an unweighted regression for four concentrations and two measurements per concentration

For the weighted regression the standard deviation was modeled as $s(x) = 100 + 5 \cdot x$; this information stems from experience with the analytical technique. Intermediate results and regression parameters are given in Tables 2.13 and 2.14. Table 2.15 details the contributions the individual residuals make.

Table 2.12. GC-Calibration (see Data File WLR.dat)

x_i	y_1	y_2
10	462.7	571.3
20	1011	1201
30	1419	1988
40	2239	2060

Table 2.13. Equations for Weighted Linear Regression

Variable	Unweighted Regression	Weighted Regression	Eq./Fig.
k	1	41'427	(2.28)
n	8	8	
x_{mean}	25.0	19.35	(2.30)
y_{mean}	1369.0	1053.975	(2.31)
S_{xx}	1000.0	865.723	(2.32)
S_{yy}	3'234'135	2'875'773.6	(2.33)
S_{xy}	54'950	48'554.81	(2.34)
V_{res}	35'772	25'422.53	(2.13)
s_{res}	±189	±159.44	(2.13)
r^2	0.9336	0.9470	(2.1)
a	−4.75	−31.6	(2.12)
rel CI(a)	±8436%	±922%	(2.15)
b	54.95	56.09	(2.11)
rel CI(b)	26.6%	23.6%	(2.14)
LOD	7.3	5.19	Fig. 2.14
LOQ	13.0	9.35	Fig. 2.14

Conclusions: the residual standard deviation is somewhat improved by the weighting scheme; note that the coefficient of determination gives no clue as to the improvements discussed in the following. In this specific case, weighting improves the relative confidence interval associated with the slope b. However, because the smallest absolute standard deviations $s(x)$ are found near the origin, the center of mass x_{mean}/y_{mean} moves toward the origin and the estimated limits of detection resp. quantitation, LOD resp.

Table 2.14. Comparison of Interpolations of Weighted and Unweighted LinReg

Interpolations	Unweighted	Weighted
$x = 10$	$Y = 545 \pm 274(50\%)$	$529 \pm 185(35\%)$
$x = 20$	$Y = 1094 \pm 179(16\%)$	$1090 + 138(13\%)$
$x = 30$	$Y = 1644 \pm 179(11\%)$	$1651 \pm 197(12\%)$
$x = 40$	$Y = 2193 \pm 274(12\%)$	$2212 \pm 306(14\%)$
$y = 544.8, M = 1$	$X = 10.0 \pm 9.8(98\%)$	$10.3 \pm 7.7(75\%)$
$y = 1094, M = 1$	$X = 20.0 \pm 9.0(45\%)$	$20.1 \pm 7.4(37\%)$
$y = 1644, M = 1$	$X = 30.0 \pm 9.0(30\%)$	$29.9 \pm 7.8(26\%)$
$y = 2193, M = 1$	$X = 40.0 \pm 9.8(24\%)$	$39.7 \pm 8.8(22\%)$

Table 2.15. Contributions of the Individual Residuals Toward the Total Variance

Unweighted			Weighted		
w_i	r_i	$\%V$	w_i	r_i	$\%V$
1	−82.0	3.1	1.84	−66.6	5.4
1	26.5	0.3	1.84	42.0	2.1
1	−83.2	3.2	1.04	−79.1	4.3
1	107	5.3	1.04	111	8.4
1	−225	23.5	0.66	−232	23.4
1	344	55.2	0.66	337	49.4
1	45.8	1.0	0.46	27.2	0.2
1	−133	8.3	0.46	−152	7.0
$\Sigma = 8$	$\Sigma = 0.0$	$\Sigma = 100.0$	$\Sigma = 8$	$\Sigma = 0.0$	$\Sigma = 100.0$

LOQ, are improved. The interpolation $Y = f(x)$ is improved for the smaller x-values, and is worse for the largest x-values. The interpolation $X = f(y^*)$, here given for $m = 1$, is similarly influenced, with an overall improvement. The largest residual has hardly been changed, and the contributions at small x have increased. This example shows that weighting is justified, particularly when the poorly defined measurements at $x = 30$ and 40 were just added to better define the slope, and interpolations are planned at low y^* levels. Note that the CL of the interpolation $X = f(y^*)$ are much larger ($m = 1$) than one would expect from Figure 2.18; increasing m to, say, 10 already brings an improvement by about a factor of 2. The same y^* values were chosen for the weighted as well as for the unweighted regression to show the effect of the weighting scheme on the interpolation. The weighted regression, especially the CL(X), give the best indication of how to dilute the samples; although the relative CL(X) for the unweighted regression are smallest for large signals y^*, care must be taken that it is exactly these signals that have the largest uncertainty: $s_y = 100 + 5 \cdot x$; if there is marked heteroscedacity, the unweighted regression is a poor model.

If it should happen that both the abscissa and the ordinate measurements suffer from heteroscedacity and the assumption of $s_y \gg s_x$ cannot be upheld, then a means must be found to introduce both weighting functions, $s_x = f(x)$ and $s_y = f(y)$. What cannot be done is to selectively use one function for the abscissa and the other for the ordinate values, because in that case it could happen that the weighted means $\mathbf{x}_{mean, w}$ and $\mathbf{y}_{mean, w}$ would combine to a pivotal point coordinate that is outside the range of measurements. See Eqs. (2.30) and (2.31). Therefore, it is suggested that the functions be combined

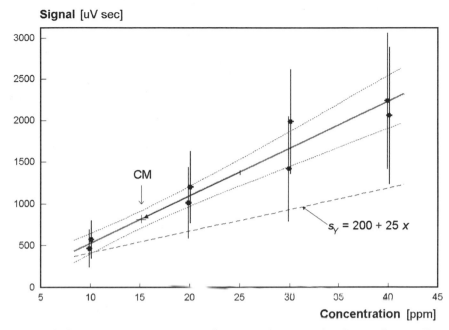

Figure 2.18. The data from the preceding example are analyzed according to Eqs. (2.27)–(2.39), (2.10)–(2.17); the effect of the weighting scheme on the center of mass and the confidence limits is clearly visible because the noise model was changed.

using Eq. (2.27) to read $s^2 = s_x^2 + s_y^2$ by transforming $s_y = f(y)$ as a function of x: $s_y = f'(x)$.

2.2.11   The Intersection of Two Linear Regression Lines

There are a number of analytical techniques that rely on finding the point on the abscissa where two straight-line segments of an analytical signal intersect, e.g., titration curves. The signal can be any function, $y = f(x)$, such as electrical potential *versus* amount added, that changes slope when a species is consumed. The evaluation proceeds by defining a series of measurements $y_{i1} \ldots y_{i2}$ before, and another series $y_{j1} \ldots y_{j2}$ after the break-point, and fitting linear regression lines to the two segments. Finding the intersection of the regression lines is a straightforward exercise in algebra. There are several models for finding the confidence limits on the intersection.[71–73] Program INTERSECT calculates the overlap of the distribution functions PD_{Y1} and PD_{Y2} as a function of x to estimate the x-range within which the intercept probably lies, see Fig. 2.19.

Conductivity

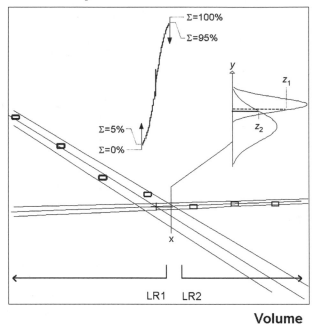

Figure 2.19. Intersection of two linear regression lines (schematic). In the intersection zone (gray area), at a given x-value two PD-curves of equal area exist that at a specific y-value yield the densities z_1 and z_2 depicted by the dashed and the full lines. The product $z_1 \cdot z_2$ is added over the whole y-range, giving the probability-of-intersection value for that x. The cumulative sum of such probabilities is displayed as a sigmoidal curve; the x-values at which 5, respectively 95% of $\Sigma_x \Sigma_y (z_1 \cdot z_2)$ is reached are indicated by vertical arrows. These can be interpreted as the 90%-CL(X_{intersec}).

2.3   NONLINEAR REGRESSION

Whenever a linear relationship between dependent and independent variables (ordinate–resp. abscissa–values) is obtained, the straightforward linear regression technique is used: the equations make for a simple implementation, even on programmable calculators.

The situation changes drastically when curvature is observed[23]:

1. Many more measurements are necessary, and these have to be carefully distributed over the x-range to ensure optimally estimated coefficients.

2. The right model has to be chosen; this is trivial only when a well-proven theory foresees a certain function $y = f(x)$. Constraints add a

further dimension.[113] In all other cases a choice must be made between approaches (3) and (4).

3. If the graph y vs. x suggests a certain functional relation, there are often several alternative mathematical formulations that might apply, e.g., $y = \sqrt{x}$, $y = a \cdot (1 - \exp(b \cdot (x + c)))$, and $y = a \cdot (1 - 1/(x + b))$: choosing one over the others on sparse data may mean faulty interpretation of results later on.[114] An interesting example is presented in Ref. 115 (cf. Section 2.3.1). An important aspect is whether a function lends itself to linearization (see Section 2.3.1), to direct least-squares estimation of the coefficients, or whether iterative techniques need to be used.

4. In some instances, all one is interested in is an accurate numerical representation of data, without any intent of physicochemical interpretation of the estimated coefficients: a simple polynomial might suffice; the approximations to tabulated statistical values in Chapter 5 are an example.

5. In analytical practice, fitting a model to data is only the first step; in analogy to Eq. (2.19) an interpolation that uses y^* to estimate $X(y^*)$ is necessary. For many functional relationships $y = f(x)$ finding an inverse, $x = f^{-1}(y)$, is difficult enough; without confidence limits such a result is nearly worthless.[114]

The linearization technique mentioned under item 3 is treated in the next section.

2.3.1 Linearization

"Linearization" is here defined as one or more transformations applied to the x- and/or y-coordinates in order to obtain a linear "y vs. x" relationship for easier statistical treatment. One of the more common transformations is the logarithmic one; it will nicely serve to illustrate some pitfalls.

Two aspects—wanted or unwanted—will determine the usefulness of a transformation:

1. Individual coordinates (x_i/y_i) are affected so as to eliminate or change a curvature observed in the original graph.

2. Error bars defined by the confidence limits $CL(y_i)$ will shrink or expand, most likely in an asymmetric manner. Since we here presuppose near absence of error from the abscissa values, this point applies only to y-transformations. A numerical example is 17 ± 1 ($\pm 5.9\%$, symmetric CL), upon logarithmic transformation becomes $1.23045 - 0.02633 \ldots 1.23045 + 0.02482$.

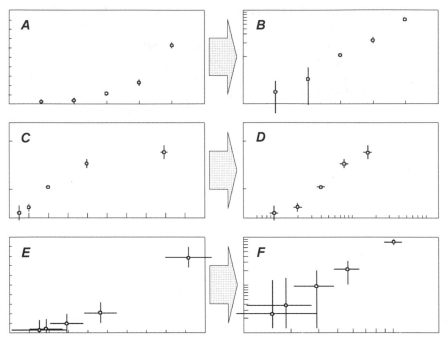

Figure 2.20. Logarithmic transformations on x- or y-axes as used to linearize data. Notice how the confidence limits change in an asymmetric fashion. In the top row, the y-axis is transformed; in the middle row, the x-axis is transformed; in the bottom row, both axes are transformed simultaneously.

Both aspects are combined in Fig. (2.20) and Table 2.16, where the linear coordinates are x resp. y, the logarithmic ones u, resp. v. Regression coefficients established for the lin/lin plot are a, b, whereas those for the transformed coordinates are p, q.

Note that the intercept p in the y-transformed graph becomes a multiplicative preexponential factor in the original non- (resp. back-) transformed graph and that functions always intersect the ordinate at $y = 10^p$. A straight line in logarithmic coordinates, if the intercept p is not exactly zero, will become an exponential function with intercept 10^p after back transformation. Since double-logarithmic transformations are often employed to compress data, such as GC-FID response over a $1:1000$ dynamic range, statistical indistinguishability of two such transformed response functions must not be interpreted as an indication of identity: for one, any straight line in a lin/lin plot takes on a slope of 1.000 in a log/log plot, and any difference between intercepts p, however small, translates into two different slopes in the original plot, while the intercept a is always zero.

Table 2.16. Lin/Log Transformations

Transformed Axis	Linear Presentation	Transformation	Logarithmized Presentation
x-axis	$y = a + b \cdot x$ $y = p + q \cdot \log_{10}(x)$	$u = \log_{10}(x) \Rightarrow$ $\Leftarrow x = 10^u$	$y = a + b \cdot 10^u$ $y = p + q \cdot u$
y-axis	$y = a + b \cdot x$ $y = 10^p \cdot (10^x)^q$	$v = \log_{10}(y) \Rightarrow$ $\Leftarrow y = 10^v$	$10^v = a + b \cdot x$ $v = p + q \cdot x$
both axes	$y = a + b \cdot x$ $y = 10^p \cdot x^q$	$u = \log_{10}(x) \Rightarrow$ $v = \log_{10}(y) \Rightarrow$ $\Leftarrow x = 10^u$ $\Leftarrow y = 10^v$	$10^v = a + b \cdot 10^u$ $v = p + q \cdot u$

2.3.2 Nonlinear Regression and Modeling

Many functional relationships do not lend themselves to linearization; the user has to choose either option 3 or 4 in Section 2.3 to continue.

Option 4 (multilinear regression, polynomials[114,116]) is uncomplicated insofar as clear-cut procedures for finding the equation's coefficients exist. Today, linear algebra (use of matrix inversion) is most commonly employed; there even exist cheap pocket calculators that are capable of solving these problems. Things do become quite involved and much less clear when one begins to ponder the question of the correct model to use. First, there is the weighting model: the least-squares approach, see Eq. (2.3),[117] is implicit in most commercially available software; weighting can be intentionally introduced in some programs, and in others there is automatic calculation of weights for grouped data (more than one calibration point for a given concentration x). Second, there is the fitted model: is one to choose a polynomial of order 5, 8, or even 15? Or, using a multi-linear regression routine, should the terms x^2, y^2, xy, or the terms x^2y and x^2y^2 be introduced[114]? If one tries several alternative fitting models, how is one to determine the optimal one? From a purely statistical point of view, the question can be answered. There exist powerful program packages that automatically fit hundreds of models to a data set and rank-order them according to goodness of fit, a sort of hit parade of mathematical functions. So far, so good, but what is one to say if, for a series of similar experiments, the proposed "best" models belong to completely different mathematical functionalities, or functions are proposed that are at odds with the observed processes (for example, diffusion is an exponential process and should not be described by fractional functions)? Practicability demands that the model be kept as simple as possible. The more terms (coefficients) a model includes, the larger the danger that a perfect fit will entice the user into perceiving more than is justified. Numerical simulation is a valuable tool for shattering these illusions: The hopefully "good" model and some sensible assumptions concerning residual variance can be used to construct several sets of synthetic data that are functionally (form of curve) and statistically (residual standard deviation) similar to the experimental data but differ stochastically (detailed numbers; Monte Carlo technique, see programs SIMGAUSS and SIMILAR); these data sets are analyzed according to the chosen fitting model(s). If the same model, say a polynomial of order 3, is consistently found to best represent the "data," the probability of a wise choice increases. However, if the model's order or one or more of the coefficients are unstable, then the simplest model that does the job should be picked. For a calculated example see Section 3.4.

Option 3 (arbitrary models) must be viewed in a similar light as option 4, with the difference that more often than not no direct procedure for estimating the coefficients exists. Here, formulation of the model and the initial conditions for a sequential simplex-search of the parameter-space are a delicate matter. The simplex-procedure[118–120] improves on a random or systematic trial-and-error search by estimating, with a minimal set of vertices (points, experiments), the direction of steepest decent (toward lower residual variances), and going in that direction for a fixed (classical simplex) or variable (Fletcher-Powell, or similar algorithms) distance, retaining the best vertices (e.g., sets of estimated coefficients), and repeating until a constant variance, or one below a cut-off criterion, is found. The system, (see Figure 3.4) is not fool-proof; plausibility checks and graphics are an essential aspect (see Section 4.2). Even if an arbitrary model is devised that permits direct calculation of its coefficients, this is no guarantee that such a model will not break down under certain conditions and produce nonsense; this can even happen to the unsuspecting user of built-in, unalterable firmware.[115]

The limit of quantitation in the nonlinear case is discussed in Ref. 121.

2.4   MULTIDIMENSIONAL DATA: VISUALIZING DATA

When confronted with multidimensional data it is easy to "plug" the figures into a statistical package and have nice tables printed that purportedly accurately analyze and represent the underlying factors. Have the following questions been asked:

- Does the model conform to the problem?
- Is the number of factors meaningful?
- Is their algebraic connection appropriate?

The point made here is that nearly any model can be forced to fit the data; the more factors (coefficients) and the higher the order of the independent variable(s) (x, x^2, x^3, etc.) the better the chance of obtaining near-zero residuals and a perfect fit. Does this make sense, statistically, or chemically?

It is proposed to visualize the data before any number-crunching is applied, and to carefully ponder whether an increase in model complexity is necessary or justified. The authors abstain from introducing more complicated statistics such as multiple linear regression (MLR), principal component analysis (PCA), or partial least squares (PLS)[122] because much experience is necessary to correctly use them. Since the day the computer invaded the laboratory, publications have appeared that feature elegant, and increasingly abstract, recipes for extracting "results" from a heap of numbers; only time will tell whether many of these concepts are generally useful. The fact that these standard methods are presented again and again, nearly always with some refinement for the particular field of application, or are critically compared, is in itself an indication that wielding high-powered, off-the-shelves tools is not without risks.[123] In an era of increasing accountability, there is no guarantee that the justification report for a particular data-evaluation scheme will not land on a lawyer's desk some day; it is hard to explain involved evaluation schemes to lay persons for whom a standard deviation is "advanced science." If schemes A and B are statistically more or less equivalent under some set of circumstances, but a given decision could go one way or the other, depending on which scheme is used, then a legal trap is set. A case in point would be a very large data set that is first reduced to a few latent variables through the use of PCA; because the MLR model built on these linear combinations does not sufficiently well map the data, the model could be expanded to include quadratic and cubic terms. The obvious question, "do you really know what you are doing?" could not be shrugged off, particularly if a reasonably similar situation was analyzed in a wholly different manner by some other company. The simple traditional models do have an advantage: their use is widespread and there is much less discussion about when and how to apply them.

Visualizing Data: the reader may have guessed from previous sections that graphical display contributes much toward understanding the data and the statistical analysis. This notion is correct, and graphics become more important as the dimensionality of the data rises, especially to three and more dimensions. Bear in mind that:

- The higher the dimensionality, the more acute the need for a visual check before statistical programs are indiscriminately applied.

- The higher the dimensionality, the harder it becomes for humans to

Table 2.17. Multidimensional Data

Method	Calculated means					Calculated SDs × 100				
	1	2	3	4	Mean	1	2	3	4	Mean
Batch										
A	**99.03**	99.16	99.27	99.41	99.22	±7	±2	±12	±2	±6.6
B	99.45	99.34	**99.06**	99.35	99.42	±2	±1	±17	±15	±10.6
C	**99.17**	99.55	99.42	99.55	99.42	±11	±7	±10	±7	±8.3
D	99.04	99.36	**98.58**	99.20	99.30	±10	±7	±7	±14	±9.3
Mean	99.17	99.35	99.33	99.38	**99.25**	±7.7	±4.7	±11.2	±10.2	**±9.5**

grasp the situation; with color coding and pseudo-three-dimensional displays, four dimensions can just be managed.

- Scientific software tends to be better equipped for handling and presenting complex data than the commonly available business graph packages, but even when one knows which options to choose, the limits that two-dimensional paper imposes remain in place.

The reader is urged to try graphics before using mathematics for reasons that will become evident in the example of Table 2.17. However, it is suggested to stick with one or two dimensions if that suffices to present the information, and to resist the urge to add pseudo-dimensions even if the illustration looks slicker that way.

Example 36: Four batches, A, B, C, and D, of an amino acid hydrochloride were investigated; four different titration techniques were applied to every sample:

1. Direct titration of $^+H^-OOC-R-N^+R_1R_2R_3 \cdot Cl^-$ with NaOH,
2. Indirect titration using an excess of NaOH and back-titration to the first,
3. Respectively second equivalence point, and
4. Titration of the chloride.

Six repeat titrations went into every mean and standard deviation listed (for a total of $4 \times 4 \times 6 = 96$ measurements). The data in Table 2.17 are for batches A, B, C, and D, and the calculated means respectively standard deviations are for methods 1, 2, 3, and 4. The means for both rows and columns are given. The lowest mean in each row is given in **bold**. The overall

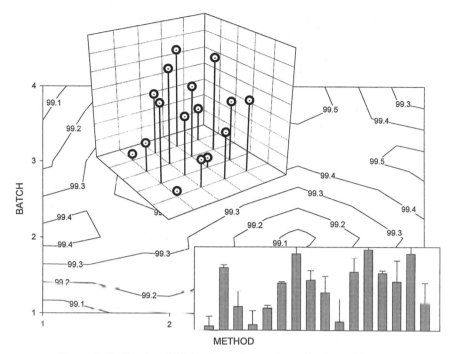

Figure 2.21. Results of 96 titrations of an amino acid. (See Table 2.17.)

mean is 99.25; relative to this, the overall SD is ± 0.245, which includes the average repeatability (±0.095) and a between-group component of ±0.226. Because $n = 6$, $t(p = 0.05, f = 5)/\sqrt{6} = 1.04$, which means the standard deviation is nearly equal to the confidence interval of the mean.

A bar chart and a pseudo-3D depiction are shown superimposed on a contour map in Figure 2.21; the same information as in Table 2.17 is given in an appealing, credible-looking manner in all three cases. The bar chart is the least confusing but misrepresents truth all the same; the other figures ought to make one think even more before calculations are started:

1. It is easy to prove that there are differences between two given means using the t-test (Case b_1 or c).

2. A one-way (simple) ANOVA with six replicates can be conducted by either regarding each titration technique or each batch as a group, and looking for differences between groups.

3. A two-way ANOVA (not discussed here) would combine the two approaches under 2.

Finding differences is one aspect of the problem, the second is to integrate problem-specific chemical know-how with statistics:

a. The chloride titration seems to give the highest values, and the direct titration the lowest. Back-titration to the first equivalent point appears to be the most precise technique.
b. The four batches do not appear to differ by much if means over methods or batches are compared.
c. Batches A, B, and C give similar minimal values.
d. The average standard deviation is $9.5/100/100 \approx 0.1\%$, certainly close to the instrumentation's limits.

By far the most parsimonious, but nonstatistical, explanation for the observed pattern is that the titrations differ in selectivity, especially as regards basic and acidic impurities. Because of this, the only conclusion that can be drawn is that the true values probably lie near the lowest value for each batch, and everything in excess of this is due to interference from impurities. A more selective method should be applied, e.g., polarimetry or ion chromatography. "*Parsimony*" is a scientific principle: make as few assumptions as possible to explain an observation; it is in the realm of wishful thinking and "fringe science" that combinations of improbable and implausible factors are routinely taken for granted.

The lessons to be learned from this example are clear:

- Most statistical tests, given a certain confidence level, only provide clues as to whether one or more elements are different from the rest.
- Statistical tests incorporate mathematical models against which reality, perhaps unintentionally or unwillingly, is compared, for example:
  — "additive difference": *t*-test, ANOVA;
  — "linear relationship." linear regression, factorial test.
- More appropriate mathematical models must be specifically incorporated into a test, or the data must be transformed so as to make it testable by standard procedures.
- A decision reached on statistical grounds alone is never as good as one supported by (chemical) experience and/or common sense.
- Never compare "apples" and "oranges" if the distinction is evident or plausible.
- Refrain from assembling incomplete models and uncertain coefficients into a spectacular "theoretical" framework without thoroughly testing the premises. "Definitive" answers so produced all too easily take on a life of their own as they are wafted through top floor corridors.

CHAPTER

3

RELATED TOPICS

Except perhaps in some academic circles, analytical chemistry is not a stand-alone affair as it is always embedded in real-world problems that need to be solved in a reasonable manner, that is legally binding, fast, precise, cheap, etc. Making a method reliably deliver the requested results at the pre-determined quality level[31] means installing a framework of equipment and procedures, and providing the necessary skill base and support functions. This is exactly what the GMPs embody; the key phrases are explained below. If the concept is adhered to, meaningful numbers will be generated even under difficult circumstances, and statisticians will be able to work the numbers; if not, the "results" might be worth less than the paper they are written on.

Statistics are employed both to check the quality and to set the requirements (synonymous with "specifications").

Chapters 1 and 2 introduced the basic statistical tools. The necessary computer can do more than just run statistics packages: in this chapter, a number of techniques are explained that tap the benefits of fast data handling, namely filtering, optimization, and simulation.

3.1 GMP BACKGROUND

With the safety of the consumer at the top of their minds, the Health Authorities of various countries have over the years established rules and guidelines to prevent the sale of harmful or ineffectual "medicines," and to establish minimal quality goals for those medications that do get their approval. As can be imagined, the variety of national givens (metric/Anglo-Saxon measuring systems; linguistic definition of terms; differences in medical practice; historical preferences; legal precedents, etc.) brought about a near-Babylonic confusion that was of no further consequence as long as everybody produced only for the local market. Today, globalization enforces harmonization against many an authority's will. Obtaining quality results is only possible if all of the manufacturer's players are included in the effort: definition of market needs, product design, production, control, and logistics. The GMP guidelines touch on all of these.

137

The pharmaceutical industry is a convenient example because the level of regulation is high and encompassing, and codification and enforcement are thorough. Other industries face similar pressures, perhaps less out of public safety needs than to play the "quality and reliability" card and stay in the game for a long time. Thus, other systems have evolved, like the EN and the ISO norm, that naturally evolved to something less specific, but very similar to the GMPs, and more generally applicable.

Of all the requirements that have to be fulfilled by a manufacturer, starting with responsibilities and reporting relationships, warehousing practices, service contract policies, airhandling equipment, etc., only a few of those will be touched upon here that directly relate to the analytical laboratory. Key phrases are underlined or are in *italics*: Acceptance Criteria, Accuracy, Baseline, Calibration, Concentration range, Control samples, Data Clean-Up, Deviation, Error propagation, Error recovery, Interference, Linearity, Noise, Numerical artifact, Precision, Recovery, Reliability, Repeatability, Reproducibility, Ruggedness, Selectivity, Specifications, System Suitability, Validation.

Selectivity and Interference: Selectivity means that only that species is measured which the analyst is looking for.[10] A corollary is the absence of *chemical interferents*. A lack of selectivity is often the cause of nonlinearity of the calibration curve. Near infra-red spectroscopy is a technology that exemplifies how seemingly trivial details of the experimental set-up can frustrate an investigator's best intentions; Ref. 124 discusses some factors that influence the result.

Linearity appeals to mind and eye and makes for easy comparisons. Fortunately, it is a characteristic of most analytical techniques to have a linear signal-to-concentration relationship over at least a certain *concentration range*;[125,126] in some instances a transformation might be necessary for one axis (e.g., logarithm of ion activity) to obtain linearity. At both ends there is a region where the calibration line gradually merges into a horizontal section; at the low-concentration end one normally finds a *baseline* given by background *interference* and measurement *noise*. The upper end of the calibration curve can be abrupt (some electronic component reaches its cut-off voltage, higher signals will be clipped), or gradual (various physical processes begin to interfere see Fig. 3.1). As long as there is no disadvantage associated with it, an analyst will tend toward using the central part of the linear portion for quantitative work. This strategy serves well for routine methods: a few calibration points will do and interpolation is straightforward.

Nonlinear calibration curves are not forbidden, but they do complicate things quite a bit: more calibration points are necessary, and interpolation from signal to concentration is often tedious. It would be improper to apply

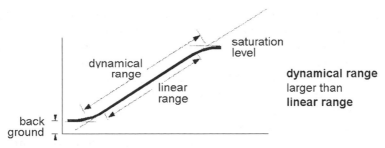

Figure 3.1. Definition of linear (LR) and dynamic (DR) ranges. The DR is often given as a proportion, i.e., 1 : 75, which means the largest and the smallest concentrations that could be run under identical conditions would be different by a factor of 75.

a regression of concentration x on signal y to ease the calculational load, cf. Section 2.2.1, because all error would be assigned to the concentrations, and the measured signals would be regarded as relatively free of uncertainty. The authors are aware of AAS equipment that offers the user the benefit of "direct concentration read-out" of unknown samples, at the price of improper statistical procedures and curve-fitting models that allow for infinite or negative concentrations under certain numerical constallations.[115] Unless there is good experimental evidence for or sound theoretical reasoning behind the assumption of a particular nonlinear model, justification is not easy to come by. A good example for the use of the nonlinear portion near the detection limit is found in an interference-limited technique, namely the ion-selective electrode. An interpolation is here only possible by the time-honored graphical method, or then by first fitting a moderately complex nonlinear theoretical model to the calibration data and then iteratively finding a numerical solution. In both cases, in linear regression, the preferred option where appropriate, and in curve-fitting, the model is best justified by plotting the residuals $r_i = y_i - Y(x_i)$ *versus* x_i (Section 2.2.3) and discussing the evidence.

Accuracy is the term used to describe the degree of *deviation* (*bias*) between the (often unknown) true value and what is found by means of a given analytical method. Accuracy cannot be determined by statistical means; the test protocol must be devised to include the necessary comparisons (blanks, other methods).

Precision: The *repeatability* characterizes the degree of short-term control exerted over the analytical method. *Reproducibility* is similar, but includes all the factors that influence the degree of control under routine and long-term conditions. A well-designed standard operating procedure permits one to repeat the sampling, sample work-up, and measurement process and repeatedly obtain very similar results. As discussed in Sections 1.1.3 and 1.1.4, the absolute or relative standard deviation calculated from experimental data is influenced by a variety of factors, some of them beyond the control of the analyst.[75,90,94,127] Thus there is no one agreed-upon relative standard deviation to judge all methods and techniques against (example: "±0.5% is 'very good', ±3% is 'acceptable', ±5% is 'insufficient'"); the analyst will have to assemble from different sources a notion of what constitutes an "acceptable" RSD for his or her particular problem. The residual curvature will also have to be judged against this value.

Reliability: This encompasses factors such as the laboratory environment and organization, equipment design and maintenance, personnel training, skills and experience, design and handling of the analytical method, etc.

Economic Considerations: Quality systems like the GMPs and the ISO norm do not require operations to be economical; it is recognized, however, that zero risk implies infinite costs, and that the type and amount of testing should be scientifically justified such that there is reasonable assurance that a product meets specifications that are in line with the associated risks and the intended use pattern, and still are affordable.

3.2   DEVELOPMENT, QUALIFICATION, AND VALIDATION

Demonstrating that an analytical procedure performs as intended is a GMP concern.[3,19,35–37,128–132] To this end, the employed equipment and the design of the method should be such that the intended goals can be met. The method validation delivers the formal proof that the outcome meets the expectations.

Installation-, Operations-, and Performance Qualification: Starting with the given task, minimal design and performance criteria are written down: this results in a target profile for the selection and purchase of the necessary equipment, and a checklist of the tests that need to be carried out. The vendor will in general provide a description of equipment performance that covers the typical uses it is put to; tests at the vendor's lab may be necessary to make certain the intended application will work. The purchasing agreement will include the necessary safeguards and a plan on how to proceed up to the customer's acceptance signature. A first check after the installation shows whether the equipment configuration is okay. Each module is then tested for correct operation, and the entire machine is challenged for absence of unintended interactions between and correct cooperation amongst the modules. Finally, equipment performance is demonstrated by checking the baseline for stability, the signal/noise ratio at various parameter settings, etc. The qualification plan and all documentation is then filed for future reference; a maintenance and an operator training plan must be set up. The equipment is then ready for use.

Method Development: An analytical method should be designed with standard sample work-up techniques[13] and average equipment performance levels in mind. The intended goal should first be put in writing ("determine A, B, and C at their typical concentrations of ... mg/ml, sample size X grams, required precision 0.7%, LOQ ≤ ... mg/ml; expected concentration ranges A_{min} ... A_{max}, ..., specification limits ...; throughput of ≥ 15 samples/hour"). The requisite calibration and control samples should be defined. A *system suitability test* should be devised that is easy to implement and that will, by reference to clearly stated criteria, show up any deficiencies that would lead to measurement problems, such as insufficient recovery, resolution, precision, or linearity. The method is completed by laying out a standard procedure that includes preparatory work, equipment settings, calibration, systems suitability decision, controls, sample sequence, and data evaluation. The exact procedure by which the raw data is transformed into results, i.e., statistical model, computer program, data entry, calculations, presentation of intermediate and final results, ... , must be defined. The calculational procedure should consider such things as calculator or software word-length, *error propagation*, and *numerical artifacts*. The decision pathways are pre-defined so that a result within certain value-range will lead to an ACCEPTED, all others to a REJECTED entry in the interpretation/comments section of the report. Certain combinations of observations (numerical result, peak-shape details, impurity profile, etc) will trigger actions that are detailed in an *"error recovery protocol,"* a standard operating procedure that tells the analyst what measures are to be taken in case of an unexpected outcome. For example, "If the above test did not yield a significant difference, continue with the measurements according to Procedure A, or else recalibrate using Procedure B." There are always things one could not possibly have thought of beforehand, so a catch-all clause, "if none of the aforementioned situations apply, notify the supervisor," is included; he will launch an investigation.

Method Validation: The signal path from detector up to the hard copy output of the final results must be perceived as a chain of error-prone components: there are errors due to conception,[115] construction, installation, calibration, and (mis-)use. Method validation checks into these aspects.

With precise expectations as to method performance set out, a test protocol is drawn up that lists the experiments that are to be carried out to challenge the method. High or low concentrations, low recovery levels, new chromatography columns, low or high temperature or pressure settings, other operators or days of the week, season, etc are brought to bear. For each experiment, the reasoning and the success criteria are written down. The intention is to prove that the method works according to expectations and does not return faulty or misleading results; the method must be shown to be rugged and reliable. The validation report contains the full documentation of these experiments and concludes with a statement that the method can or cannot be employed.

Ruggedness is achieved when small deviations from the official procedure, such as a different make of stirrer or similar-quality chemicals from another supplier are tolerable and when the stochastic variation of instrument readings does not change the overall interpretation upon duplication of an experiment. A non-rugged method leaves no margin for error and thus is of questionable value for routine use in that it introduces undue scatter and bias in the results (Section 4.1). Ruggedness is particularly important when the method must be transferred to other hands, instruments, or locations. This is no idle question: an involved protocol is not likely to be followed for very long, especially if it calls for unusual, tricky, or complicated operations, unless these can be automated. Unrecognized systematic errors,[17] if present, can frustrate the investigator or even lead to totally false results being taken for true.[17,24,90] The ruggedness of an analytical method thus depends on the technology employed and, just as important, on the reliability of the people entrusted to apply it.

Recovery is a measure of the efficiency of an analytical method, especially the sampling and sample work-up steps, to recover and measure the analyte *spiked* into a blank matrix (in the case of a pharmaceutical dosage form containing no active principle, the blank matrix is called a "*placebo*"). Note: even an otherwise perfect method suffers from lack of credibility if the recovered amount is much below what was added; the rigid connection between content and signal is then severed. *Recovery* is best measured by adding equal and known amounts of analyte to (a) the solvent and (b) the *blank matrix* from which the analyte is to be isolated and determined. The former samples establish a reference calibration curve. The other samples are taken through the whole sample work-up procedure and yield a second calibration curve. If both curves coincide recovery is 100% and interference is negligible. A difference in slope shows the lack of extractive efficiency or selectivity. Problems crop up, however, if the added standard does not behave the same way as the analyte that was added at the formulation stage of the product. Examples of such unexpected "matrix-effects" are the (1) slow adsorption of the analyte into packaging material,[11] (2) the redistribution of an analyte between several compartments/phases in the course of days or months, or (3) the complexation of the analyte with a matrix component under specific conditions during the manufacturing process that then leave the analyte in a "frozen equilibrium" from which it cannot be recovered. A series of schematic graphs shows some of the effects that are commonly observed (combinations are possible). (See Fig. 3.2.)

Calibration and *control* means appropriate standards as well as positive

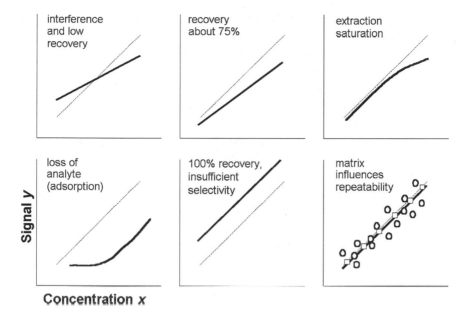

Figure 3.2. Schematic representation of some of the effects on the calibration curve observed during validation.[173,174] The dotted line represents the reference method or laboratory, the solid line is the test method or new laboratory.

and negative control samples are available, and their use to secure control over the methodology is established.[13,26] Control samples are blinded, if possible, and pains are taken to achieve stability over months to years so that identical control samples can be run with calibrations that are far removed in time or location from the initial method development site.[9–11] Control chart techniques are used to supervise both the standards and the analytical method. (See Section 1.8.4.) The linking of calibration factors across time and space is discussed in Section 4.32.

As far as the bench-chemist is concerned, the following nonexhaustive list of points should be incorporated into the experimental plan:

- Ensure that the actual instrument configuration conforms to what is written under "Experimental": supplier, models, modifications, consumables (HPLC or GC columns, gaskets, etc.), and software for the main instrument, peripherals (injectors, integrators, computers, printers, plotters, etc.), and ancillary equipment (vortexer, dispensers, balances, centrifuges, filters, tubing, etc.).

- For all critical equipment log books are available that show the installation and performance qualification checks, maintenance, validation, and calibration history, and status.

- All reagents are traceable to certificates of analysis, are used within the posted shelf-life, and are from known suppliers.

- Standard substances of defined purity must be available for major components and the main impurities (criteria: mole fraction, toxicity, legal requirements, etc.).[9]

- Solvents and sample matrices must be available in sufficient amounts and appropriate purity (e.g., pooled blood plasma devoid of analytes of interest).

- An internal standard (IS) that is physicochemically reasonably similar to the main analyte must be available.

- The analytical method must have been run using a variety of compounds that might turn up as impurities (synthesis byproducts, degradation products) to prove that there is sufficient selectivity. Materials in contact with the sample (tubing, filters, etc.) that might interfere with the analytes of interest must be proven to be innocuous.

- A series of calibration standards (CS) is made up that covers the concentration range from just above the limit of detection to beyond the highest concentration that must be expected (extrapolation is not accepted). The standards are made up to resemble the real samples as closely as possible (solvent, key components that modify viscosity, osmolality, etc.).

- A series of blinded standards is made up (usually low, medium, high; the analyst and whoever evaluates the raw data should not know the concentration). Aliquots are frozen in sufficient numbers so that whenever the method is again used (later in time, on a different instrument or by another operator, in another laboratory, etc.), there is a measure of control over whether the method works as intended or not. These so-called "QC-standards" (QCS) must contain appropriate concentrations of all components that are to be quantified (main component, e.g., drug, and any impurities or metabolites).

- During the method validation phase, the calibration, using the CS solutions, is repeated each day over at least one week to establish both the within-day and the day-to-day components of the variability. To this end, at least 6 CS, evenly spread over the concentration range, must be repeatedly run ($m \approx 8$–10 is usual), to yield $n \approx 50$ measurements per day. If there are no problems with linearity and heteroscedacity, and if the precision is high (say, $CV \leq 2$–5%, depending on the context), the number of repeats m per concentration may be reduced from the second day onwards ($m = 2 - 3$ is reasonable). The reasoning behind the specifics of the validation/calibration plan is written down in the method justification document for later reference.

- After every calibration, repeat QC-samples are run ($m = 2 - 3$)
- At the end of the validation phase, an overall evaluation is made and various indicators are inspected, such as
  - Variability and precision of calibration line slope.
  - Variability and significance of the intercept.
  - Linearity, usually assessed by plotting the residuals *vs.* the concentration.
  - Back-calculated CS concentrations $\pm$ CL(X).
  - Interpolated QCS concentrations $\pm$ CL(X).
  - Residual standard deviation: within-day effects and for pooled data.
  - Correlation coefficient r or coefficient of determination r^2 (these indicators are not overly useful, see Section 2.1, but so well-known that some bureaucrats are unhappy if they do not have them, so they were included here but rounded to four or five decimal places).
  - For each of the CS and the QC concentrations the overall mean and standard deviation are compared to the daily averages and SDs; from this, variance components for the within-day and day-to-day effects are estimated by subtraction of variances.
- If the analytical method survives all of the above criteria (suitably modified to match the situation), it is considered to be "under control."
- Changing major factors (instrument components, operators, location, etc.) means revalidation, generally along the same lines.

3.3 DATA TREATMENT SCHEME

Data acquisition is not treated in this book. The most common technique is to convert a physicochemical signal into a voltage by means of a sensor, and feed the electrical signal into a digital volt meter or a chart recorder. With today's instrumentation this is no longer the problem it used to be.

Acceptance Criteria: *System suitability tests* are to be conducted, if necessary, every time an analysis method is installed, in order to ensure that meaningful results are generated. Criteria are in place to supply the necessary information for a go/no go decision.

Data Assembly and Clean-Up: Section 4.31 provides an example of how tables often look when data was compiled from a multitude of sources: formats might differ, data quality is uneven, and comments in text format were smuggled in to further qualify the entries. Since vital background information will be missing unless the statistician has chemical or similar training, the interpretation can suffer.

Data evaluation used to be a question of a graduated ruler, a pencil, and either a slide-rule or a calculator. Today's equipment takes the drudge out of such mundane jobs, but even for such widely employed operations such as the calculation of the area under a peak, the availability of an "integrator" is no guarantee for correct results. The algorithms and the hardware built into such machines were designed around a number of more or less implicit assumptions, such as the form of chromatographic signals.[15,17] It should not come as a surprise then, that analytical methods that were perfectly well behaved on one instrument configuration will not work properly if only one element, such as an integrator, is replaced by one of a different make. The same is true, of course, for software packages installed on PCs (e.g., use of single- or double-precision math as a hidden option, or the default parameters that set, for example, horizontal or valley-to-valley baseline options). (See Figure 3.3.)

Statistical and algebraic methods, too, can be classed as either rugged or not: they are rugged when algorithms are chosen that on repetition of the experiment do not get derailed by the random analytical error inherent in every measurement,[107,133] that is, when similar coefficients are found for the mathematical model, and equivalent conclusions are drawn. Obviously, the choice of the fitted model plays a pivotal role. If a model is to be fitted by means of an iterative algorithm, the initial guess for the coefficients should not be too critical. In a simple calculation a combination of numbers and truncation errors might lead to a division by zero and crash the computer. If the data evaluation scheme is such that errors of this type could occur, the validation plan must make provisions to test this aspect.

An extensive introduction into robust statistical methods is given in Ref. 134; a discussion of non-linear robust regression is found in Ref. 135. An example is worked in Section 3.4.

Presentation of Results: The whole derivation of the final result must remain open to scrutiny. *Raw data* is whatever is noted when reading off a gauge, displayed by an instrument, printed on paper, or dumped into a file (an unadulterated hardcopy no longer qualifies, according to the FDA). The raw data must be signed and dated by the operator (validated electronic systems can incorporate an "electronic signature and date stamp"). This typically includes the print-out of a balance (tare, net, total), the $y - t$ chart of a titrator, an IR spectrum, or a chromatogram. Computerized instruments will provide much more, like data tables, date and time, instrument type and identification. Any printout that can fade is best photocopied to ensure legibility after several years; it is even better to avoid such printers altogether, if a legal challenge is likely.

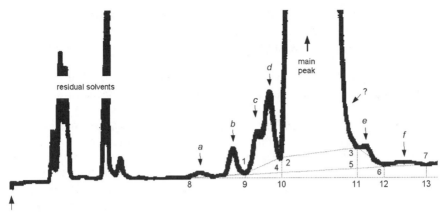

Figure 3.3. Peak integration options. The thick curve is an actual HPLC-trace from a purity-profile run. The residual solvent peaks are disregarded because this is information that is better acquired on a capillary-column GC. Besides the main peak, which is far off-scale due to the enormous signal-expansion chosen, one can detect at least six impurities ($a - f$); other chromatograms indicate a further impurity in the tailing flank of the main peak (?), which seems to have just about disappeared here. Its peak area is probably similar to that of peak e, but because of marginal resolution, it remains unresolved in many runs. This product poses the typical industrial quandary. Due to very similar chemical structures, all impurities cluster close to the main compound, and optimizing for one invariably destroys the resolution achieved for one or more other impurities. Chromatography systems offer a variety of integration options that generally are designed with the integration of the main peak in mind. The baseline (BL) can be set 8–13 (flat BL), 8–7 (sloping BL), 8–1–2–3–6–7 (valley-to-valley BL), or ?–3–6–7 (parabolic or exponential fit BL). The peak areas can then be assigned such that verticals are dropped (11–3–6–12–11 belongs to e) or 1–2–4 belongs to the main peak and the area of peaks c and d is either separated using a vertical or by slicing c off of d. Every such model introduces assumptions that might be fine for one situation but distort the truth for some other peak. Statistical evaluation of such data may require fine adjustments in chromatographic and integration parameters from run to run, so as not to introduce further artifacts into the data base. If baseline resolution of adjacent peaks is not possible, one should neither tinker with the integration parameters from one run to the next, nor simply accept default settings and hope a little bit of chemometrics would paper over the inconsistencies.

All formulas are to be written out by hand with the specific numbers in the right places on a sheet provided for the purpose, and the calculations are to be done by calculator. *Caution*: Write the numbers exactly as printed, do not round any digits, or the quality assurance unit (a sort of corporate vice squad) will not approve the report out of fear that someone could have cheated. A validated program can be used. While an Excel spreadsheet as such needs no validation, a simple cell-formula calls for extensive tests and documentation and proof that the sheet is password protected against fraudulent manipulation. On top of that, the analyst's supervisor is required to confirm the calculation and sign off on it, which in effect reduces the computer to an expensive typewriter. The recalculation requirement is relaxed if one works with a LIMS or a similarly tested and secured system. The authorities want to make sure that a decision can be revisited, and fraud or incompetence exposed.

The final result is written in a predefined manner: number of decimals, units, number of repeats, relevant test criteria, and a decision statement (fails/passes).

Specifications: At least in the pharmaceutical industry, specification limits are usually given by default values which need not be defended, such as the ±5% limits on the assay in Europe. If there is any reason for wider limits, solid scientific evidence and reasoning must be brought forward if an approval is to have much of a chance. Since a basic demand is that state-of-the-art methodology and equipment be used, there is some pressure that the generous specifications granted years ago (because analytical technology was limiting the attainable precision) be tightened up after a grace period; a good example is the transition from quantitative TLC to HPLC in the 1980s, with a concomitant shift from ±10% to ±5% limits.

If one is less restrained in setting specification limits, a balance can be struck between customer expectations and the risk and cost of failure; a review of available data from production and validation runs will allow confidence limits to be calculated for a variety of scenarios (limits, analytical procedures, associated costs; see Fig. 2.15 for an example).

Records Retention: Product design, manufacturing, and control documentation must remain available for many years after the product has been sold. For pharmaceuticals, the limit can be as low as one year after expiration, and as high as 30 years after taking the product off the market, for instance, if a patient suffered side-effects that might be related to the drug. This clearly points out the care that must be invested in making documents readable and self-explanatory. Electronic records pose a special risk because there is no guarantee that the particular hard- and software under which the records were generated will still be available in 5, 10, or more years.

3.4 EXPLORATORY DATA ANALYSIS (EDA)

Natural sciences have the aura of certainty and exactitude. Despite this, every scientist has experienced the situation of being befuddled:

- Analytical projects must often be initialized when little or nothing is known about the system to be invesigated. Thus data are generated under conditions believed to be appropriate, and after some numbers have accumulated, a review is undertaken.

- Well-known processes sometimes produce peculiar results; data covering the past 20 or so batches are collated for inspection. Frequently, one has no idea what hypotheses are to be tested.

Both situations lend themselves to exploration: The available data are assembled in a data file along the lines "one row per batch, one column per variable," or an equivalent organization. In a first step the data has to be critically reviewed and cleaned up. (See Section 4.31.) A little detective work almost always turns up transcription or calibration errors, or even calculational mistakes (someone hit the wrong button on the calculator and got ($\div$3) instead of ($\times$3); at first glance the result looks plausible and is jotted down because the exponent that is off by -1 or $+1$ is overlooked). Programs are then used to plot data (e.g., Figures 4.41 and 4.42) and search for correlations (program CORREL) or choose any three variables for display in a pseudo three-dimensional format (programs XYZ or XYZCELL). Combinations of variables are tried according to intuition or any available rule of thumb. Perceived trends can be modeled, that is, a mathematical function can be fitted that is then subtracted from the real data. (See program TESTFIT.) What ideally remains are residuals commensurable in size with the known analytical error and without appreciable trend.

Exploratory data analysis is a form of a one-man brain-storming session; the results must be accordingly filtered and viable concepts can then be turned into testable hypotheses. The filtering step is very important, because inexplicable and/or erroneous data turn up in nearly every data set not acquired under optimally controlled conditions. Global figures of merit, such as the correlation coefficient (see Section 2.1), if not supported by visual trend analysis, common sense, and plausibility checks, may foster wrong conclusions. An example is worked in Section 4.11, "Exploring a Data Jungle," and in Section 4.22.

The EDA technique cannot be explained in more detail because each situation needs to be individually appraised. Even experienced explorers now and then jump to apparently novel conclusions, only to discover that they are the victims of some trivial or spurious correlation.

3.5 OPTIMIZATION TECHNIQUES

After the first rounds of experimentation, both the analytical method (equipment and procedure) and the data evaluation scheme are rarely as good as they could be. Assuming the equipment is a given, the procedure, the various settings, and the model can be subjected to an investigation that closes gaps and polishes the rough edges, figuratively speaking. One should of course

Table 3.1. Design of an Experiment

Classical experiment	$\Leftarrow ? \Rightarrow$	Statistically guided experiment
Full analysis after all experiments are finished	$\Leftarrow ? \Rightarrow$	On-the-run analysis
Functional relationships	$\Leftarrow ? \Rightarrow$	Correlations
Orthogonal factors, simple mathematical system	$\Leftarrow ? \Rightarrow$	Correlated factors, complex mathematics

first know what one wants to improve: speed, consumption of reagents, precision, accuracy, costs, etc. Once this is set, the goal can be approached from different sides, all with their pros and cons:

- Small improvements based on experience: This is sure to work, however, the gains may be small.

- Search for the overall optimum within the available parameter space: Factorial, simplex, regression, and brute-force techniques. The classical, the brute-force, and the factorial methods are applicable to the optimization of the experiment. The simplex and various regression methods can be used to optimize both the experiment and fit models to data.

- Simulation of the system: This is fine if the chemistry and physics are well known.

3.5.1 Full Factorial vs. Classical Experiments

In planning experiments and analyzing the results, the experimenter is confronted with a series of decisions (see Table 3.1).

What strategy should one follow? In the classical experiment, one factor is varied at a time, usually over several levels, and a functional relationship between experimental response and factor level is established. The data analysis is carried out after the experiment(s). If several factors are at work, this approach is successful only if they are more or less independent, that is, do not strongly interact. The number of experiments can be sharply increased as in the brute-force approach, but this might be prohibitively expensive if a single production-scale "experiment" costs five- or six-digit dollar sums. Figure 3.4 explains the problem for the two-factor case.

As little as three factors can confront the investigator with an intractable situation if he chooses to proceed classically. One way out is to use the factorial approach, which can just be visualized for three factors. An example from process optimization work will illustrate the concept. Assume that temperature, the excess concentration of a reagent, and the pH have been iden-

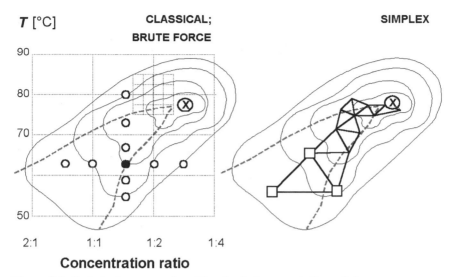

Concentration ratio

Figure 3.4. Optimization approaches. The classical approach fixes all factors except one, which is systematically varied (rows of points in left panel); the real optimum (x) might never be found this way. The brute-force approach would prescribe experiments at all grid points (dotted lines), and then further ones on a finer grid centered on 80/1:2, for example.

A problem with the simplex-guided experiment (right panel) is that it does not take advantage of the natural factor levels, e.g., molar ratios of 1:0.5, 1:1, 1:2, but would prescribe seemingly arbitrary factor combinations, even such ones that would chemically make no sense, but the optimum is rapidly approached. If the system can be modeled, simulation[136] might help. The dashed lines indicate ridges on the complex response surface. The two figures are schematic.

tified as probable factors that influence yield; a starting value (40°C, 1.0% concentration, pH 6) and an increment (10, 1, 1) is decided upon for each; this gives a total of eight combinations that define the first block of experiments. (See Figure 3.5.) Experiment 1 starts off with 40°C, 1% concentration and pH 6, that is, all increments are zero. Experiments 2–8 have one, two or three increments different from zero. The process yields are given in Table 3.2, second row. These eight combinations, after proper scaling, define a cube in 3-dimensional space.

Program FACTOR8 in Section 5.2.3 lists the exact procedure, which is only sketched here. Table 3.2, for each factor and interaction, lists the observed effect, and the specific effect, the latter being a slope or slopes that make(s) the connection between any factor(s) and the corresponding effect. A t-test is conducted on the specific effects, i.e., $t = E_j/(\Delta x \cdot s_E)$, where E_j is the observed effect $y_j - y_1$, s_E its (estimated) standard deviation, and Δx the change in the factor(s) that produced the effect. If t is larger than the

Table 3.2. Results of a Factorial Experiment[a]

Item	Factor	1	a	b	c	ab	ac	bc	abc
Measurement	$y()$	51	62	54	46	68	59	52	62
Obs. effect	$E()$	0	12.0	4.5	-4.0	0.0	-0.5	0.0	-1.5
Spec. effect	$F()$	±0.0	+1.2	+4.5	-4.0	+0.0	-0.1	+0.0	-0.6
Model	$M()$	—	+1.2	+4.5	-4.0	—	—	—	-0.6
Estimate	$Y()$	51.25	61.75	54.25	45.75	67.75	59.25	51.75	62.25
Deviation	$y()-Y()$	-.25	+.25	-.25	+.25	+.25	-.25	+.25	-.25

[a]See data file FACTOR.dat.

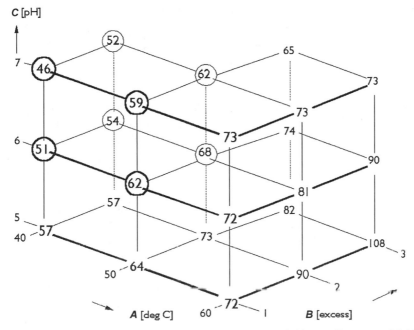

Figure 3.5. Factorial space. Numbers in circles denote process yields actually measured (initial data set); all other numbers are extrapolated process yields used for planning further experiments (assuming that the repeatability $s_y = \pm 0.1$; all values are rounded to the nearest integer); the "estimated yield" of 108% shows that the simple linear model is insufficient.

critical $t(p, f)$, the specific effect F_j is taken to be significant, otherwise it is not included in the model; a model effect M_j for a combined factor, e.g., ac, indicates an interaction between factors a and c. Specific effects found to be nonsignificant are given as a dash (—). In Fig. 3.5, the yields measured for a reaction under different conditions of pH, temperature, and excess of a reagent are circled. The extrapolation to the adjacent cube is straightforward, the geometrical center of the original cube serving as a starting point; in this fashion the probable direction of yield-increase is easily established. Note that the reduction of a complex reality to a linear three-factor model can result in extrapolations that do not make sense: a yield cannot exceed 100%.

Example 38: For $T = 55°C$, 1.7% excess, and pH 7.4, the coordinates relative to the center of the cube defined by the first eight experiments are $a = 55 - (40 + 50) \div 2 = 10$; $b = 1.7 - (1 + 2) \div 2 = 0.2$; $c = 7.4 - (6 + 7) \div 2 = 0.9$; estimated yield is $Y = (51 + 62 + 54 + 46 + 68 + 59 + 52 + 62)$

$\div 8 + 1.2 \cdot a + 4.5 \cdot b - 4.0 \cdot c + 0 \cdot a \cdot b + 0 \cdot a \cdot c + 0 \cdot b \cdot c - 0.6 \cdot a \cdot b$
$\cdot c$; $Y = 56.75 + 12 + 0.9 - 3.6 - 1.08 = 64.97$. Thus, the yield is expected
to increase in this direction; after further extrapolations, the vicinity of the
coordinate 60°C, 2% excess, and pH 6 appears to be more promising.

The course of action is now to do either of two things:

1. Continue with measurements so that there are always eight points as
   above by adding four coplanar points parallel to the side of the existing
   cube, e.g., at 60°C, and replacing the extrapolations (72, 81, 73, resp.
   73) by real measurements. This choice would have the advantage of
   requiring only minimal mathematics, but one is limited by having to
   choose cubically arranged lattice points.

2. Program FACTOR8 uses a brute-force technique (cf. Figure 3.4) to
   estimate the changes necessary in each of the three factors in order to
   move closer to the local maximum. Since the precise direction is sensitive to the noise superimposed on the original eight measurements,
   these "suggested changes" should only be taken as indications to plan
   the next two to four experiments, whose coordinates do not necessarily
   lie on the previously established lattice. All available results are entered
   into a multilinear regression program, e.g., Ref. 122 (programs of this
   type are available in many program packages, even on handheld computers, so that further elaboration is not necessary here). In essence,
   the underlying model is an extension of the linear regression model
   explained in Chapter 2, in that more than one independent variable x_i
   exists:

$$Y = a_0 + a_1 \cdot x_1 + a_2 \cdot x_2 + a_3 \cdot x_3 + \ldots \tag{3.1}$$

Because the preceding factor experiment suggests a, b, c, and abc as independent variables, cf. bottom row in Table 3.2, the data table would take on
the form:

Table 3.3. Data Matrices for MLR

a_1	b_1	c_1	$a_1 b_1 c_1$	y_1
a_2	b_2	c_2	$a_2 b_2 c_2$	y_2
.	.	.	.	.
.	.	.	.	.
.	.	.	.	.
a_n	b_n	c_n	$a_n b_n c_n$	y_n

Some experimenting might be necessary if it turns out that quadratic terms such as b^2 improve the fit between the model and the data. However, the relevant point is that such a model is only a means to refine and speed up the process of finding optimal conditions. For this purpose it is counter-productive to try for a perfect fit, it might even be advantageous to keep the model simple and throw out all but the best five to 10 experiments, choose new conditions, and then return to the work bench.

The factorial experiment sketched out above is used in two settings (cf. Ref. 137 for a tutorial):

1. If very little is known about a system, the three factors are varied over large intervals; this maximizes the chances that large effects will be found with a minimum of experiments, and that an optimal combination of factors is rapidly approached (for example, new analytical method to be created, no boundary conditions to hinder investigator).

2. If much is known about a system, such as an existing production process, and an optimization of some parameter (yield, purity, etc.) is under investigation, it is clearly impossible to endanger the marketability of tons of product. The strategy is here to change the independent process parameters (in random order) in small steps within the operating tolerances (cf. Fig. 1.8). It might be wise to conduct repeat experiments, so as to minimize the overall error. In this way a process can evolve in small steps, nearly always in the direction of an improvement. The prerequisites are a close cooperation between process and analytical chemists, a motivated staff willing to exactly follow instructions, and preferably, highly automated hardware. The old-style and undocumented "lets turn off the heat now so we can go to lunch" mentality introduces unnecessary and even destructive uncertainty

The choice of new vertices should always take into consideration the following aspects:

- A new point should be outside the lattice space already covered.
- The expected change in the dependent (= target) variable (effect) should be sufficient to distinguish the new point from the old ones.
- The variation of a factor must be physically and chemically reasonable.
- Extrapolations should never be made too far beyond *terra cognita*.

The optimization technique embodied in Program FACTOR8 could easily be expanded from three purported factors to four or more, or to three, four, or even five levels per factor. Mathematically, this is no problem, but (1) the

Table 3.4. Number of Necessary Experiments

	With L = Number of Levels per Factor f = Number of Factors, Eq. (2.36)			
$n = L^f$	$L = 2$	3	4	5
Number of Factors				
$f = 3$	$n = 8$	27	64	125
4	16	81	256	625
5	32	243	1024	3125

number of experiments rises sharply, and (2) improper design can lead to false conclusions, as is elegantly explained in Ref. 138. The most economical approach is then to pick out three factors, study these, and if one of them should prove to be a poor choice, replace it by another. The necessary number of experiments for the first round of optimization is as shown in Table 3.4; see also Figure 3.5.

3.5.2 Simplex-Guided Experiments

In three-dimensional space the simplex consists of a triangle as shown in Figure 3.4; the concept is readily expanded to four dimensions (tetragon-shaped simplex), or higher. Because a large number of experiments needs to be run, simplex optimization is best used in situations where an experiment does not cost much, or, preferably, can be done on-line by changing some settings and waiting a few seconds until the signal has stabilized. A good example would be the optimization of an AAS instrument configuration with the factors burner gas-flow, oxidant gas-flow, source intensity, horizontal and vertical source positions; the target variable could be the signal-to-noise ratio for a standard sample. Each of the five parameters is changed by turning a knob and reading the appropriate gauge. The parameter values and the measurement are entering into the computer, which then does the calculation and suggests a new combination. Here it is even more important than with the factorial experiments that the factors be continuous variables, particularly in the region of the optimum, where small steps will be taken. The choice of initial conditions is crucial: Since the simplex-program does not know its chemistry and human imagination fails in four-space and higher, it is hard to tell whether one was led off towards a secondary maximum (see Section 4.2) even if simplex algorithms employ some sophisticated tricks to automatically adjust the step size and direction.

3.5.3 Optimization of the Model: Curve Fitting

Once a mathematical model has been chosen, there is the option of either fixing certain parameters (see Section 4.10) or fixing certain points, e.g., constraining the calibration line to go through the origin.[63–66,74,113]

When one tries to fit a mathematical function to a set of data one has to choose a method or algorithm for achieving this end, and choose a weighting model with which to judge the goodnss of fit.

As with the method, four classes can be distinguished:

- Graphical fit.
- Closed algebraic solution.
- Iterative technique.
- Brute force approach.

Each of these four classes has its particular advantages.

The *graphical* fit technique makes use of a flexible ruler (or a steady hand): through a graphical representation of the data a curve is drawn that gives the presumed trend. While simple to apply, the technique is very much subject to wishful thinking. Furthermore, the found "curve" is descriptive in nature, and can only be used for rough interpolations or estimation of confidence intervals. A computerized form of the flexible ruler is the *spline* function[139,140] that exists in one- and two-dimensional forms. The gist is that several successive points (ordered according to abscissa value) are approximated by a quadratic or a cubic polynomial; the left-most point is dropped, a further one is added on the right side, and the fitting process is repeated. These local polynomials are subject to the condition that at the point where two of them meet, the slope must be identical. Only a very small element out of each constrained polynomial is used, which gives the overall impression of a smooth curve. The programs offer the option of adding "tension," which is akin to stiffening a flexible ruler. The spline functions can be used wherever a relatively large number of measurements is available and only a phenomenological, as opposed to a theoretical, description is needed. Splines can, by their very nature, be used in conditions where no single mathematical model will fit all of the data, but, on the other hand, the fact that every set of coefficients is only locally valid, means that comparisons can only be done visually.

The *algebraic* solution is the classical fitting technique, as exemplified by the linear regression (Chapter 2). The advantage lies in the clear formulation of the numerical algorithm to be used and in the uniqueness of the solution. If one is free to choose the calibration concentrations and the number of

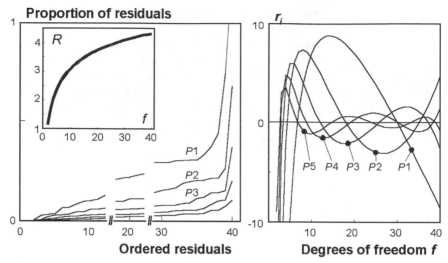

Figure 3.6. Approximating a given function using polynomials of order 1 through 5.

repeats, tips on how to optimize are found in Ref. 141. The disadvantages that polynomials can have are quite evident in Figure 3.6, where polynomials P of order 1 ... 5 were imposed on the Studentized range function from Figure 1.2. The data D is displayed in the inset, while the %-deviation $100 \cdot (P - D)/D$ is given at right; the alternating stretches of hi-low-hi ... bias are evident. If the absolute residuals $|P - D|$ are sorted according to size, the left figure is obtained, which shows that at least $\frac{3}{4}$ of all estimates found for a $P2$ model are larger than all but the worst one or two estimates found with the $P5$ model. There is no point in going to higher polynomials, because $P5$, with a total of six coefficients, is already down to $39 \div 6 = 6.5$ points per coefficient.

The *iterative* method encompasses those numerical techniques that are characterized by an algorithm for finding a better solution (set of model coefficients), starting from the present estimate[142]: Examples are the *regula falsi*, Newton's interpolation formula, and the simplex techniques.[118,129] The first two are well known and are amenable to simple calculations in the case of a single independent variable. Very often though, several independent variables have to be taken account of; while two variables (for an example see Section 4.2) barely remain manageable, further ones tax imagination. A simplex-optimization program (available at computing centers, as PC software, and also in the *Curve ROM* module of the HP-71 handheld computer) works in a $(k + 1)$-dimensional space, k dimensions being given by the coefficients (a_1, a_2, ... a_k), and the last one by χ^2. The global minimum in the $\chi^2 = f(a_1, a_2, \ldots$

a_k) function is sought by repeatedly determining the "direction of steepest descent" (maximum change in χ^2 for any change in the coefficients a_i), and taking a "step" to establish a new vertex. A numerical example is found in Table 1.26. An example of how the simplex method is used in optimization work is given in Ref. 143.

Here a problem enters that is already recognizable with Newton's formula: If pure guesses instead of graphics-inspired ones are used for the initial estimate, the extrapolation algorithm might lead the unsuspecting user away from the true solution, instead of toward it. The higher the number of dimensions, the larger the danger that a local instead of the global minimum in the goodness-of-fit measure is found. Under the keyword "ruggedness" a technique is shown that helps reduce this uncertainty. Simplex methods and their ilk, because of their automatic mode of operation, demand thoughtful use and critical interpretation of the results. They can be used to either fit a model to data or guide an experiment.

The *brute force* method depends on a systematic variation of all involved coefficients over a reasonable parameter space. The combination yielding the lowest goodness-of-fit measure is picked as the center for a further round with a finer raster of coefficient variation. This sequence of events is repeated until further refinement will only infinitesimally improve the goodness-of-fit measure. This approach can be very time-consuming and produce reams of paper, but if carefully implemented, the global minimum will not be missed, cf. Figures 3.4 and 4.4.

The algebraic/iterative and the brute force methods are numerical respectively computational techniques that operate on the chosen mathematical model. Raw residuals r are weighted to reflect the relative reliabilities of the measurements.

The weighting model with which the *goodness-of-fit* or figure-of-merit (GOF = $\Sigma(u_i)$) is arrived at can take any of a number of forms. These continuous functions can be further modified to restrict the individual contributions u_i to a certain range, for instance r_i is minimally equal to the expected experimental error, and all residuals larger than a given number r_{max} are set equal to r_{max}. The transformed residuals are then weighted and summed over all points to obtain the GOF. (See Table 3.5.)

A good practice is to use a weighting model that bears some inner connection to the problem and results in GOF figures that can be physically interpreted. A function of the residual standard deviation, s_{res}, which has the same dimension as has the reproducibility, s_y, might be used instead of χ^2.

The chosen weighting model should also be applied to a number of repeat measurements of a typical sample. The resulting GOF figure is used as a benchmark against which those figures of merit resulting from parameter fitting operations can be compared. (See Table 1.26.) The most common

Table 3.5. Weighting Schemes

$Y = f(x/a \dots)$	Y designates the model, with the independent variable(s) x, and parameters $a \dots$
y_i	Measured value
$r_i = y_i - Y(x\|a\dots)$	Residual
$w_i = w(x)$	Weighting function, e.g. reciprocal variance, cf. Section 2.2.10
$u_i = w_i \cdot g(r_i)$	Statistically and algebraically weighted residual
	Particular examples:
$u_i = w_i \cdot \text{ABS}(r_i)$	Linear
$u_i = w_i \cdot (r_i)^2$	Quadratic, absolute (least squares)
$u_i = w_i \cdot (r_i/Y(x\|a\dots))^2$	Quadratic, relative (least squares)
$u_i = w_i \cdot (r_i)^2/Y(x\|a\dots)$	Intermediate (χ^2)

situation is that one compares the residual standard deviation against the known standard deviation of determination (reproducibility or repeatability).

Once a fitted model is refined to the point where the corresponding figure of merit is smaller than the benchmark (Table 1.26), introducing further parameters or higher dimensions is (usually) a waste of time and only nourishes the illusion of having "enhanced" precision.

If several candidate models are tested for fit to a given data set, it need not necessarily be that all weighting models $w(x) \cdot g(r_i)$ listed as examples above would indicate the same model as being the best (in the statistical sense), nor that any given model is the best over the whole data range.

It is evident that any discussion of the results rests on three premises: Constraints, the fitted model, and the weighting model. Constraints can be boon or bane, depending on what one intends to use the regression for.[64] The employed algorithm should be of secondary importance, as long as one does not get trapped in a local minimum or by artifacts. While differences among fitted models can be (partially) judged by using the goodness-of-fit measure (graphical comparisons are very useful, especially of the residuals), the weighting model must be justified by the type of the data, by theory, or by convention.

3.5.4   Computer Simulation

"What if?" This question epitomizes both serious scientific thinking and children's dreams. The three requirements are (a) unfettered fantasy, (b) mental discipline, and (c) a thinking machine. The trick to get around the apparent

EXPERIMENT   SIMULATION

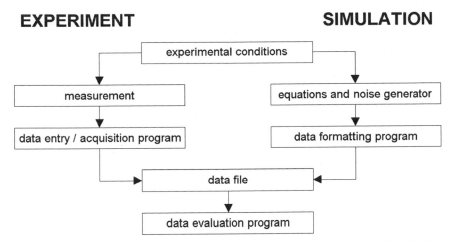

Figure 3.7. Schematic depiction of the relation between experiment and simulation. The first step is to define the experimental conditions (concentrations, molecular species, etc.), which then form the basis either for the experiment or for simulation. Real data are manually or automatically transferred from the instruments to the data file for further processing. Simulated values are formatted to appear indistinguishable from genuine data.

contradiction between points (a) and (b) is to alternately think orderly and freely. The "unimpeded" phase is relaxing: Think of all those "impossible" situations that were never dealt with in school because there are no neat and simple answers: That is precisely why there is a problem waiting to be solved. Then comes the bridled phase: Systematically work through all those imagined combinations. This can be enormously tiring, can tax memory and common sense, and demands precise recording. For this reason a computer is a valuable assistant (c).

What are the mechanics of simulation[42,136,144]? In Figure 3.7 the classical sequence experiment/raw-data/evaluation is shown. A persistent but uninspired investigator might experiment for a long time without ever approaching an understanding of the system under scrutiny, despite his attempts to squeeze information from the numbers. A clever explorer will with a few experiments gather some fundamentals facts, e.g., noise and signal levels, limits of operation of the instrumentation and the chemistry (boiling points, etc.), and combine this knowledge with some applicable mathematical descriptions, like chemical equilibria, stoichiometry, etc. This constitutes a rough model; some parameters might still have to be estimated. The model is now tested for fit against the available experimental data, and refinements are applied where necessary. Depending on the objective of the simulation procedure, various strategies can be followed: What is common to all of them is that the experiment is replaced by the mathematical model. Variation of the model parameters and/or assumptions allows "measurements" to be simulated (calculated) outside the experimentally tested parameter domain. The computer is used to sort through the numbers to either find conditions that should permit pivotal experiments[136] to be conducted, or to determine critical points. In either case, experiments must be done to verify the model. The circle is now closed, so that one can continue with further refinements of the model.

What are the advantages to be gained from simulation? Some are given in Section 1.4 and others are listed here:

- Refined understanding of the process under scrutiny.

- Optimization of the experimental conditions, so that reliable data can be acquired, or that data acquisition is more economical.

- The robustness of assumptions and procedures can be tested.

- If a well-developed theory is available, the limits of the experimental system (chemistry, physics, instrumentation) can be estimated.

- Very complex systems can be handled.

- Deterministic and random aspects can be separately studied.

A typical problem for simulation is the investigation of a reaction mechanism involving coupled equilibria, such as a pH-dependent dissociation in the presence of metal ions, complexing agents, and reagents. Some of the interactions might be well known, and for others bounds on association constants and other parameters can at least be roughly estimated. A model is then set up that includes the characteristics of UV-absorption and other observables for each species one is interested in. Thus equipped, one begins to play with the experimentally accessible concentrations to find under what conditions one of the poorly known values can be isolated for better characterization. In the end, various combinations of component concentrations can be proposed that with high probability will permit successful experiments to be conducted, in this way avoiding the situation where extensive laboratory effort yields nothing but a laconic "no conclusion can be drawn."

Here again, no precise instructions can be given because each situation will demand a tailored approach. (*Note*: Numerical simulation in many ways resembles the "what-if" scenario technique available in spreadsheets programs. Several programs supplied with this book allow the reader to play with functions and noise levels.)

3.5.5 Monte Carlo Technique (MCT)

In Section 1.4, the MCT was introduced in general terms as an important numerical tool for studying the relationship of variables in complex systems of equations. Here, the algorithm will be presented in detail, and a more complex example will be worked.

The general idea behind the MCT is to use the computer to roll dice, that is to generate random variations in the variables, which are then inserted into the appropriate equations to arrive at some result. The calculation is repeated over and over, the computer "rolling the dice" anew every time. In the end, the individual results average to that obtained if average variables had been inserted into the equations, and the distribution approximates what would have been found if the propagation-of-errors procedure had been applied. The difference between classical propagation of errors and MCT is that in the former the equation has to be in closed form (i.e., result $= f(x, y, z, \ldots)$, the result variable appearing only once, to the left of the equals sign), and the equation has to be differentiable. Also, propagation of errors assumes that there is a characteristic error Δx, symmetrical about x_{mean}, as in mean/standard deviation, that is propagated. Often one of these conditions is violated, either because one does not want to introduce simplifying assumptions in order to obtain closed solutions (e.g., n is assumed to be much larger than 1 to allow the transition from "$n - 1$" to "n" in the denominator), or because differentiation is impossible (step-functions, iterative algorithms, etc.).

MCT allows one to choose any conceivable error distribution for the variables, and to transform these into a result by any set of equations or algorithms, such as recursive (e.g., root-finding according to Newton) or matrix inversion (e.g., solving a set of simultaneous equations) procedures. Characteristic error distributions are obtained from experience or the literature, e.g., Ref. 95.

In practice, a normal distribution is assumed for the individual variable. If other distribution functions are required, the algorithm $z = f(CP)$ in Section 5.1.1, respectively the function FNZ() in Table 5.16 has to be appropriately changed.

The starting point is the (pseudo-) randomization function supplied with most computers; it generates a rectangular distribution of events, that is, if called many times, every value between 0 and 1 has an equal probability of being hit. For our purposes, many a mathematician's restraint regarding randomization algorithms (the sequence of numbers is not perfectly random because of serial correlation, and repeats itself after a very large number of cycles,[145] $c \approx 10^9$ for a PC) is irrelevant, as long as two conditions are met:

Table 3.6. Generation of Normally Distributed Random Numbers

1	R = random number $0 \leq R \leq 1$	
2	if $R \leq 0.5$ then CP = R	sign $w = -1$
	if $R > 0.5$ then CP = $1 - R$	sign $w = +1$
3	Calculate $u = \log_{10}(\text{CP})$	
4	Calculate $v = f(u)$	
5	Calculate $z = w \cdot v$	
6	Calculated ND-variable = $x_{\text{mean}} \pm z \cdot s_x$	

1. Fewer than that very large number of cycles are used for a simulation (no repeats), and
2. The randomization function is initialized with a different seed every time the program is run.

The rectangular distribution generated by the randomize function needs to be transformed into the appropriate distribution for each variable, generally a normal distribution. Since the ND-transformation function is symmetrical, the algorithm can be compressed by using only one quadrant: R-values larger than 0.5 are reflected into the 0 to 0.5 range, and the sign of the result z is accordingly changed. Note that the $z = f(CP)$ function described in Sections 1.2.1 and Table 5.16 is used. Here, x_{mean} and s_x are parameters of the MC subroutine; $f(u)$ is the second polynomial given in Section 5.1.1. Fig. 3.8 and Tables 3.6 and 3.7 illustrate the function of the algorithm.

Simple examples for the technique are provided in Figures 1.9, 1.10, and 1.19 and in program SIMGAUSS. Additional operations are introduced in Figures 1.2, 1.3, and 1.24, namely the search for the extreme values for a series. A series of interesting applications, along with basic facts about MCT is to be found in Ref. 145.

In the following, an example from Chapter 4 will be used to demonstrate the concept of statistical ruggedness, by applying the chosen fitting model to data purposely "corrupted" by the Monte Carlo technique. The data are normalized TLC peak heights from densitometer scans. (See Section 4.2):

- An exponential and several polynomial models were applied, with the χ^2 measure of fit serving as arbiter.
- An exponential function was fitted using different starting points on the parameter space (A, B). (See Figure 4.4.)

For the purpose of making the concept of ruggedness clear, a polynomial of order 2 suffices. The fitted parabola $y = 4 + 25 \cdot x - 1.12 \cdot x^2$ was subtracted from the data and the residuals were used to estimate a residual stan-

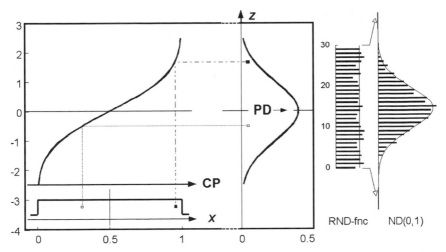

Figure 3.8. The transformation of a rectangular into a normal distribution. The rectangle at the lower left shows the probability density (idealized observed frequency of events) for a random generator *versus* x in the range $0 \le x \le 1$. The curve at the upper left is the cumulative probability CP *versus* deviation z function introduced in Section 1.2.1. At right, a normal distribution probability density PD is shown. The dotted line marked with an open square indicates the transformation for a random number smaller or equal to 0.5, the dot-dashed line starting from the filled square is for a random number larger than 0.5.

Table 3.7. Four Examples for $x_{mean} = 1.23$, $s_x = 0.073$

R	0.0903	0.7398	0.995	0.999
sign w	-1	$+1$	$+1$	$+1$
CP	0.0903	0.2602	0.005	0.001
u	-1.0443	-0.5847	-2.3010	-3.0000
v	$+1.3443$	$+0.6375$	$+2.5761$	$+3.0944$
z	-1.3443	$+0.6375$	$+2.5761$	$+3.0944$
$x_{mean} + s_x \cdot z$	1.1320	1.2770	1.418	1.456
Uncertainty	±0.0004	±0.0004	±0.000	±0.007

dard deviation: $s_{res} = 6.5$. Then, a Monte Carlo model was set up according to the equation

$$Y = (A = 4) + (B = 25) \cdot x + (C = -1.12) \cdot x^2 \pm ND(0, 6.5^2) \qquad (3.2)$$

A y_i-value was then simulated for every x_i-value in Table 4.5. This new, synthetic data set had statistical properties identical (n, S_{xx}), or very similar $(s_{xy}, s_{yy}, s_{res})$ to those of the measured set, the difference residing in

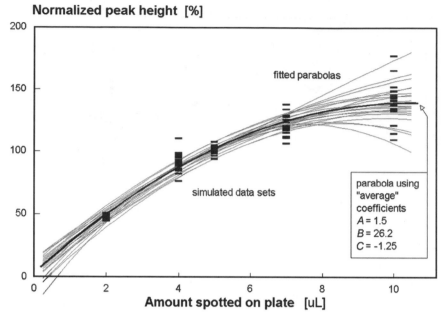

Figure 3.9. Demonstration of ruggedness. Ten series of data points were simulated that all are statistically similar to those given in Table 4.5. (See program SIMILAR.) A quadratic parabola was fitted to each set and plotted. The width of the resulting band shows in what x-range the regression is reliable, higher where the band is narrow, and lower where it is wide. The bars depict the data spread for the ten statistically similar synthetic data sets.

the stochastic variations of the y_i' around the estimate Y, see program SIM-ILAR. The regression coefficients were calculated as they had been for the measured data, and the resulting parabola was plotted. This sequence of steps was repeated several times. This resulted in a family of parabolas (each with its own set of coefficients similar to those in Eq. (3.5); a graph of coefficient a_1 (≈ 25) vs. b (≈ -1.12) yields a highly elliptical patch of points) that very graphically illustrates the ruggedness of the procedure: any individual parabola that does not markedly deviate from the "confidence band" thus established must be regarded as belonging to the same population. By extension, the simulated data underlying this nondeviant curve also belongs to the same population as the initial set. The perceived confidence band also illustrates the limits of precision to which interpolations can be carried, and the concentration range over which the model can safely be employed. An "average" set of coefficients would be found near the middle of the above-mentioned elliptical patch, see Figures 3.9 and 3.10; these must be investigated for dispersion and significance: The standard deviation helps to define

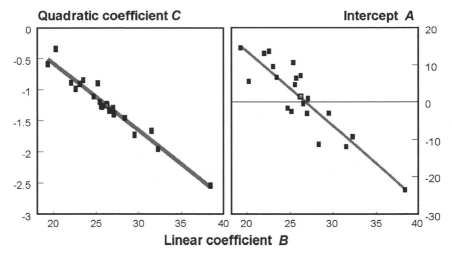

Figure 3.10. The relationship between the three coefficients A, B, and C for the curves shown in Figure 3.9; the quadratic and the linear coefficients are tightly linked. The intercept suffers from higher variability because it carries the extrapolation penalty discussed in Section 2.2.5 and Figure 2.8.

the number of significant digits, and a t-test carried out for H_0: $\mu = 0$ reveals whether a coefficient is necessary at all. The CVs are 730% (A), 16.4% (B), resp. 38.1% (C); with $n = 20$, the t-factor is approximately 2 for $p = 0.05$, so B and C are significant, while A is not.

3.6   SMOOTHING AND FILTERING DATA

All techniques introduced so far rely on recognizable signals; what happens if the signal more or less disappears in the baseline noise?

If the system is <u>static</u>, repeating the measurement and averaging according to Eq. (1.1) will eventually provide a signal-to-noise ratio high enough to discriminate signal from background. If the system is <u>dynamic</u>, however, this will only work if the transient signals are captured and accumulated by a computer, as in FT-NMR. A transient signal that can be acquired only once can be smoothed if a sufficiently large number of data points is available.[85] It must be realized that the procedures that follow are cosmetic in nature and only serve to enhance the presentability of the data. Distortion of signal shape is inevitable.[28] An example will illustrate this: in Figure 3.11 three Gaussian signals of different widths [full width at half maximum (FWHM)] are superimposed on a sloping background. With Monte Carlo simulation noise is added and

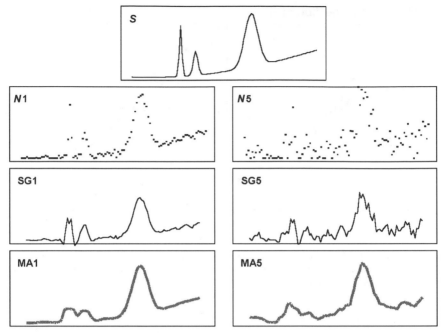

Figure 3.11. Smoothing a noisy signal. The synthetic, noise-free signal is given at the top. After the addition of noise by means of the Monte Carlo technique, the panels in the second row are obtained (little noise, left, five times as much noise, right). A seven-point Savitzky-Golay filter of order 2 (third row) and a seven-point moving average (bottom row) filter are compared.

individual "measurements" are generated. How is one to extract the signals? There are various digital filtering techniques[130,146–152] such as box-car averaging, moving average, Savitzky–Golay filtering, Fourier transformation,[153,154] correlation analysis,[155] and others.[156–158] The first three of these are related, differing only in the number of calculations and the weighting scheme. The filter width must be matched to the peak-widths encountered.

Box-car averaging is the simplest approach. The name can be traced to an image that comes to mind: Imagine a freight train (a locomotive and many boxcars) running across the page. All measurements that fall within the x-range spanned by one freight car (one "bin") are averaged and depicted by a single resulting point in the middle of this range.

The moving average can be explained as before. After having obtained the first set of averages, the "train" is moved by one point and the operation is repeated. This simply provides more points to the picture. Wider bins result in more averaging. There is a trade-off between an increase of the signal-to-noise ratio and distortion of the signal shape. Note that narrow signals that are defined by only a few measurements become distorted or even disappear.

Savitzky-Golay filtering[159–163] operates by the same mechanism, the difference being that instead of just averaging the points in a bin, a weighting function is introduced that prefers the central relative to the peripheral points. There are Savitzky-Golay filters that just smooth the signal, and there are others that at the same time calculate the slope[164–166] or the second derivative.[164] Depending on the coefficient sets used, "tension" can be introduced that suppresses high-frequency components. The ones that are presented here have a width of seven points, see Figure 3.11, Ref. 28, and use the coefficients −2, 1, 6, 7, 6, 1, and −2. Program SMOOTH incorporates a SG filter (restricted to smoothing polynomials of order 2 and 3, filter width 3 ... 11) that provides filtered output over the whole data range; BC and MA filters are also available.

An extension to two- resp. multidimensional filters is presented in Refs. 114, 167, 168. Matched filters[114] are the most elegant solution, provided one knows what form the signal has. The choice between different filter types is discussed in Ref. 114.

The Cusum technique (Section 1.8.5, program CUSUM) can also be regarded as a special type of filtering function, one that detects changes in the average y.

Every electronic instrument (sensor, amplifier) filters the true signal, usually by suppression of the high-frequency components through the action of resistive and capacitive elements. Whether by intention or not, this serves to smooth the signal and stabilizes any signal-processing done downstream. If the signal is digitized, all of the filters discussed here can be applied.

3.7 ERROR PROPAGATION AND NUMERICAL ARTIFACTS

The flowsheet shown in the introduction and that used in connection with a simulation (Section 1.4) provide insights into the pervasiveness of errors: at the source, random errors are experienced as an inherent feature of every measurement process. The standard deviation is commonly substituted for a more detailed description of the error distribution (see also Section 1.2), as this suffices in most cases. Systematic errors due to interference or faulty interpretation cannot be detected by statistical methods alone; control experiments are necessary. One or more such primary results must usually be inserted into a more or less complex system of equations to obtain the final result (for examples, see Refs. 23, 91–94, 104, 105, 142. The question that imposes itself at this point is: "how reliable is the final result?"[39] Two different mechanisms of action must be discussed:

1. The act of manipulating numbers on calculators of finite accuracy leads to numerical artifacts. (See Table 1.1.)[15,16]

2. The measurement uncertainty is transformed into a corresponding uncertainty of the final result due to algebraic distortions and weighting factors, even if the calculator's accuracy is irrelevant.

An example of mechanism (1) is given in Section 1.1.2: Essentially, numerical artifacts are due to computational operations that result in a number, the last digits of which were corrupted by numerical overflow or truncation. The following general rules can be set up for simple operations:

- Ten nearly identical numbers of s significant digits each, when added, require at least $s + 1$ significant digits for the result. Note that when the numbers are different by orders of magnitude, more significant digits are needed. Even the best calculators or software packages work with only so many digits. For 12-digit arithmetic, cf. Table 1.1, adding the squares of five-digit numbers already causes problems.

- Two numbers of s significant digits each, when multiplied, require $2 \cdot s$ digits for the result.

- The division of two numbers of s significant digits each can yield results that require nearly any number of digits (only a few if one number is a simple multiple of the other, many if one or both numbers are irrational); usually the number of bytes the computer displays or uses for calculation sets an upper limit.

- If the rounding procedures used for the display did not mask the effect, the math packs that are in use would come up with seemingly nonsensical results now and then, e.g., instead of a straight 6 one might obtain 5.9999999999999 or 6.0000000000001; those who do their own programming in BASIC, for example, are aware of this.

- Differences of nearly identical sums are particularly prone to truncation. [See Eqs. (1.3b) and (2.7)–(2.9).] A few examples will illustrate these rules (See Table 3.8.).

Classical error propagation (2) must not be overlooked: if the final result R is arrived at by way of an algebraic function

$$R = f(x_1, x_2, \ldots, x_k, A, B, \ldots) \tag{3.3}$$

with $x_1 \ldots x_k$ variables and $A \ldots$ parameters, the function f must be fully differentiated with respect to every x:

Table 3.8. Common Mathematical Operations and the Number of Significant Digits that are Involved

Calculation	Number of Significant Digits		
(a) (operation) (b) = (c)	a	b	c
1.2345678 + 9.8765432 = 11.1111111	8	8	9
1.23457 · 8.56789 = 10.5776599573	6	6	12
3.333333333333333 · 3 = 10	∞	1	2
0.0333/0.777777 = 0.42814322529 ...	3	6	∞
1024/256 = 4	4	3	1

$$d_i = \left\{ \frac{\partial f}{\partial x_i} \right\}, \qquad i = 1 \ldots k \tag{3.4}$$

and typical errors e_i must be estimated for all variables.[19,95] The linear terms of the full expansion of $(\delta f / \delta x)$ are then used to estimate the error in R. The assumption is that all higher-order terms are much smaller because the individual errors are uncorrelated and thus cancel each other:

$$e_R^2 = \Sigma \, (d_i \cdot e_i)^2 \tag{3.5}$$

If, as is usual, standard deviations are inserted for e_i, e_R has a similar interpretation. Examples are provided in Refs. 23, 75, 89, 93, 142, 169–171 and in Section 4.17. In complex data evaluation schemes, even if all inputs have Gaussian distribution functions, the output can be skewed,[172] however.

3.8 PROGRAMS

Computer programs have become a fixture of our lives. The following comments apply to using and writing programs.

Using Program Packages

Many fine program packages for the statistical analysis of data, and untold numbers of single-task programs are available for a variety of computers ranging from main-frames to programmable pocket calculators. The commercial packages from reputed software houses can generally be considered to be validated, and can be used in GMP areas. However, see Presentation of Results in the preceding, the validation status becomes questionable as soon as the user starts writing instructions or macros.

Even the best software contains errors:

- Software can be poorly designed: data format specifications are perhaps incompatible, intermediate results are inaccessible, all data are lost if an input error is committed, or results and data are not transferable to other programs. The division into tasks (modules, menu positions) might reflect not human but computer logic, and the sequence of entries the user is forced to follow might include unnecessary or redundant steps.
- Software designed for general or specific use might be useless in the intended application: sociologists, mathematicians, clinical researchers, and physicists, just to name a few user categories, all gather and analyze data, to be sure, but each one has particular data structures and hypotheses. Tests must be selected to fit the application.
- The software's author, as is often the case, is unfamiliar with the particulars of analytical chemistry or unaware of what happens with the numbers after the instrument has issued a nice report. Is it surprising then that an unknown's concentration is given without confidence limits? Most chromatography software might be fine for manipulating chromatograms acquired in a short time-span, but the comparison of stacks of chromatograms is accomplished only if there are no nonproportional shifts in retention times, a phenomenon often observed when HPLC columns age.
- Instrument suppliers are well acquainted with the design, construction, promotion, and sale of their products. The analytical problem-solving capabilities thereof are more often than not demonstrated on textbook variety "problems" that are only remotely indicative of a machine/software combination's usefulness. If software is tailored toward showroom effectiveness, the later user suffers.
- Software is rarely completely free of coding errors. While manifest errors are eliminated during the debugging stage, the remaining ones crop up only when specific data patterns or sequences of instructions are encountered.

There is a two-stage procedure to help judge the veracity of a program's output:

1. Validated data are input, and the results are compared on a digit-by-digit basis with results from a reference program. If possible, the same is done with intermediate results; see the example in Table 1.1.

2. The above data base is systematically modified to force the program to its limits. For example, the number of "measurements" is multiplied by entering each datum several times instead of only once, etc. That this step is necessary is shown by the authors' occasional experience that commercial programs as simple as a sorting routine would only work with the supplied demonstration data file.

Note Added in Proof

Two very interesting references in this regard are:

M. G. Cox and P. M. Harris, Design and use of reference data sets for testing scientific software, *Analytica Chimica Acta*, **380**, 339–351 (1999).

B. P. Butler, M. G. Cox, S. L. R. Ellison, and W. A. Hardcastle, *Statistics Software Qualification (Reference Data Sets)*, The Royal Society of Chemistry, ISBN 0-85404-422-1.

COMPLEX EXAMPLES

Life is a complicated affair; otherwise none of us would have had to attend school for so long. In a similar vein, the correct application of statistics has to be learned gradually through experience and guidance. Often enough, the available data does not fit simple patterns and more detective work is required than one cares for. Usually a combination of several simple tests is required to decide a case. For this reason, a series of more complex examples was assembled. The presentation closely follows the script as the authors experienced it; these are real-life examples straight from the authors' fields of expertise with only a few of the more nasty or confusing twists left out.

4.1 TO WEIGH OR NOT TO WEIGH

Situation and Design   A photometric assay for an aromatic compound prescribes the following steps:

1. Accurately weigh about 50 mg of compound
2. Dissolve and dilute to 100.0 ml (master solution MS)
3. Take a 2.00-ml aliquot of MS
4. Dilute aliquot to 100.0-ml (sample solution SS)
5. Measure the absorbance A of SS

The operations are repeated for samples 2 and 3; the final result is the mean of the three individual results per batch.

The original method uses the classical "weigh powder, dilute volumetrically" scheme. In order to improve the efficiency and at the same time guarantee delivery of results of optimal quality, an extensive review was undertaken.

In principle, diluter technology accurate to 10 μl or less would have been an option, and appropriately smaller sample amounts could have been taken at the third step (a reduction at step 1 would have entailed another proof of homogeneity and representativity if the number of crystals per sample became too small; if unlimited material is available, reducing the sample size does not save money). High-precision diluters require rinsing cycles to eliminate carry-over if the sample solution is aspirated and dispensed, which increases cycle-time and creates the need for validation. If corrosive solvents or solutes are involved, the machine needs to be serviced more often.

The other option is weighing. Since low-tech glassware or disposable pipettes could continue to be used, obtaining a tare and a total weight on a top-loading balance would add only a few seconds for each sample, but inherently be much more accurate and precise than any operation involving just pipettes and graduated flasks.[25]

Consideration   Any change in procedure would have to be accepted by the clients, who would have to adapt their raw materials testing procedures, something that could take months to negotiate and accomplish. If any client used the material in a GMP-regulated process, a minor change, especially if it increases reliability, could simply be communicated to the authorities, whereas a major change would result in an expensive revalidation and reapproval of the whole process. Also, if some clients accepted the change and others did not, the manufacturer would be stuck with having to repeat the analysis twice, the old way and the new way.

The plan, then, was to do the next 20 analyses according to a modified protocol that incorporated a weighing step after every volumetric operation. The evaluation of the absorbance measurements (data files VVV.dat, VWV.dat, and WWW.dat) was carried out according to three schemes: for the aforementioned steps 2 to 4, to use either the volumetric or the gravimetric results, or use a combination thereof:

W: weight solution to five significant digits, in grams
V: volumetrically dispense solution

The results are presented in Figs. 4.1 and 4.2.

Questions

1. Which scheme is best, which is sufficient?
2. What part of the total variance is due to the analytical methodology, and how much is due to variations in the product?
3. Can an assay specification of "not less than 99.0% pure" be upheld?
4. How high must an in-house limit be set to in order that at least 95% of all analyzed batches are within the foregoing specification?

Example 39: The variances were calculated according to Tables 1.13 and

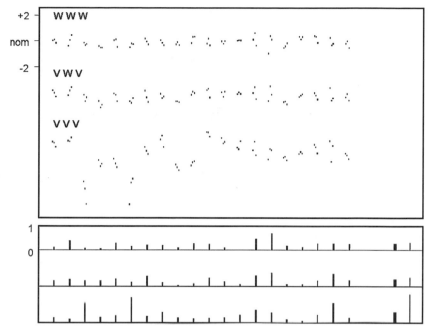

Figure 4.1. Assay results calculated according to three schemes WWW (top), VWV (middle), and VVV (bottom). The raw values (top panel, scale: ±2%) are plotted chronologically. In the bottom panel the standard deviations for the triplicate determinations are shown (scale: 1%); the bold bar at right signals the average within-group SD, and the thin line besides it the overall SD. The VVV scheme does look inferior in this metric, but the raw data graph is much more powerful in conveying the idea.

1.14, respectively (Eq. 1.3). Use program MULTI to obtain Tables 4.1 and 4.2. The reason that H_1 was found for WWW is that the intrinsic standard deviation has become so small that two values now fall out of line: 0.015 and 0.71; if these are eliminated, the quotient $s_{x,\text{before}}/s_{x,\text{after}}$ is about 46!

Evidently, replacing volumetric work by weighing improves the within-group variance from 0.192 (VVV) to 0.0797 (WWW, factor of 2.4), and the standard deviation from ±0.44 to ±0.28 (±0.28 is a good estimate of the analytical error); much of the effect can be achieved by weighing only the aliquot (VWV), at a considerable saving in time. The picture is much more favorable still for the between-groups variance: the improvement factor can be as large as 34; a look at Figure 4.1 shows why this is so: the short-term reproducibility of the fixed-volume dispenser used for transferring 2.0 ml of solution MS cannot be much inferior to that of the corresponding weighing step W because the within-group residuals are similar for WWW and VVV.

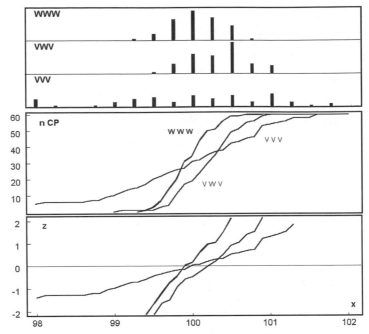

Figure 4.2. The top panel gives the histograms for the three sets of results calculated from Fig. 4.1, and two derivatives, the cumulative number of points (middle), respectively the nonlinear NPS-transform. The VVV-outliers on the low side are easily discerned.

However, some undetermined factor produces a negative bias, particularly in samples nos. 3, 6, 9, and 10; this points to either improper handling of the fixed-volume dispenser, or clogging. The reduction of the within-group variance from VVV to VWV is, to a major part, due to the elimination of

Table 4.1. Decomposition of Variance by ANOVA

	WWW	VWV	VVV	Direction of Calculation	VVV $\cdot f$	f
Mean	99.92	100.19	99.81			
VARIANCE:						
Within groups[*]	0.0797	0.0884	0.192	→	7.68	40
Between groups[**]	0.111	0.235	3.82	→	72.58	19
Total	0.0899	0.136	1.36	←	80.26	59

([*]) Analytical repeatability, ([**]) made up of analytical artifacts and/or production reproducibility; f: degrees of freedom; the arrows indicate in which direction the calculation proceeds.

Table 4.2. Confirmed Hypotheses (MRT: Multiple Range Test)

Test	VVV	VWV	WWW
Bartlett	H_0	H_0	H_1
ANOVA	H_1	H_1	n.a.
MRT[#of groups]	10	6	n.a.

the large residuals associated with samples #3 and #6. The extreme standard deviations in WWW are 0.015 and 0.71, $F > 2100$, which means the ANOVA and the MRT tests cannot be carried out.

The total variance (corresponding standard deviations ±0.3, ±0.37, resp. ±1.17) is also improved by a factor of 15, which means the specifications could be tightened accordingly.

The process mean with the VVV scheme was 99.81, that is essentially indistinguishable from the nominal 100%. The proposed SL is "more than 99.0%"; the corresponding z-value (see Section 1.2.1) is $(99.81-99.00)/\sqrt{0.192} = 1.85$, which means that about 3.2% [use $CP = f(z)$ in Section 5.1.1] of all measurements on good batches are expected to fall below the specification limit (*false negative*). The mean of three replicates has a z-value of $1.85 \cdot \sqrt{3} = 3.20$ (use Eq. (1.5)), giving an expected rejection rate of 0.13%. The corresponding z-values and rejection rates for the WWW scheme are only minimally better; however, the reliability of the WWW and VWV schemes is vastly improved, because the occurrence of systematic errors is drastically reduced. The same calculation could have been carried out using the t- instead of the normal distribution; because the number of degrees of freedom is relatively large ($f = 19$), virtually the same results would be obtained.

False negative responses of 0.13–3.2% are an acceptable price. What are the chances of *false positives* slipping through? Four alternative hypotheses are proposed for the VVV scheme (compare μ to SL = 99.0, with $p = \beta$, the type II error !):

Table 4.3. Power of Discrimination for a Single Measurement

H_n	μ	z	p	Power
H_1:	98.8	$(99.8 - 99.0)/\sqrt{0.192} = 0.46$	0.323	$100 \cdot (1 - 0.323) \approx 68\%$
H_2:	98.5	$(98.5 - 99.0)/\sqrt{0.192} = 1.14$	0.127	$100 \cdot (1 - 0.127) \approx 87.3\%$
H_3:	98.0	$(98.0 - 99.0)/\sqrt{0.192} = 2.28$	0.0113	$100 \cdot (1 - 0.0113) \approx 98.87\%$
H_4:	97.7	$(97.7 - 99.0)/\sqrt{0.192} = 2.97$	0.0015	$100 \cdot (1 - 0.0015) \approx 99.85\%$

Interpretation   While good batches of the quality produced (= 99.81%) purity) have a probability of being rejected (false negative) less than 5% of the time, even if no replicates are performed, false positives are a problem: an effective purity of $\mu = 98.5\%$ will be taxed "acceptable" in 12.7% of all cases because the found x_{mean} is 99% or more. Incidentally, plotting $100 \cdot (1-p)$ *versus* μ creates the so-called power-curve, see file POWER.xls and program HYPOTHESIS.exe.

If the much better and still reasonably workable VWV scheme (V = 0.0884) is chosen, the probabilities for false positives (cf. above) drop to 0.25, 0.046, 0.0005, and 0.0005; thus $\mu = 98.5\%$ will be found "acceptable" in less than 5% of all trials. The WWW scheme would bring only a small improvement at vastly higher cost: for the above case 0.046 would reduce to 0.038; the corresponding power-curves are much steeper, indicating a "sharper" cut-off and better decision-making properties.

Conclusion   The VWV scheme will have to be implemented to make precision commensurable with the demanding limits proposed by the Marketing Department. If the Production Department can fulfill its promise of typically 99.8% purity, all is well. For the case of lower purities, Quality Assurance will have to write a SOP that contains an action Limit for x_{mean} at AL $\approx$ 99.5%, and an error Recovery Procedure that stipulates retesting the batch in question so as to sufficiently increase n and decrease V_x. The Production Department was exonerated by the discovery that the high between-groups variances were analytical artifacts that could be eliminated by the introduction of a weighing step. The problem remains, however, that there is no margin for error, negligence, or lack of training.

4.2   NONLINEAR FITTING

Initial Situation   A product contains three active components that up to a certain point in time were identified using TLC. Quantitation was done by means of extraction/photometry. Trials to circumvent the time-consuming extraction steps by quantitative TLC (diffuse reflection mode) had been started but were discontinued due to reproducibility problems. The following options were deemed worthy of consideration:

1. HPLC (ideal because separation, identification and quantitation could be accomplished in one step).
2. Diode-array UV-spectrophotometer with powerful software (although the spectra overlapped in part, quantitation could be effected in the

first derivative mode. The extraction/separation step could be circumvented.)

3. Conventional UV/VIS spectrophotometer (manual measurements from strip-chart recorder traces and calculations on the basis of fitted polynomials; the extraction/separation step remains).

4. Quantitative TLC (an additional internal standard would be provided; the extraction/separation step could be dropped).

Options 1 and 2 had to be eliminated from further consideration because the small company where this was conducted was not producing sufficient cash-flow. Despite the attractiveness of photometric quantitation, option 3 did not fulfill its promise; careful revalidations revealed a disturbing aspect: The extractions were not fully quantitative, the efficiency varied by several percent depending on the temperature and (probably) the origin of a thickening agent. Thus, one had the choice of a multi-step extraction, at further cost in time, or a large uncertainty that obviated the advantage of the photometric precision. Option 4 was still beset by "irreproducibility": Table 4.4. contains reflection-mode densitometer peak-height measurements for one of the three components from 7 TLC plates; cf. Fig. 4.3.

Data Normalization   Since the general trend for every set of numbers is obviously similar, a simple normalization was applied: for each plate, the mean over all height measurements at 5 μl/spot was set equal to 100%, yield-

Table 4.4. Peak Heights Taken from Densitometric Scans for Various Amounts of a Given Sample Applied onto Seven TLC Plates Taken from the Same Package

Peak-Height [mm]	TLC Plate						
	1	2	3	4	5	6	7
2 μl/spot				42.5	31.5	39.5	
2				44.5	33.2	39.5	
4	62	67	74			75	
4	61	68	69			75	
5	63	80	85.5	92	67	74.8	93.5
5	71	80.5	83	99	75.5	75.5	95
7	87						109
7	81						114
10				124	95	110.8	
10				153	97	112.8	

Peak height

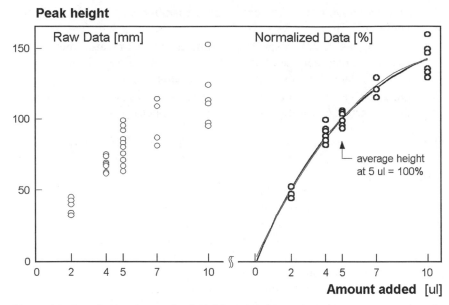

Figure 4.3. Raw (left) and normalized (right) peak heights *versus* the amount spotted on the plates. The average of all 5 μl spots per plate was set to 100% (right); the improvement due to the normalization is evident. The thick curve is for Eq. (4.1). The fine curve is for the best quadratic parabola, cf. Fig. 3.9.

ing the reduced height values given in bold numbers in Table 4.5 (see data file TLC.dat). This data normalization reduces the plate-to-plate variability from ±11.6 to ±6.4 residual standard deviation and permits a common calibration curve to be used, see Fig. 4.3.

Curve Fitting   It is immediately apparent that a straight line would be a poor approximation, but a polynomial of order 2 or 3 would be easy to fit; this was done in the example in Section 3.4. From a theoretical perspective, a disadvantage of polynomials, especially the quadratic one, is their nonasymptotic behavior: while one can easily impose the restriction $Y(0) = 0$ for the left branch, the strong curvature near the focus normally falls into a region that physically does not call for one, and at larger concentrations the peakheight would decrease. Also, while each x_i can unambiguously be assigned a single Y_i, the reverse is not necessarily true.[114] An equation more suited to the problem at hand is $Y = A \cdot (1 - \exp(B \cdot X))$ in accord with the observation that calibration curves for TLC measurements are highly nonlinear and tend to become nearly horizontal for larger amounts of sample spotted onto the plate. The disadvantage here is that there is no direct method for obtaining

Table 4.5. Data from Table 4.4 Normalized to the Average of the 5 µl Spots (Bold) on a Plate-by-plate Basis; a Quadratic Regression was Applied to These Data, Yielding Eq. (3.5)

Peak-Height [mm]	TLC Plate							
	1	2	3	4	5	6	7	
2 µl/spot				44.5	44.2	52.6		
2				46.6	46.6	52.6		
4	92.5	83.5	87.8			99.8		
4	89.6	84.7	81.9			99.8		
5	**94.0**	**99.7**	**101.5**	**96.3**	**94.0**	**99.5**	**99.2**	average
5	**106.0**	**100.3**	**98.5**	**103.7**	**106.0**	**100.5**	**100.8**	= 100%
7							115.6	
7							121.0	
10				129.8	133.3	147.4		
10				160.2	136.1	150.1		

the coefficients a and b (if $Y(x = \infty)$ were precisely known, this would not be a problem).

Three paths can be advanced: (1) expansion, e.g., Taylor series; (2) trial and error, e.g., generating curves on the plotter; and (3) simplex optimization algorithm. (See Section 3.1.)

1. An expansion brings no advantage in this case because terms at least quartic in coefficient b would have to be carried along to obtain the required precision.

2. If a graphics screen or a plotter is available a fairly good job can be done by plotting the residuals $r_i = y_i - Y$ *versus* x_i, and varying the estimated coefficients a and b until suitably small residuals are found over the whole x-range; this is accomplished using program TESTFIT.

3. If iterative optimization software is available, a result of sufficient accuracy is found in short order. This is demonstrated in the following discussion.

In Fig. 4.4, the starting points (initial estimates) are denoted by circles in the A, B plane; the evolution of the successive approximations is sketched by arrows that converge on the optimum, albeit in a roundabout way. The procedure is robust: no local minima are apparent; the minimum, marked by the square, is rapidly approached from many different starting points within the given A, B plane (black symbols); note that a B-value ≥ 0, an A-value ≤ 0, and combinations in the lower left corner (gray symbols) pose problems

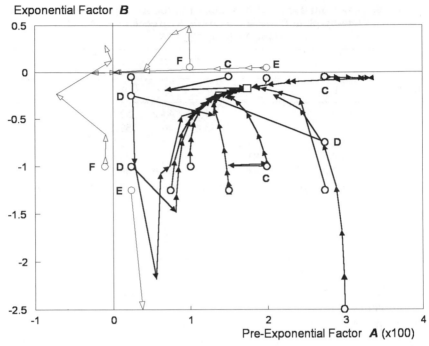

Figure 4.4. Optimization of parameters. The exponential equation (4.1) was fitted to the normalized data from Table 4.5.

for the algorithm or lead to far-off relative minima. Starting points that first lead off in one direction, only to turn around, are marked "C"; starting points that lead to traces that cross others are marked "D". Those that start at "E" or "F" lead to other minima. Despite the fact that a unique solution was found (Eq. (4.1)), good practice is followed by graphing the reduced data together with the resulting curve (Fig. 4.3); the residuals were plotted separately. A weighting scheme could have been implemented to counter heteroscedacity, but this, in the light of the very variable data, would amount to statistical overkill.

The equation, fitted to data in Table 4.5, is

$$Y = A \cdot (1 - \exp(B \cdot x)), \qquad \text{with } A = 176.65, B = -0.16 \qquad (4.1)$$

In this specific case, the predictive power of the polynomial P^2 (see Fig. 3.9) and the exponential function are about equal in the x-interval of interest. The peak height corresponding to an unknown sample amount would be divided by that of the standard (5 μl/spot) run on the same plate. If possible, each sample and the standard would be spotted several times on the same plate, e.g., in alternating sequence, and the mean of the sample respectively the standard peak heights would be used for improved precision. The resulting quotient is entered on the ordinate and the unknown amount is read off the abscissa after reflection. Despite the normalization, the results are not exceptional in terms of reproducibility; thus it is perfectly acceptable to use a graphical approximation by constructing upper and lower limits to the calibration points with the help of a flexible ruler in order to find confidence limits for X.

As long as the health authorities accept 90–110% specification limits on the drug assay, the normalization method presented above will barely suffice for batch release purposes. Since there is a general trend toward tightening the specification limits to 95–105% (this has to do with the availability of improved instrumentation and a world-wide acceptance of GMP-standards), a move toward options 1 (HPLC) and 2 (DA-UV) above is inevitable.

For an example of curve fitting involving classical propagation of errors in a potentiometric titration setting, see Ref. 142.

4.3   UV-ASSAY COST STRUCTURE

Problem   The drug substance in a pharmaceutical product is to be assayed in an economical fashion. The following constraints are imposed:

- UV-absorbance measurement at a given wavelength, either on a photometer, which is known to be sufficiently selective, or on a HPLC
- The extraction procedure and the solvent
- The number of samples to be processed for every batch produced: six samples of 13 tablets each are taken at prescribed times after starting the tablet press (10 tablets are ground and well mixed (= compound sample), two average aliquots are taken, and each is extracted); the additional three tablets are used for content uniformity testing; this gives a total of $6 \cdot (2 + 3) = 30$ determinations that have to be performed.

Available Lab Experience

- The relative standard deviation of the determination was found to be ±0.5% (photometer) resp. ±0.7% (HPLC[175]) for samples and references.
- The relative content varies by nearly ±1% due to inhomogeneities in the tablet material and machine tolerances.

- Duplicate samples for the photometer involve duplicate dilutions because this is the critical step for precision; for the HPLC the same solution is injected a second time because the injection/detection/integration step is more error-prone than the dilution.

Requirements

- The mean content must be 98–102% for nine out of 10 compound samples
- The individual contents must be 95–105% for nine out of 10 tablets and none outside 90–110%
- The linearity of the method must be demonstrated

Options The analyst elects to first study photometry and place the three reference concentrations symmetrically about the nominal value (= 100%). The initial test procedure consists of using references at 80, 100, and 120% at the beginning of the series, and then a 100% reference after every fifth determination of an unknown sample.

Known Cost Factors for the UV-Method

- A technician needs an average of 10 minutes to prepare every photometer sample for measurement.
- The necessary reference samples are produced by preparing a stock solution of the pure drug substance and diluting it to the required concentrations. On the average, 12 minutes are needed per reference.
- The instrument-hour costs 20 currency units.
- A lab technician costs 50 currency units per hour (salary, benefits, etc). The price levels include the necessary infrastructure.
- The instrument is occupied for 5 minutes per determination, including changing the solutions, rinsing the cuvettes, etc.
- The technician is obviously also occupied for 5 minutes, and needs another 3 minutes per sample for the associated paperwork.

First Results The confidence interval from linear regression CI(X) was found to be too large; in effect, the plant would have to produce with a 0% tolerance for drug substance content; a tolerance of about ±1% is considered necessary, cf. Fig. 2.12 and Section 4.24 (Fig. 4.35).

Refinements The confidence interval must be small enough to allow for some variation during production without endangering the specifications, so the additivity of variances is invoked: Since 90% of the results must be within ±2% of nominal, this can be considered to be a confidence interval

$$\tfrac{1}{2}\text{CI} = t\text{-factor} \cdot \sqrt{V_{\text{prod}} + V_{\text{analyt}}} \leq 2\%.$$

Example 40: Trial calculations are done for $V_{\text{analyt}} = (0.5)^2$ and $V_{\text{prod}} = (0.9)^2 \ldots (1.1)^2$; the required t-factors for $p = 0.1$ turn out to be 1.94 ... 1.66, which is equivalent to demanding $n = 7 \ldots 120$ calibration samples. Evidently, the case is critical and needs to be underpinned by experiments. Twenty or 30 calibration points might well be necessary if the calibration scheme is not carefully designed.

Solution 20–30 calibration points are too many, if only for reasons of expended time. The analyst thus searches for a combination of perhaps $n = 8$ calibration points and $m = 2$ replications of the individual samples. This would provide the benefit of a check on every sample measurement without too much additional cost. An inspection of the various contributions in Eq. (2.17) toward the $\text{CI}(X)$ in Table 2.9 reveals the following for $n = 8$ and $m = 2$:

about 70.9% contribution due to the $1/m$ term,

about 17.7% contribution due to the $1/n$ term, and

about 11.3% contribution due to the $1/S_{xx}$ term ($S_{xx} = 1250$, $\Delta x = 10$).

Various calibration schemes similar to those given in Section 2.2.8 were simulated. The major differences were (1) the assumption of an additional 100% calibration sample after every fifth determination (including replications) to detect instrument drift, and (2) the cost structure outlined in Table 4.6, which is summarized in Eq. (4.2) below. The results are depicted graphically in Figure 4.5, where the total cost per batch is plotted against the estimated confidence interval $\text{CI}(X)$. This allows a compromise involving acceptable costs and error levels to be found.

Interpolations at 110% of nominal were simulated; if the interpolations are envisaged closer to the center-of-mass, the third term will diminish in importance, and the sum of the first two terms will approach 100%.

Besides the photometric method, a HPLC method could also be used. (Cf. the first constraint under the earlier head "PROBLEM." The HPLC has a higher relative standard deviation for two reasons: The transient nature of the signal results in a short integration time, and the short pathlength makes for a very small absorption. The available HPLC is equipped with an auto-sampler, so that a different calibration scheme can be implemented: Instead of a complete calibration curve comprising n points at the beginning of the series, two symmetrical calibration samples, e.g., 80/120% are run at the beginning, and after five to 10 determinations a double calibration is again performed, either at the same two or at two different concentrations. In this way, if no instrument drift or change of slope can be detected during the run, all calibration results can be pooled. Overall linearity is thus established. Also, S_{xx} increases with every calibration measurement. In the photometer case above, the additional 100% of nominal measurement after every 5–10 determinations does not contribute toward S_{xx}. The number of replicate sample injections, m, was taken to be 1, 2, 3, 4, or 5. Preparing HPLC samples takes longer because vials have to be filled and capped.

Known Cost Factors for the HPLC Method

- The technician needs 11 minutes to prepare every sample solution and 13 minutes to prepare a calibration solution. A vial is filled in 1 minute.
- The instrument costs 30 currency units per hour.
- The instrument is occupied for 15.5 minutes per determination, including 30 seconds for printing the report.
- The technician needs 1 minute per determination to load the sample into the autosampler, type in sample information, start the machine, etc., and 20 minutes for preparing the eluent per batch.
- The technician needs only 5 minutes per batch for evaluation, including graphics and tables because the computerized instrument and the integrator do the rest.

The HPLC instrument is more expensive and thus costs more to run; because it is used 12 to 16 hours per day, *versus* only 2 to 5 hours/day for the photometer, the hourly rates are not proportional to the initial investment and servicing needs. The cost structure given in Table 4.6 can now be derived.

The total costs associated with the 30 determinations are calculated to be

$$5.83 \cdot n + 500.0 \cdot m + 10.0 \cdot k \qquad \text{(photometer)} \qquad (4.2)$$
$$8.33 \cdot n + 232.5 \cdot m + 10.87 \cdot k + 301.89 \qquad \text{(HPLC)} \qquad (4.3)$$

The number of man-hours expended is equal to the sum over all items in the columns marked "operator".

Table 4.6. Cost Components for UV and HPLC Measurements; the Numbers Give the Necessary Time in Hours; amortization of the Equipment and Infrastructure is Assumed to be Included in the Hourly Rates

Item	Photometry		HPLC	
	Machine	Operator	Machine	Operator
Costs	20	50	30	50
Prepare calibration solutions	—	$k/5$	—	$k/4.6$
Measure calibration points	$n/12$	$n/12$	$n/4$	$n/60$
Prepare samples	—	$30 \cdot m/6$	—	$30/5.5$
Measure samples	$30 \cdot m/12$	$30 \cdot m/12$	$30 \cdot m/4$	$30/60$
Evaluate results	—	$30 \cdot m/20$	$30 \cdot m/120$	$1/12$

n: number of calibration points; m: number of replicates per sample; k: number of calibration solutions prepared; costs: currency units per hour.

A few key results are given in Fig. 4.5. The arrows give the local tangents.

It is interesting to see that the two curves for the photometer and the HPLC nearly coincide for the above assumptions, HPLC being a bit more expensive at the low end but much cheaper for the best attainable precisions. Note the structure that is evident especially in the photometer data: This is primarily due to the number m of repeat determinations that are run on one sample

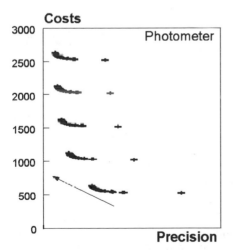
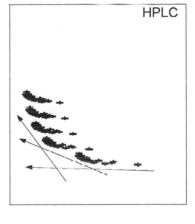

Figure 4.5. Estimated total analytical cost for one batch of tablets *versus* the attained confidence interval CI(X). 640 (UV) resp. 336 (HPLC) parameter combinations were investigated (some points overlap on the plot).

($m = 1$ at bottom, $m = 5$ at top). One way for reducing the photometer costs per sample would be to add on an autosampler with an aspirating cuvette, which would allow overnight runs and would reduce the time the expensive technician would have to spend in front of the instrument (a time-interval of less than about 5 minutes between manual operations cannot be put to good use for other tasks). Roughly, three cost regimes can be distinguished: Dramatic improvements in precision are to be had practically free of charge (nearly horizontal arrow); in order to reduce the confidence interval of the result from $\pm 1\%$ to $\pm 0.6\%$ relative, HPLC costs would about double on a per batch basis (middle arrow); if even better precision is targeted, cost begins to soar. This simulation shows that the expected analytical error of $\pm 0.7\%$ (HPLC) will only be obtained at great cost. Experiments will be necessary to find out whether this assumption was overly optimistic or not. A trade-off between tolerance toward production variability and analytical error might be possible and should be investigated to find the most economical solution.

4.4 PROCESS VALIDATION

Situation   During the introduction of a new tablet manufacturing process, the operation of a conditioner had to be validated; the function of this conditioner is to bring the loaded tablets to a certain moisture content for further processing.

Question   Does the conditioner work in a position-independent mode, that is, all tablets in one filling have the same water content no matter into which corner they were put, or are there zones where tablets are dried to a larger or lesser extent that the average?

Experiment   Ten different positions within the conditioner representing typical and extreme locations relative to the air inlet/exhaust openings were selected for analysis. Eight tablets were picked per position; their water content was accurately determined on a tablet-to-tablet basis using the Karl Fischer technique. Table 4.7 gives an overview of all results:

Data Evaluation   The Bartlett test (Section 1.7.3; cf. program MULTI using data file MOISTURE.dat) was first applied to determine whether the within-group variances were homogeneous, with the following intermediate results: $A = 0.1719$, $B = -424.16$, $C = 1.4286$, $D = 70$, $E = 3.50$, $F = 1.052$, $G = 3.32$.

Table 4.7. Residual Moisture in Tablets

Tablet					Position					
	1	2	3	4	5	6	7	8	9	10
1	1.149	1.082	1.847	1.096	1.109	1.181	1.191	1.175	1.152	1.169
2	1.020	1.075	1.761	1.135	1.084	1.220	1.019	1.245	1.073	1.126
3	1.106	1.112	1.774	1.034	1.189	1.320	1.198	1.103	1.083	1.050
4	1.073	1.060	1.666	1.173	1.170	1.228	1.161	1.235	1.076	1.095
5	1.108	1.068	1.779	1.104	1.252	1.239	1.183	1.101	1.021	1.041
6	1.133	1.003	1.780	1.041	1.160	1.276	1.109	1.162	1.051	1.065
7	1.108	1.104	1.816	1.141	1.148	1.232	1.158	1.171	1.078	1.127
8	1.026	1.049	1.778	1.134	1.146	1.297	1.140	1.078	1.030	1.023
mean	1.090	1.069	1.775	1.107	1.157	1.249	1.145	1.159	1.071	1.087
std. dev.	0.047	0.034	0.052	0.049	0.051	0.045	0.059	0.062	0.040	0.051
n	8	8	8	8	8	8	8	8	8	8
Σx	8.723	8.553	14.20	8.858	9.258	9.993	9.159	9.270	8.564	8.696
$\Sigma(x - x_{mean})^2$	.0155	.0081	0.190	.0168	.0181	.0144	.0240	.0266	.0114	.0180

Table 4.8. Variance Decomposition

$S_1 = 0.1719$	$f_1 = 70$	$V_1 = 0.00246$	Variance within groups
$S_2 = 3.2497$	$f_2 = 9$	$V_2 = 0.36108$	Variance between groups
$S_T = 3.4216$	$f_T = 79$	$V_T = 0.043311$	Total variance

Example 41: For $f = k - 1 = 9$ degrees of freedom the χ^2-value comes to 3.50 (uncorrected) resp. 3.32 (corrected); this is far below the critical χ^2 of 16.9 for $p = 0.05$. Thus the within-group variances are indistinguishable.

Because of the observed homoscedacity, a simple ANOVA-test (see Table 4.8) can be applied to determine whether the means all belong to the same population. If there was any indication of differences among the means, this would mean that the conditioner worked in a position-sensitive mode and would have to be mechanically modified.

Interpretation   Because the calculated F-value ($0.36108/0.00246 = 148$) is much larger than the tabulated critical one (2.04, $p = 0.05$), the group means cannot derive from the same population.

The multiple range test was used to decide which means could be grouped together. First, the means were sorted and all possible differences were printed, as in Table 4.9.

Next, each difference was transformed into a q-value according to Eq. (1.25). With $70 - 9 = 61$ degrees of freedom, the critical q-value for the longest diagonal (adjacent means) would be 2.83, that for the top right corner (eight interposed means) 3.33, see Table 1.11. For this evaluation sep-

Table 4.9. Table of Differences Between Means (Rounding Errors May Occur, e.g., $1.070500 - 1.069125 = 0.001375$ is Given as $1.071 - 1.069 = 0.001$ in the Top Left Cell)

Ordered Means	1.071	1.087	1.090	1.107	1.145	1.157	1.159	1.249	1.775
1.069	0.001	0.018	0.021	0.038	0.076	0.088	0.090	0.180	0.706
1.071		0.017	0.020	0.037	0.074	0.087	0.088	0.179	0.705
1.087			0.003	0.020	0.058	0.070	0.072	0.162	0.688
1.090				0.017	0.055	0.067	0.068	0.159	0.685
1.107					0.038	0.050	0.052	0.142	0.668
1.145						0.012	0.014	0.104	0.630
1.157			differences $\Delta x_{ji}\ldots$				0.002	0.092	0.618
1.159								0.090	0.616
1.249									0.526

arate tables would have to be used for different error probabilities p. The conversion to reduced q-values eliminated this inconvenience with only a small risk (see Section 1.5.4); this was accomplished by dividing all q-values by $t(61, 0.05) \cdot \sqrt{2} = 1.9996 \cdot \sqrt{2} = 2.828$; cf. Tables 1.12 and 4.9, and file Qred_tbl.dat, yielding reduced critical q-values in the range 1.00 ... 1.18. (See Table 4.10.)

The critical reduced q-values pertinent for each diagonal are given in the right-hand column: For the diagonal 0.43–0.74–1.17–1.35–1.04–2.11–12.5, the tabulated reduced critical q_c is 1.09. The number of decimal places was reduced here so as to highlight the essentials.

Which means could be grouped together? A cursory glance at the means would have suggested (values rounded) 1.07 ... 1.16 for one group, 1.25 resp. 1.78 as second resp. third one.

An inspection of the bottom diagonal 0.03 ... 10.6 shows that the two largest means differ very much ($10.6 \gg 1.83 > 1.1$), whereas each of the other means could be grouped with at least its nearest neighbor(s). The staircase line distinguishes significant from nonsignificant reduced q-values: The mean 1.249 is larger than 1.159, and the means 1.145, 1.157, and 1.159 are indistinguishable among themselves, and could be grouped with 1.107, but not with 1.087 (e.g., $1.35 > 1.09$). The values 1.069–1.107 form a homogeneous group, with the value 1.107 belonging both to this and the next group, see Table 4.11.

On the evidence given here, the tablet conditioner seemed to work well, but the geometrical positions associated with the means (1.249, 1.775) differ from those with the means (1.069 ... 1.107); indeed, five of these positions were near the entry port of the controlled-humidity airstream, and the other two were situated in corners somewhat protected from the air currents. The 1.159 group marks the boundary of the acceptable region.

Tablet samples were pulled according to the same protocol at different times into the conditioning cycle; because the same pattern of results emerged repeatedly, enough information has been gained to permit mechanical and operational modifications to be specified that eliminated the observed inequalities to such a degree that a more uniform product could be guaranteed. The groups are delineated on the assumptions that the within-group distributions are normal and the between-group effects are additive. The physicochemical reasons for the differentiation need not be similarly structured.

4.5 REGULATIONS AND REALITIES

SITUATION: Low limits on impurities are a requirement that increasingly is being imposed on pharmaceuticals and high-quality chemicals. A specification that still commonly applies in such cases is the "2% total, 0.5%

Table 4.10. Similarity Determined Using q-Factors

Ordered Means	1.071	1.087	1.090	1.107	1.145	1.157	1.159	1.249	1.775	Reduced Critical $q^{*)}$
1.069	0.03	0.36	0.43	0.77	1.53	1.78	1.81	3.64	14.3	1.18
1.071	—	0.33	0.40	0.74	1.51	1.76	1.79	3.62	14.3	1.17
1.087		—	0.07	0.41	1.17	1.42	1.45	3.28	13.9	1.16
1.090			—	0.34	1.10	1.35	1.38	3.21	13.9	1.15
1.107				—	0.76	1.01	1.04	2.87	13.5	1.13
1.145					—	0.25	0.28	2.11	12.8	1.11
1.157						—	0.03	1.86	12.5	1.09
1.159							—	1.83	12.5	1.05
1.249								—	10.6	1.00

reduced q_{ij}

Table 4.11. Sample Assignment to Groups

Group				Sorted Means					
1	1.069	1.071	1.087	1.090	1.107				
2					1.107	1.145	1.157	1.159	
3									1.249
4										1.775

individual" rule, which is interpreted to mean that any one single impurity must remain below 0.5%, while the sum of all impurities shall not exceed 2.0% w/w. Conversely, the purity of the investigated compound must come to 98.0% or higher. In practice, this numerical reconciliation, also known as "mass balance," will not work out perfectly due to error propagation in the summation of the impurities, and the nonzero confidence interval on the >98% purity figure; thus it is considered improper to measure impurity by way of the (100 minus %-purity) difference. Because for the HPLC data both the individual peak areas and the result of the addition (100%) are known ("closure,"[176]), not all information can be used in one calculation. Also, the sum of area-% is not proportional to the sum of weight-% because no two chemicals have exactly the same absorptivity at the chosen wavelength.

The Case   In 1986 there were semi-official indications that a major registration authority might tighten the above requirement on pharmaceutical grade chemicals by a factor of 4 to "0.5 weight-% total and an individual upper limit on every impurity found in an 'Accepted Sample'". The plant manager, after hearing this, asks the analyst to report whether a certain product characterized by a large number of small impurities would comply.

Review of Data   The analyst decides to review the critical HPLC and GC impurity methods by retrieving from his files the last 15 or so sets of chromatograms. On the average, 14.5 (range: 11–22) HPLC impurity peaks ranging from 0.004 to 0.358 area-% are found per chromatogram. Some components are more precisely analyzed by GC, so these results are used. The fact that most of these impurities are either insufficiently characterized because of the small amounts found, or no clean reference samples exist, means that area-% (integrated absorbance) instead of weight-% (integrated concentration) are used for the evaluation; this is accepted by the authorities. Also, the numbers bear out a large variability for any one impurity from batch to batch; a typical impurity might yield $0.1 \pm 0.02\%$ ($n = 16$). The "purity" given by the main peak averages to $98.7 \pm 0.14\%$. From these first results the operations manager draws the conclusion that an additional crystalliza-

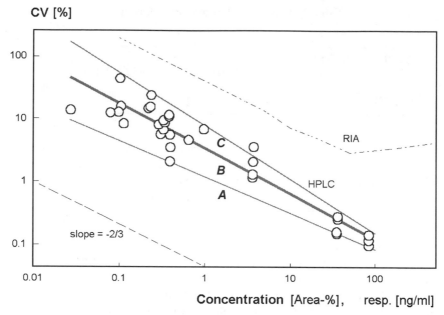

Figure 4.6. Trend of the CV.

tion step should push the two critical impurities below the 0.1% limit, and the sum of impurities to just below 0.5%.

Experience The analyst is skeptical, and for good reason: Because of the low average levels, a higher relative standard deviation must be expected for the small peaks than for the large ones, that is, as each impurity is reduced, the analytical uncertainty would increase even more and levels, as delineated by the CL_U, might not appear to become smaller at all ("false positive"). This would force the analyst to make multiple purity determinations to keep confidence intervals within reasonable limits, and thus incur higher analytical overhead, and/or force the plant to aim for average impurity levels far below the present ones, also at a cost. The analyst then assembles Figure 4.6: The abscissa is the logarithm of the impurity concentration in %, while the ordinate gives the logarithm of the CV on the impurity, i.e., $\log_{10}(100 \cdot s_x/c_x)$. For comparison, RIA data is shown as a dashed line: The slope is about the same!

The trend $\log_{10}(CV)$ vs $\log_{10}(c)$ appears reasonably linear (compare this with Ref. 177; some points are from the method validation phase where various impurities were purposely increased in level). A linear regression line (*B*) is used to represent the average trend (slope = −0.743). The target level for any given impurity is estimated by a simple model. Because the author-

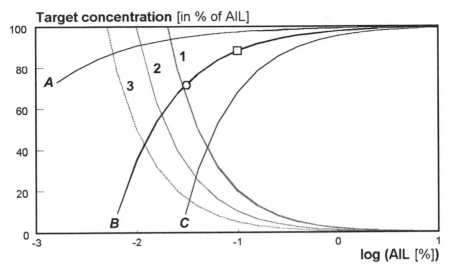

Figure 4.7. Consequences for the case that the proposed regulation is enforced: The target level for an impurity is shown for several assumptions in percent of the level found in the official reference sample that was accepted by the authorities. The curves marked A (pessimistic), B, and C (optimistic) indicate how much the detected signal needs to be below the approved limits for assumptions concerning the signal-to-noise relationship, while the curves marked 1–3 give the LOQ in percent of this limit for LOQs of 0.02, 0.01, resp. 0.005. The circle where curves B and 1 intersect points to the lowest concentration of impurity that could just be handled, namely 0.031%. The square is for an impurity limit of 0.1%, for which the maximal signal ($\approx 0.087\%$) would be just a factor of ≈ 4.4 above the highest of these LOQs.

ity thought it would allow a certain level for each impurity in an Accepted Sample (usually that which was used for the first clinical trial if it is a drug), the impurity levels in all following batches could not be higher. The model assumes a target level TL that is below the allowed impurity level AIL by an amount that ensures the risk of being higher is less than $p = 0.05$. The curves shown in Figure 4.7 demonstrate that the TL rapidly drops to the LOD as the AIL goes below about 0.1%. Even under the most optimistic assumptions, the intersection does not drop below 0.01%.

$$\text{target level TL} = \text{allowed impurity level AIL} - t_{(\alpha = 0.1, f)} \cdot s_x \qquad (4.4)$$

This amounts to stating "the analytical results obtained from HPLC-purity determinations on one batch are not expected to exceed the individual limit AIL more than once in 20 batches." Since a one-sided test is carried out here, the $t(\alpha = 0.1, f)$ for the two-sided case corresponds to the $t(\alpha/2 = 0.05, f)$ value needed. The target level TL is related to the AIL as is the lower end

Table 4.12. Accepted and Target Impurity Concentrations (Target Concentrations for Impurities, Under Assumption of the Regression Line in Fig. 4.7 (B: $a = 0.92$, $b = -0.743$, $m = 1$); If the LOQ of the Method were 0.03%, the Target Concentration in the Last Line (0.011) Would be Inaccessible to Measurement!

Accepted Impurity Level (%)	Target Concentration (%)	Target Concentration (% of AIL)
10.0	9.85	98.5
5.0	4.87	97.5
3.0	2.89	96.3
2.0	1.90	95.0
1.0	0.917	91.7
0.5	0.430	86.1
0.3	0.239	79.7
0.2	0.145	72.5
0.1	0.054	54.0
0.05	0.011	23.0

of the "production tolerance" range to the lower specification limit for the "probably inside" case in Fig. 2.12 or batch C to the upper SL in Fig. 2.13.

Example 42: Assume that a certain impurity had been accepted by the authorities at the AIL = 0.3% level; what level must one now aim for to assure that one stays within the imposed limit? Eq. (4.4) is rewritten with both TL and CI/2 in % of AIL, $x = \log(\text{AIL})$, $y = \log((100 \cdot t \cdot s_y)/\text{AIL})$, and the right-hand expression is replaced by the linear estimate from Fig. 4.6.

$$
\begin{aligned}
t \quad &= t(f = n - 1, p = 0.1) & &\text{Student's } t \\
y[\%] \quad &= 100 \cdot \text{TL}/\text{AIL} \\
&= 100 - 100 \cdot t \cdot s_x/\text{AIL} & &\text{linear estimate} \\
&= 100 - 10\^\{a + b \cdot \log_{10}(\text{AIL})\} & &\text{substitution} \\
&= 100 - 10\^\{0.92 - 0.743 \cdot \log_{10}(0.3)\} & &\text{insert linear regression} \\
& & &\text{parameters} \\
&= 79.7\% & &\text{result}
\end{aligned}
$$

$$(4.5)$$

For repeat measurements ($m = 2$, etc.), one has to subtract the logarithm of the square root of m from the sum ($a + b \cdot \log_{10}(c)$). For $m = 2$ resp. $m = 3$, the result 79.7% would change to 85.7% resp. 88.3%. (See Table 4.12.)

Consequences   While this may still appear reasonable, lower accepted impurity limits AIL quickly demand either very high m or then target levels TL below the LOQ, as is demonstrated in Fig. 4.7. If several impurities are involved, each with its own TL and AIL, the risk of at least one exceeding its AIL rapidly increases (joint probabilities, see Section 4.24). For k impurities, the risk is $[1 - (1 - 0.05)^k]$, that is for $k = 13$, every other batch would fail!

Actually, it would be reasonable for the authorities to replace by 0.1% the individual limit concept for all impurities lower than about 0.1% in the accepted sample, provided that toxicity is not an issue, because otherwise undue effort would have to be directed at the smallest impurities. Various modifications, such as less stringent confidence limits, optimistic estimates (line (A) in Fig. 4.6), etc. somewhat alleviate the situation the plant manager is in, but do not change the basic facts.

The effect of such well-intentioned regulations might be counterproductive: Industry could either be forced to withdraw products from the market despite their scientific merits because compliance is impossible, or they might dishonestly propose analytical methods that sweep all but a scapegoat impurity below the carpet.

That these are not idle thoughts is demonstrated by the following example that involves a process intermediate that was offered by two specialty chemicals houses; about half a dozen synthesis steps had been performed up to this point, all involving solvents of similar polarity. The two steps that would lead to the final product would be conducted under somewhat different conditions, but not different enough to ensure that the previous process impurities would all be extracted. So, it was important to select a precursor that would not emburden the process from the beginning. (See Fig. 4.8.) Vendor A might have had a better price, but any advantage there could have been eaten up by more waste or a poor impurity profile in the final product, an uncertainty that could only be laid to rest by testing the two vendor's samples in the synthesis and re-running the analysis to see how well the individual impurity was eliminated.

4.6   DIFFUSING VAPORS

Situation   Two different strengths of plastic foil are in evaluation for the packaging of a moisture-sensitive product. Information concerning the diffusion of water vapor through such foils is only sparsely available for realistic conditions. To remedy this lack of knowledge, samples of the product are sealed into pouches of either foil type and are subjected to the following tests:

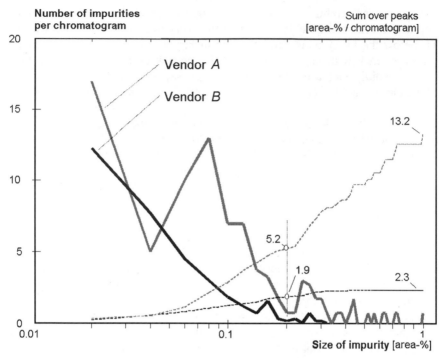

Figure 4.8. Comparison of impurity profiles for the same chemical intermediate from two different suppliers. The impurity peak-areas for each chromatogram were tallied in 0.02 area-% bins for each vendor, the data was normalized by dividing by the number of chromatograms. Vendor *A*'s material has many more peaks in the 0.05–0.2% range, which drives the total impurity level to ≈5.2% (*vs.* ≈ 1.9 for Vendor *B*) for ≤0.2%; the number of excess peaks above 0.2% does not appear as dramatic, but greatly adds to the total impurity level: ≈ 13.3 *vs.* ≈ 2.3%!

- Normal storage (results expected after some months to years)
- Elevated temperature and humidity (accelerated test, approximate results expected after 2 weeks)

Experimental   Weight gain of hygroscopic tablets due to moisture absorption is taken as a surrogate indicator in lieu of direct moisture transmission tests. The pouches are individually weighed with a resolution of 0.1 mg every few days and the mean weight gain is recorded *versus* time. Because the initial value is a measurement (as opposed to a theoretical assumption), there are $n = 5$ coordinates. Subtraction of the initial weight corresponds to a translation in weight-space, with no loss of information. Empty pouches serve as

Table 4.13. Weight Gain for two Different Thicknesses of Foil

| Day | Weight Gained in mg | | Ratio |
	Thin Foil	Thick Foil	
0	0	0	
2	2.0	1.4	0.70
5	4.1	3.4	0.83
8	6.5	5.4	0.83
12	10.2	8.1	0.79
			0.79 ± 0.06

controls (these absorbed such small quantities of water that the water content of the plastic foil could safely be ignored). (See Table 4.13.)

Analysis From the graph of the stress test results, linear regression is seen to be appropriate (this in effect means that at these small quantities of water vapor the hygroscopic tablets are far from thermodynamic saturation). The slopes, in mg of water per day per pouch, are compared:

Results The uncertainties associated with the slopes are very different and $n_1 = n_2$, so that the pooled variance is roughly estimated as $(V_1 + V_2)/2$, see case c in Table 1.10; this gives a pooled standard deviation of 0.020: a simple t-test is performed to determine whether the slopes can be distinguished. $(0.831 - 0.673)/0.020 = 7.9$ is definitely larger than the critical t-value for $p = 0.05$ and $f = 3$ (3.182). Only a test for H_1: $t > t_c$ makes sense, so a one-sided test must be used to estimate the probability of error, most likely of the order $p = 0.001$ or smaller.

Example 43: Single-sided t-tables are often not available and many of the two-tailed t-tables do not provide values for $p < 0.01$ (corresponding to $p < 0.005$ for the single-sided case), however, so that interpolation becomes necessary: $\log(t)$ vs. $\log(p)$ is reasonably linear over the $p = 0.1 \ldots, p = 0.001$ interval (check by graphing the data, see program LINREG and data file INTERPOL1.dat). For $f = 4$, one finds $t = 4.604$ at $p = 0.01$, resp. $t = 8.610$ at $p = 0.001$; linear interpolation for $y = 7.9$ yields $p \approx 0.0014$ (two-sided), respectively $p \approx 0.0007$ (one-sided); the same result is obtained using the t-approximations for $p = 0.001$ and 0.002 in program CALCVAL, option ⟨Student's t⟩. The ratio is $0.673/0.831 = 0.81$, which means the thicker foil will admit about 80% of what the thinner one will. The potentially extended shelf-life will have to be balanced against higher costs and alternative designs. A very similar result is found if the regression lines are forced through the origin (*a*

Table 4.14. Comparison of Moisture Transmission through Foils.

	Regression with $a \neq 0$	Regression with $a = 0$
b_{thin}:	0.831 ± 0.028	0.871 ± 0.088
b_{thick}:	0.673 ± 0.003	0.688 ± 0.012

= 0): The ratio of the slopes is 0.79, but due to the high standard deviation on the first one (±0.088), the ratios cannot be distinguished.

Comment   The linear regression technique here serves to average the two trends before comparing them. Under these ideal conditions it would have been just as correct to calculate the ratio of weight gains at every point in time, see Table 4.13, last column, and then averaging over all ratios. However, if for one foil a nonlinear effect had been observed, the regression technique, employing only the linear portion of those observations, would make better use of the data. For exploratory investigations like this the results need only suggest one course of action amongst alternatives because the only facts a regulatory authority will accept are real-time stability data in the proposed marketing put-up. (See Section 4.7 and Ref. 178.)

4.7 STABILITY *à la carte*

Pharmaceutical products routinely undergo stability testing; the database thus accumulated serves to substantiate claims as to shelf life and storage conditions. Every year, sufficient material of a production batch is retained to permit regular retesting.[178] In the following case of Fig. 4.9 that occurred many years ago in an obscure and old-fashioned company, stability data pertaining to two batches of the same product manufactured 27 months apart were compared. The visual impression gained from the graph was that something was out of control. Two lab technicians and the hitherto uncontested production manager were confronted, and reached as follows:

1. "Technician *B* fabricated data to make himself look more competent."
2. "Technician *A* works sloppily."
3. "The QC manager ought to keep out of this, everything has worked so well for many years."

Being 26 months into the program for the later batch (53 months for the earlier one), the method was checked and found to be moderately tricky, but workable. Technician *B* (bold circles) did the final tests on both batches. When the

Assay [% nom]

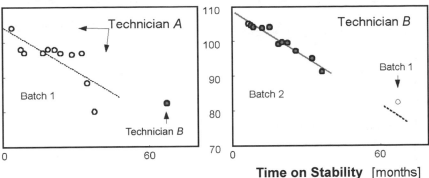

Time on Stability [months]

Figure 4.9. Product deterioration according to technicians *A* (left) and *B* (right) using the same analytical method. Technician *A*'s results are worthless when it comes to judging the product's stability and setting a limit on shelf life. The bold circles indicates the batch 1 result obtained by technician *B*; this turns out to be close to the linear regression established for batch 2, suggesting that the two batches degraded at the same rate.

same slope is used for both sets of data, technician *A*'s residual standard deviation (combined error of the method and that of the operator) is several-fold higher than that for technician *B*. This, together with other information, convinced the new head of quality control that technician *A* had for too long gone unsupervised, and that statistics and graphics as a means of control were overdue. The production manager, who, in violation of the GMP spirit, had simultaneously also served as head of QC for many years, had the habit of glancing over tabulated data, if he inspected at all. The evident conflict of interest and the absence of an appropriate depiction of data inhibited him from finding irregularities, even when a moderately critical eye would have spotted them.

4.8 SECRET SHAMPOO SWITCH

The quality control unit in a cosmetics company supervised the processing of the weekly batch of shampoo by determining, among other parameters, the viscosity and the dry residue. Control charts showed nothing spectacular. (See Fig. 4.10, top.) The cusum charts were just as uneventful, except for that displaying the dry residue (Fig. 4.10, middle and bottom): The change in trend in the middle of the chart was unmistakable. Since the analytical method was very simple and well-proven, no change in laboratory personnel had taken place in the period, and the calibration of the balances was done on a weekly basis, suspicions turned elsewhere. A first hypothesis,

Dry Residue [%]

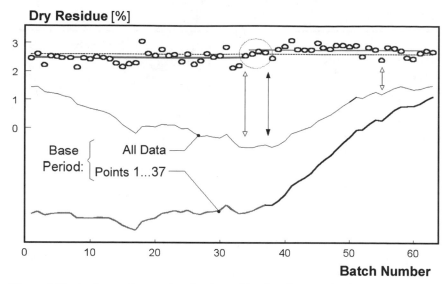

Batch Number

Figure 4.10. At the top the raw data for dry residue for 63 successive batches is shown in a standard control chart format. The fact that as of batch 34 (arrow!) a different composition was manufactured can barely be discerned, see the horizontals that indicate the means DR_{1-33} resp. DR_{34-63}. A hypothesis that a change occurred as of batch 37 would find support, though. Cusum charts for base period 1 ... 63 resp. base period 1 ... 37 make the change fairly obvious, but the causative event cannot be pinpointed without further information. Starting with batch 55 (second arrow!), production switched back to the old composition.

soon dropped, stated that the wrong amount of a major component had been weighed into the mixing vat; this would have required a concomitant change in viscosity, which had not been observed. Other components that do not influence the viscosity could have inadvertently been left out, but this second hypothesis also proved nonviable because further characteristics should have been affected. Finally, a check of the warehouse records showed that the Production Department, without notifying Quality Control, had begun to use a thickening agent from a different vendor; while the concentration had been properly adjusted to yield the correct viscosity, this resulted in a slight shift in dry residue (+0.3%), which nicely correlated with what would have been expected on the basis of the dry residues of the pure thickening agents. 21 batches later, the Production Department canceled the experiment (the trend is again nearly parallel to that on the left half of the chart). This case very nicely shows the superiority of the cusum chart over simple control charts for picking up small changes in process mean.

That improper weighing (first hypothesis) is not a farfetched thought is demonstrated by observations of how workers interact with balances. It does not matter whether digital or analog equipment is available: Repetitive work breeds casual habits, such as glancing at only a part of a number, or only the angle of an indicator needle, instead of carefully reading the whole display. A "1" and a "7", or a "0" and an "8," do not differ by more than a short horizontal bar on some LCD displays. On a particular model of multi-turn scales the difference between 37 and 43 kg resides in a "36" or a "42" display in a small window; the needle in both cases points to the "2 o'clock" position because 1 kg corresponds to 1/6 of a full turn. Mistaking the 4 with the "10 o'clock" position is also not unknown; the angle of the needle with respect to the vertical is the same in both cases. There must be a reason for a GMP requirement that stipulates a hard-copy printout of the tare and net weights and signatures by both the operator and an observer!

4.9   TABLET PRESS WOES

Initial Situation   An experimental granulation technique is to be evaluated; a sample of tablets of the first trial run is sent to the analytical laboratory for the standard batch analysis prescribed for this kind of product, including content uniformity (homogeneity of the drug substance on a tablet to tablet basis, see USP Section ⟨905⟩[43]), tablet dissolution, friability (abrassion resistance), hardness, and weight. The last two tests require little time and were therefore done first. (Note: Hardness data is either given in [kg-force] or [N], with 1 kg ≈ 9.81 Newton).

Results

Hardness: 6.9, 6.1, 6.5, 7.6, 7.5, 8.3, 8.3, 9.4, 8.6, 10.7 kg
Weight: 962, 970, 977, 978, 940, 986, 988, 993, 997, 1005 mg

Conclusion   The hardness results are very variable (CV: ± 17%), while the weight scatters only about 2%, so a request for more tablets is issued:

Results

Hardness: 7.3, 6.2, 8.4, 8.9, 7.3, 10.4, 9.2, 8.1, 9.3, 7.5 kg
Weight: 972, 941, 964, 1008, 1001, 988, 956, 982, 937, 971 mg

Conclusion   No improvement, something is wrong with either the product or the analytical procedure; measuring hardness and weighing are such simple procedures, however, that it is hard to place the blame on the very reliable laboratory technician.

Strategy   Since the tablet press is still in operation, an experiment is devised

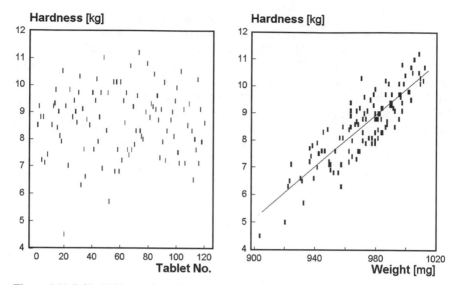

Figure 4.11 (left). Tablet hardness found for the sequence of 120 tablets. On the suspicion that there is a dependency by stamp, the data are grouped by stamp (Fig. 4.12 left). **(right).** Correlation between tablet hardness and weight. For the residuals that are obtained after correction for the tablet weight, see Fig. 4.12 right.

to test the following question: could the granulate form soft clumps and result in uneven flow from the hopper to the cavity? Good flow properties are a prerequisite for untroubled production. A suitable plastic tube is connected to the exit chute of the tablet press: the tablets accumulate at the lower end of the tube strictly in the sequence they come off the press.

Results   The hardness and the weight are measured (for 10 tablets from each of the 12 stamps, see HARDNESS.dat, use programs XYZ, XYZCELL, LINREG) and are graphed (Fig. 4.11, left).

Conclusion   The spread in both measures is large, and the correlation strengthens the notion that the hardness-variability is due to weight-variability. In particular, there are some very light and soft tablets in the lower left corner of Fig. 4.11 (right), and a cluster of very heavy ones near the limit of hardness such a formula will support. It is known from experience that higher weight yields higher hardness values because of the constant tablet volume to which the granulation compressed.

Plan   Obtain the intrinsic s_H by subtracting the weight-dependent part and taking the standard deviation of the residuals. Check the hardness values for

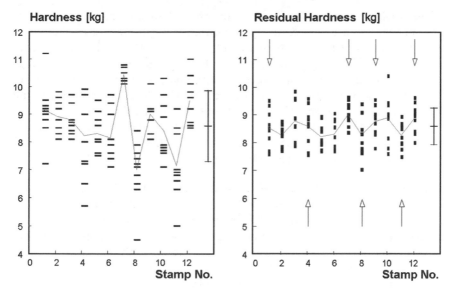

Figure 4.12 (left). The hardness data is grouped by stamp **(right).** By subtracting the weight-dependent portion of the hardness, the residuals are seen to cluster much more closely, particularly for those stamps marked with arrows. Because of the limited hardness resolution (0.1 kg), the symbols for two or more data points can overlap. The vertical bar indicates $\pm s_H$.

any connection to stamp number. The linear regression is calculated and the hardness values are corrected through:

$$H'[\text{kg}] = H_i - (W_i - W_{mean}) \cdot 0.047417[\text{kg} \cdot \text{mg}^{-1}].$$

Data Reduction and Interpretation   With the weight-dependent part of the hardness subtracted, see Fig. 4.12 (right), a residual standard deviation $s_{H,res}$ = ±0.64 (kg) is obtained, being somewhat high, but still reasonable in view of the preliminary nature of the experiment. Thus it is improbable that the granulation is fully at fault.

The stamp associated with the extreme hardness values (number 7, Fig. 4.12, left; use STAMP.dat with program MULTI) is the next suspect: It is identified and inspected on disassembly of the tablet press: Due to mechanical wear, the movement of the stamp assembly is such that an above-average amount of granulate drops into cavity number 7, and is thus compressed to the limiting hardness supported by the granulate. The hardness for stamps 4, 8, and 11 tends to be low, and high for 12, but the data is insufficient to come to definite conclusions for these tools. The tablets from these stamps "contaminated" what would otherwise been a fairly acceptable product. Because of the small IPC sample size ($n = 10$), the chances of spotting aberrant tablets are not that good, unless curiosity is aroused through hints or suspicions.

The analysis of the 12 sets of weight data yields H_0 for the Bartlett test, H_1 for the ANOVA test, and the multiple range test proposes five groups (two that greatly overlap and three smaller ones, with Stamp #7 almost in a group of its own). The analysis of the hardness data fails with H_0 in the Bartlett test because set #7 values cluster so tightly, probably a result of being at the hardness limit. If this set's data is manipulated to produce a SD similar to that of others (use program DATA, option ⟨Edit⟩), then one ends up with one central group and four fringe groups.

Conclusion The problem areas are tentatively identified; the formulations department is asked to improve the flow properties of the granulate and thus decrease the weight dispersion. The maintenance department will now have to find a proposal for countering the excessive wear on one stamp. *Note*: On more modern, heavily instrumented and feed-back controlled tablet presses, the described deviations would have become more readily apparent, and mechanical damage could have been avoided.

4.10 SOUNDING OUT SOLUBILITY

Situation A poorly soluble drug substance is to be formulated as an injectable solution. A composition of 2% w/v is envisaged, while the solubility in water at room temperature is known to be around 3% w/v. This difference is insufficient to guarantee stability under the wide range of temperatures encountered in pharmaceutical logistics (chilling during winter transport). A solubility-enhancing agent is to be added; what is the optimal composition? Two physicochemical effects have to be taken into account:

- At low concentrations of the enhancer a certain molar ratio enhancer/drug might yield the best results.
- At the limiting concentration the enhancer will bind all available water in hydration shells, leaving none to participate in the solution of the drug substance.

Mathematical Modeling A function $v = g(u)$ (Fig. 4.13, right) is found in the literature that roughly describes the data $y = f(x)$ but does not have any physicochemical connection to the problem at hand (Fig. 4.13, left); since the parameter spaces x and y do not coincide with those of u and v, transformations must be introduced:

Since u must be between zero and one, experimental data x from Fig. 4.13 would have to be compressed and shifted along the abscissa:

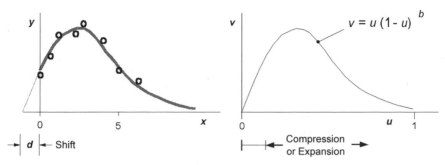

Figure 4.13. The solubility function expected for the combination of the above-mentioned physicochemical effects (left), and a similar mathematical function (right).

$$u = c \cdot (x + d) \qquad v = u \cdot (1 - u)^b \qquad y = a \cdot u \cdot (1 - u)^b \qquad (4.6)$$

Three parameters thus need to be estimated, namely the scalar factor a, the compression factor c, and the shift d. Parameter b was dropped for two reasons: (1) the effect of this exponent is to be explored, so it must remain fixed during a parameter-fitting calculation, and (2) the parameter estimation decreases in efficiency for every additional parameter. Therefore the model takes on the form

$$y( ) = f(x( ), a, c, d \,|\, b),$$

where $y( )$ and $x( )$ contain the experimental data, a, c, and d are to be optimized for every b, and b is to be varied in the interval 1 ... 6. This function was appropriately defined in a subroutine which was called from the optimization program. The optimization was carried out for the exponents $b = 1$ to 6. Four of the resulting six curves are practically equivalent in the x-interval 0 ... 7, and the other two clearly do not fit the data.

Contrary to what is suggested in Section 2.3.2, not the simplest model of those that well represented the data was chosen, but one with a higher exponent. (See Fig. 4.14.) The reason becomes apparent when the curves are compared at $x > 8$: For the model with $b = 1$ a sharp drop in drug solubility at $x = 10$ is predicted, whereas the $b = 6$ model suggests a more gradual reduction in solubility, more in accord with what is expected from the involved chemistry. The issue could only have been decided with an experiment in the $10 \le x \le 12$ region, which was not carried out. The point of interest was the prediction that maximal solubility would be attained for 2 to 3% of the enhancer, and not for 5%, as had originally, and somewhat naïvely, been extrapolated from experiments with a drug concentration of 0.5%.

**Solubility of Drug
Substance [% w/v]**

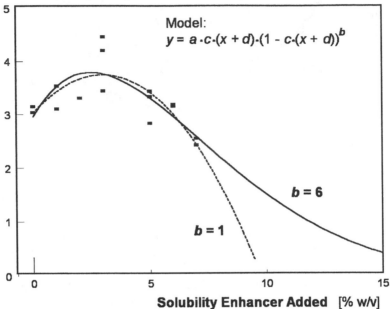

Solubility Enhancer Added [% w/v]

Figure 4.14. Solubility data and fitted models for parameters $a = 14.8$, $b = 1$, $c = 0.0741$, $d = 3.77$, resp. $a = 66.1$, $b = 6$, $c = 0.0310$, and $d = 2.14$.

Commercial program packages that either propose phenomenological models or fit a given model to data are easily available; such equations, along with the found coefficients can be entered into program TESTFIT. It is strictly forbidden to associate the found coefficients with physicochemical factors *unless* there is a theoretical basis for a particular model.

4.11 EXPLORING A DATA JUNGLE

As a new product is taken from the laboratory, through the pilot, and into the production plant, a lot of information will accumulate that helps to better understand the product and the associated production process. In general, the major characteristics are soon evident and will be exploited to optimize the production. Fine-tuning the process can only be done, however, if hitherto hidden aspects are revealed.

Situation   A case in point is a purification/drying step before the sample

Table 4.15. Extract of Data Table for 43 Batch Analyses see JUNGLE4.dat.

	1	2	3	4	5	6	7	8	9
i	Solv A	Solv B	Solv C	Solv D	Solv E	Other Impur.	Assay HPLC	Assay Titr.	Purity HPLC
1	4.1	6.9	11.0	13	23	0.6	98.0	98.0	98.7
2	5.1	6.2	11.3	15	34	0.8	94.9	97.0	98.1
3	6.3	7.0	13.3	18	46	0.7	94.0	94.0	98.0
.	.	.	.	.	.	.	.	.	.
.	.	.	.	.	.	.	.	.	.
43	6.1	7.39	13.2	16	30	0.2	95.5	95.6	99.4
x_{mean}	5.94	7.39	13.33	16.1	31.26	0.40	95.52	96.09	98.95
s_x	±1.2	±.77	±1.55	±2.7	±14.0	±.25	±2.30	±2.61	±0.56

for in-process control (IPC) is taken. Some of the analytical parameters that were determined were water content in wt-% (Karl Fischer), various residual solvents in ppm (GC), the sum of other impurities in area-% (HPLC), the content of the major compound (HPLC and titration, as determined by comparison with an external standard), and its purity as determined from area-% figures (HPLC). The severity of the drying process had varied somewhat over 43 batches, but was likely to be on the mild side, so that an interesting question was to determine the ultimate quality that was attainable (too much heat would have provoked decomposition), see file JUNGLE4.dat; an extract is given in Table 4.15.

Data Analysis Because of the danger of false conclusions if only one or two parameters were evaluated, it was deemed better to correlate every parameter with all the others, and to assemble the results in a triangular matrix, so that trends would become more apparent. The program CORREL described in Section 5.2 retains the sign of the correlation coefficient (positive or negative slope) and combines this with a confidence level (probability p of obtaining such a correlation by chance alone).

Prior to running all the data through the program, a triangular matrix of *expected effects* was set up, see Fig. 4.15, by asking such questions as "if the level of an impurity X increases, will the assay value Y increase or decrease?" (it is expected to decrease). Furthermore, the probability of this being a false statement is expected to be low; in other words, a significant negative event is expected (=). The five solvents are expected to be strongly correlated because heating and/or vacuum, which would drive off one, would also drive off the others. The titration is expected to strongly correlate with the HPLC content; because titration is less specific (more susceptible to interferences),

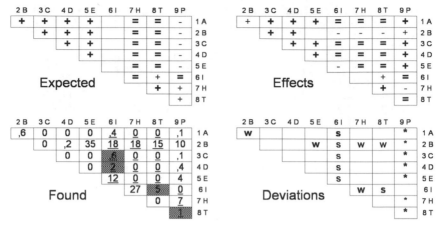

Figure 4.15. Expected and found correlations among nine quality indices determined for 43 batches of a bulk chemical; for details, see text.

the titration results are in general a bit higher. The HPLC purity of the major component and the sum of the other impurities, together with the residual solvents, should add to 100%, and a strong negative correlation is expected. A word of caution must be added here: The common practice of normalizing chromatograms to 100% peak area redistributes noise from large to small peaks because the absolute noise is relevant, not the S/N ratio; this can distort conclusions drawn from the analysis of small peaks, cf. "closure"[179]; it would be better to do the calculations with peak areas, i.e., dimension [mV · sec]. The *found correlations* are given as percent probability of chance occurrence. The *effects* figure records the probability in terms of the (= − ++) scale. The *deviation* figure compares expectations with effects and shows whether one's thinking was off. A somewhat weaker ("*w*") or stronger ("*s*") correlation is immaterial. Much weaker or much stronger correlations cannot be ignored; faulty thinking, flagged by asterisks, should set one's mind ticking.

So much for theory: The interesting thing is to ponder what went contrary to expectations, and to try to find explanations. It is here that graphical presentations are helpful to judge the plausibility of any effect: A correlation, is only a correlation, and can come about because of a single point that bucks the trend.

Interpretation   It seems as if the assay figures (columns 7, 8) were strongly negatively correlated with the presence of residual solvents (rows 3–5, not as expected) and to a lesser degree with water. The up to 8% water have a direct effect on the assay, unless compensated for, while even 50 ppm of E

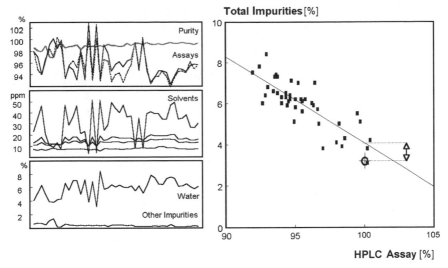

Figure 4.16 (left). Trendlines for the various components. The three scales are different %, ppm, resp. %). **(right).** Total impurities (columns 1–6, including water of crystallization, *versus* the HPLC assay of the major compound (column 7). The circle marks the hypothetically pure compound: 3.2% water of crystallization, but no other impurities. The arrow indicates the percentage of impurities expected (for this simple linear model) to remain in the product after all solvents and excess water have been driven off.

are not expected to have an effect on the assay itself; the correlation must therefore be due to the process chemistry. Surprisingly, purity is positively and not negatively correlated to the solvents. Absolute values of $p = 5\%$ and less were taken to mean "significant effect" or even "highly significant effect"; an absolute p larger than about 20% indicates that this might well be a chance result. Three correlations that were plotted are shown in Figs. 4.16 (right) and 4.17 for illustrative purposes.

The fact that in HPLC only UV-active components are registered, whereas in titration all basic functional groups are detected constitutes a difference in specificity (quality) and sensitivity (quantity) of these two methods relative to a given impurity. See Fig. 4.17 (left). [Solvent *A* (water) behaves differently from the other four as can be seen from Fig. 4.17 (right). The material was known to exist in a crystal modification that theoretically contains 3.2% water, and moderate drying will most likely drive off only the excess: Indeed, the best-dried batches are all close to the theoretical point (circle, arrow in Figs. 4.16–17), and not near zero. This is only partly reflected in Table 4.15, column A; for this reason tabular and graphic information has to be combined. Solvent *B*, which is an alcohol, behaves more like water

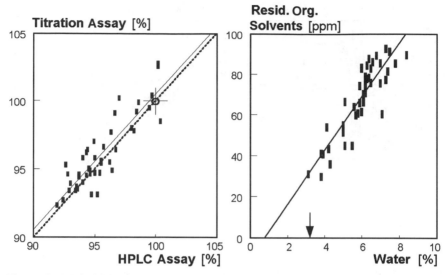

Figure 4.17 (left). HPLC assay (column 7) and titration assay (column 8) are compared: Evidently titration yields the higher values (solid line vs. dashed theoretical line); the reason is that one of the major impurities is of basic nature. The circle denotes the pure compound and perfect selectivity for both techniques. **(right).** The sum of all organic solvents (columns B–E in Table 4.15) is plotted *versus* the residual water (column A). The drying step obviously drives off organic solvents and water to a large degree, depending on the severity of the conditions. Organic solvents can be brought down to 30 ppm, while only the excess water can be driven off, the remainder being water of crystallization (arrow: theoretically expected amount: 3.2%).

(A) than the apolar solvents (C–E). All three correlation graphs demonstrate that careful drying drives off solvents and excess water, and in the process improves purity.

This technique is also used to ferret out correlations between impurities within the same HPLC chromatogram: If several reaction pathways compete for reagent, each with its own characteristic impurity profile, any change in conditions is likely to affect the relative importance of the individual pathways. Finding which impurities move in concert helps to optimize a process. (See Fig. 4.18). Data file PROFILE.dat contains another example: 11 peak areas were determined in each chromatogram for nine production runs. Impurities 5, 6, and 8 appear to be marginally correlated to the others, if at all, while the product strongly correlates with impurities 1–4, 7, the solvent, and the reagent. Since impurities 5, 6, and 8 are far above the detection limit, analytical artifacts cannot be the reason. The correlation graph is depicted in Fig. 4.28. These examples show that unless the interdependencies between various parameters are clearly reflected in the measurements, some interpre-

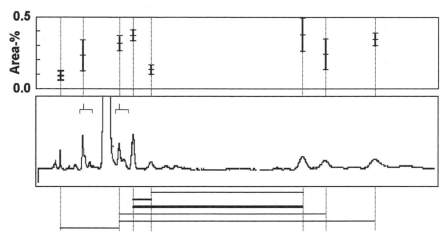

Figure 4.18. Peak-size correlation in an HPLC-chromatogram. The impurity profile of a chemical intermediate shown in the middle contains peaks that betray the presence of at least two reaction pathways. The strength of the correlation between peak areas is schematically indicated by the thickness of the horizontal lines below the chromatogram. The top panel gives the mean and standard deviation of some peak areas ($n = 21$); the two groups of peaks immediately before and after the main peak were integrated as peak groups.

tations may (apparently) contradict others. This should be taken as a hint that the type of analysis possible with program CORREL is of exploratory nature and should be viewed as food for thought.

4.12 SIFTING THROUGH SIEVED SAMPLES

Situation   There are two vendors for a particular bulk chemical who meet all written specifications. The products are equally useful for the intended reaction as far as the chemical parameters are concerned; both comply in terms of one physical parameter, the size distribution of the crystals, but on the shop floor the feeling prevails that there is a difference. Because the speed of dissolution might become critical under certain combinations of process variables, the chemical engineers would favor a more finely divided raw material. On the other hand, too many fine particles could also cause problems (dust, static charging).

Questions   Are the materials supplied by the two vendors systematically different? How could such a difference be quantified?

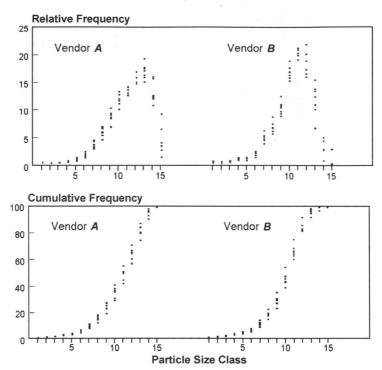

Figure 4.19. Relative (top panel) and cumulative weight of material per size fraction for two different vendors (bottom panel; *A*: left; *B*: right). The cumulation helps reduce noise.

Course of Action   A laser-light scattering apparatus LLS is used to measure the size distributions. The results are given in weight-% per size class under the assumption that all material falls into the 15 classes between 5.8 and 564 μm. For the seven most recent deliveries from each of the two vendors samples are obtained from the retained sample storage. The 14 size distributions are measured and the average distribution for each vendor is calculated. (See Fig. 4.19 and Table 4.16, right-hand columns.) The Euclidean distance for each sample is given relative to the Group Average *A* (*), respectively *B* (**) (564–261 μm in bin 15, 7.2–5.8 μm in bin 1, logarithmic classification) for samples 1–7 of each of two vendors.

As it turns out, one vendor's material contains almost no particles (0.5%) in the 261–564 μm class (bin #15); this means that the %-weight results accurately represent the situation. The other vendor's material, however, contains a sizable fraction (typically 5%, maximally 9%) in this largest size class; this implies that 1–5% "invisible" material is in the size class >564 μm. Evidently then, the size distribution curve for this second material is accurate only on

Table 4.16. Sieve Analysis on Powder Samples from Two Vendors; for Details, see Text

Vendor A							Vendor B							Group Mean		Bin
1	2	3	4	5	6	7	1	2	3	4	5	6	7	A	B	No.
9.2	1.5	6.5	3.5	4.1	2.8	6.5	0.0	2.9	0.1	0.0	0.0	0.1	0.3	4.87	0.49	15
16.0	10.8	12.4	12.0	15.7	12.2	12.6	0.9	5.0	2.7	2.7	2.8	2.9	4.4	13.10	2.99	14
17.5	16.3	15.0	17.1	19.3	17.7	15.9	6.7	10.1	10.9	15.4	13.8	13.4	12.4	16.97	11.81	13
15.0	15.8	14.7	6.9	15.9	16.8	15.0	16.5	16.0	17.3	21.8	20.1	18.9	16.0	15.73	18.09	12
12.7	14.2	12.6	13.1	13.6	14.1	14.1	20.9	19.9	20.9	21.2	19.9	20.3	19.1	13.49	20.31	11
10.1	13.3	11.1	11.7	11.7	12.8	12.0	18.9	16.3	16.9	15.3	15.8	16.8	17.6	11.81	16.80	10
6.9	10.3	9.1	9.4	7.0	8.7	8.5	12.5	10.7	10.9	8.8	9.7	9.3	10.4	8.56	10.33	9
4.6	7.0	6.7	5.9	4.6	5.4	6.0	8.7	6.8	7.4	5.7	6.7	6.1	6.7	5.74	6.87	8
3.0	4.3	4.5	3.8	3.2	3.4	3.7	6.2	4.5	5.1	3.9	4.5	4.8	5.3	3.70	4.90	7
1.5	2.0	2.5	2.1	1.3	1.6	1.8	2.5	2.4	2.4	1.4	1.8	2.2	2.5	1.83	2.17	6
0.9	1.1	1.4	1.3	0.8	1.0	1.0	1.4	1.4	1.4	0.9	1.2	1.2	1.3	1.07	1.26	5
0.6	0.9	0.9	0.9	0.7	0.9	0.8	1.4	1.0	1.2	1.0	1.2	1.1	1.1	0.80	1.14	4
0.5	0.6	0.6	0.5	0.5	0.6	0.5	0.9	0.8	0.7	0.6	0.6	0.7	0.8	0.54	0.73	3
0.4	0.5	0.5	0.4	0.4	0.5	0.4	0.6	0.7	0.6	0.4	0.5	0.6	0.6	0.44	0.57	2
0.5	0.6	0.6	0.6	0.5	0.6	0.5	0.8	0.7	0.7	0.5	0.6	0.6	0.7	0.56	0.66	1
6.0	5.0	3.4	2.4	4.2	2.9	2.3	20.4	13.6	16.2	15.9	14.8	14.9	13.8	*)		
20.3	11.7	15.5	14.2	18.1	13.7	14.9	6.9	4.2	1.7	6.0	3.2	2.2	3.0		**)	

a relative basis, but not absolutely; distorted data are a poor foundation for a statistical analysis. Thus, there are three ways to continue:

1. Use another methodology to determine the amount above 564 μm.
2. Proceed as if the data were not distorted and carry in mind that any result so obtained is biased.
3. Employ a model that mathematically describes a size distribution of this type, adjust the model parameters for best fit, and estimate the missing fraction above 564 μm; after correcting the observed frequencies, continue with a correct statistical analysis.

The first option is unworkable, though, because this other technology is unlikely to have (*A*) the same cut-off characteristics around 564 μm, or (*B*) measure the same characteristics (e.g., volume instead of length). The third options falls out of favor for the simple reason that such a model is not available, and if it were, errors of extrapolation would be propagated into any result obtained from the statistical analysis.

   Example 44: It appears that one distribution is sharper than the other; a χ^2 test is applied to the group means to confirm the difference [Eq. (1.50)]: $\chi^2 =$ 20.6 or 95.7 is found, depending on which distribution is chosen as reference, cf. Table 3.5. Since there is no theoretical distribution model to compare against, the choice of reference is arbitrary. The critical $\chi^2(p = 0.95, f = 14)$ is 23.7, which means that H_1 could have to be rejected under one perspective. The above-mentioned distortion of the data from the coarser material might have tipped the scales; this is a classical case where the human eye, used to discriminating patterns, sees something that a statistical test did not. A disturbing aspect is the fact that the individual results scatter so much so as to obscure any difference found between the means. (See Fig. 4.19.) The situation can be improved by regarding the cumulative (Fig. 4.19, bottom) instead of the individual frequencies, because through the summation the signal/noise ratio is improved.

Euclidean Distance   On the basis of the given evidence, the size distributions are different, but this is not fully borne out by the statistical test employed. To overcome the impasse, another technique is employed that allows each sample to be judged according to its proximity to given points. Cluster analysis (finding and comparing "distances" in 15-dimensional space) shows a difference between the products; the disadvantage is that cluster analysis strains the imagination. (*Cluster analysis* would allow any number of vendors to be subjected to a simultaneous comparison, each providing as many batches for size distribution analysis as he wanted.) A further problem is that at the moment no one knows what causes the process to run astray; in such a situation it would be inopportune to confront vendors with highly abstract statistical analyses and ask them to comply with undefined specifications. A simple one- or two-dimensional depiction that allows the statement "A is OK, and your product is different; batches 9 and 14 are closest, batch 8 is the worst" would help. In order to rectify this, the Euclidean distances separating every point from both the average of its group and that of the other group are projected into two dimensions. Four things are necessary:

1. For each of the two groups, the mean over every one of the 15 dimensions (classes, bins) is calculated (columns A and B in Table 4.16).
2. Corresponding elements in the two vectors of means are subtracted, and the differences are squared and added. The square root of the sum (15.21) is equal to the Euclidean distance in 15 dimensions separating the two points that represent the group means. This distance forms the base line in Fig. 4.20.
3. Corresponding elements in the vector representing one particular sample and in the appropriate vector of means are worked up as in 2) to find the Euclidean distance between point i and its group mean (see lines marked with an asterisk (*) in Table 4.16); this forms the second side of the appropriate triangle in Fig. 4.20.
4. In order to find the third side of the triangle, proceed as under 3) by replacing one vector of averages by the other. (See lines marked with double asterisks (**) in Table 4.16.)

Example 45: For sample 1, vendor A:

For point 1:
   Sample data: 9.20, 16.00, 17.50, ... 0.50
   Vector of means 1: 4.87, 13.10, 16.97, ... 0.56
   Vector of means 2: 0.49, 2.99, 11.81, ... 0.66

The baseline is $b = 15.2$, with $b^2 = (4.87 - 0.49)^2 + ... (0.56 - 0.66)^2$ (see preceding item 2); side a is $a = 6.0$, with $a^2 = (9.20 - 4.87)^2 + ... (0.50 - 0.56)^2$ (see preceding item 3); side c is $c = 20.3$, with $c^2 = (9.20 - 0.49)^2 + ...$ $(0.50 - 0.66)^2$ (see preceding item 4)

The corresponding Cartesian coordinates are $x = 12.3$ and $y = 3.7$ if the group averages are set to $x_A = -15.21/2$ and $x_B = +15.21/2$, and $y_A = y_B = 0$.

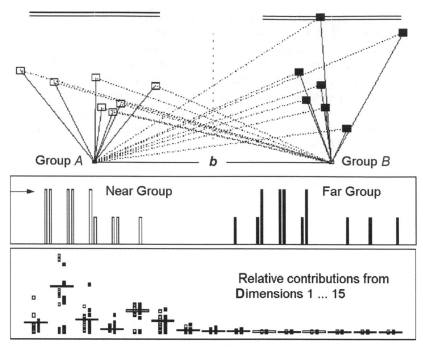

Figure 4.20. Euclidean distances of the 14 samples from their group means; the individual points can be clearly assigned to either the left (white) or the right (black) group by testing against the dashed separation line. The double bars at the top give the group means and SDs relative to the baseline *b* separating the group centers. The Euclidean distances were ordered in cumulative fashion in the middle panel and distinguished as being either near or far group. The separation is indisputable (zero distance at left, maximal distance at right; the colors black and white do not pertain to groups *A* and *B*!) The bottom panel gives the relative contributions of dimensions 1 (left) ... 15 (right) towards b^2; the horizontal bars represent the averages, and the squares give the spread for the individual points.

This illustrative technique suffers from a lack of a statistically objective measure of probability. The comparison is done visually by judging the distance of the center of a group of seven points from the center line and taking into account the diameter of a group of points, or by using the middle panel in Fig. 4.20 or the lines marked with an asterisk and a double asterisk in Table 4.16, and looking for an overlap in near- and far-group Euclidean distances in less than, say, one sample out of 10 (the smallest far-group ED = 11.7, the largest near-group ED = 6.9, so there is no overlap in this particular case).

Interpretation Using Euclidean distances, the difference between the vendor's samples shows up nicely. (See data file SIEVE1.dat; if some samples of groups *A* and *B* had been similar, an overlapping of the two distributions would have been seen.)

Instead of comparing each sample against the group averages, long-term-averages or even theoretical distributions could have been employed. Program EUCLID provides additional information: The %-contribution of each bin to the total Euclidean distance for both the group averages (horizontal lines) and the individual samples (squares). These eight dimensions that show both the lowest contributions and the smallest variability are natural candidates for elimination if the number of dimensions is to be reduced. This can be tested by deleting the last eight rows from SIEVE1.dat (available as SIEVE2.dat) and reanalyzing.

In this particular example, the individual bins all carry the same dimension and are mutually coupled through the conditions Σ(bin contents) = 100%. If unrelated properties were to be used in a comparison, of all the employed results must be numerically similar if each property is to contribute to the Euclidean distance in a roughly comparable manner. As an illustration, calculations involving group means of, say, 8.3 area-%, 550 ppm, and 0.714 AU would yield wholly different s than if 0.083, 0.052%, and 714 mAU had been used. For this reason, unrelated vectors are first normalized so that the overall mean over groups *A* and *B* is 1.00 for each bin. An example is provided in data file EUCLID.dat, where two groups of data (n_A = 6, n_B = 5) are only marginally different in each of eight dimensions, but can be almost perfectly separated visually. Normalization is achieved by using program DATA, option ⟨Modify⟩.

4.13   CONTROLLING CYANIDE

Situation and Criteria   A method was to be developed to determine trace amounts of cyanide (CN^-) in waste water. The nature of the task means precision is not so much of an issue as are the limits of detection and quantitation (LOD, LOQ), and flexibility and ease of use. The responsible chemist expected cyanide levels below 2 ppm.

Experimental   A photometric method was found in the literature which seemed to suit the particular circumstances. Two cyanide stock solutions were prepared, and an electromechanical dispenser was used to precisely prepare solutions of 20, 40, ... , 240 respectively 10, 30, 50, ... , 250 μg CN^-/100 ml. 10 ml of each calibration solution were added to 90 ml of the color-forming reagent solution and the absorbance was measured using 1-cm cuvettes. (See Table 4.17 (left and middle panels) and data file CYANIDE.dat.)

Table 4.17. Absorption Measurements on Cyanide Calibration Solutions

No.	Conc.	Absorb.	No.	Conc.	Absorb.	No.	Conc.	Absorb.**
1	0	0.000	15	0	0.000	28	0	0.0000
2	10	0.049	16	20	0.099	29	2	0.0095
3	30	0.153	17	40	0.203	30	4	0.0196
4	50	0.258	18	60	0.310	31	6	0.0298
5	70	0.356	19	80	0.406	32	8	0.0402
6	90	0.460	20	100	0.504	33	10	0.0501
7	110	0.561	21	120	0.609			
8	130	0.671	22	140	0.708			
9	150	0.761	23	160	0.803			
10	170	0.863	24	180	0.904			
11	190	0.956	25	200	0.997			
12	210	1.053	26	220	1.102			
13	230	1.158	27	240	1.186			
14	250	1.245						

0.00501	±1.0%		0.00497	±1.0%		0.00504	±1.7%	slope b
0.00588	±123%		0.00611	±113%		−0.00033	±160%	intercept a
±0.0068			±0.0060			±0.00027		res. std. dev. s_{res}
0.9997			0.9998			0.9998		coeff. determ. r^2
14			13			6		n
1.4			1.4			0.1		LOD $\mu g/100$ ml
3.0			3.0			0.2		LOQ $\mu g/100$ ml

Legend: No: number of measurement, Conc: concentration in μg CN$^-$/100 ml; Absorb: absorbance [AU]; slope: slope of regression line $\pm$ t · CV; intercept: see slope; res. std. dev.: residual standard deviation s_{res}; n: number of points in regression; LOD: limit of detection; LOQ: limit of quantitation; **: measurements using a 2-fold higher sample amount and 5-cm cuvettes—i.e., measured absorption 0 ... 0.501 was divided by 10.

Data Analysis   The results were plotted; at first glance a linear regression of absorbance *versus* concentration appeared appropriate. The two dilution series individually yielded the figures of merit given in Table 4.17, bottom. The two regression lines are indistinguishable, have tightly defined slopes,

Table 4.18. Regression Coefficients for Linear and Quadratic Model

Linear	Quadratic	Item
0.005843	−0.002125	Constant term
0.004990	+0.005211	Linear term
	−0.0000009126	Quadratic term
±6.6 mAU	±4.5 mAU	Res. std. dev.

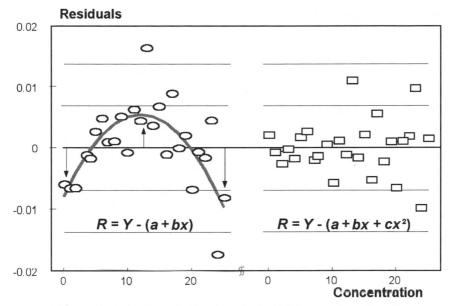

Figure 4.21. Residuals for linear (left) and quadratic (right) regressions; the ordinates are scaled ±20 mAU. Note the increase in variance toward higher concentrations (heteroscedacity). The gray line was plotted as the difference between the quadratic and the linear regression curves. Concentration scale: 0–25 μg/ml, final dilution.

and pass through the origin. The two data sets are thus merged and reanalyzed. (See Table 4.18, left column.) This rosy picture dissolves when the residuals (Fig. 4.21) are inspected: The residuals No. 8 (1.3 μg/ml) and 27 (2.4 μg/ml) are far from the linear regression line. The question immediately arises, "are these points outliers?" One could then drop one or both points from the list, repeat the regression analysis, and determine whether (a) the residual standard deviation had become smaller, and (b) whether these points were still outside the CL(y). On the other hand, the residuals in the middle of the concentration range are positive, those at the ends negative; this, together with the fact that a photometer is being used, should draw attention to the hypothesis H_1: "curved calibration function." (Note to the nonchemist: Stray light in the photometer dominates at high absorbances, which contributes to lower slopes at high concentrations. The chemical work-up can also produce lower yields of the chromophore, the light-absorbing part of the molecule, at higher concentrations). The curvature is quite evident (cf. Fig. 4.21), which makes H_1 all the more probable. A quadratic regression is applied to the merged data set and the residuals are again plotted. The quadratic regression $Y = a + b \cdot x + c \cdot x^2$ is a straightforward extension of the linear regression concept,[64,114] with programs available at all computing centers, for PCs, and even for some hand-held computers. (See Table 4.18, right column.)

Example 46: There seems to be a clear reduction in the residual standard deviation, and the F-test supports this notion: $F = (6.6/4.5)^2 = 2.15$, with $F(26, 26, 0.05) = 1.93$. Point No. 8 (see Fig. 4.21) is now only 0.011 AU above the parabola, which means it is barely outside the $\pm 2 \cdot s_{res}$ band; all other residuals are smaller. From the practical point of view there is little incentive for further improvement: the residual standard deviation ± 4.5 mAU is now only about twice the experimental standard deviation (repeatability), which is not all that bad when one considers that two dilutions and a derivatization step are involved. The scatter appears to increase towards higher concentrations: Indeed, this may be so, but to underpin the case statistically one would have to run at least eight repeats at a low and another eight at a high concentration if $s_{high} = 2 \cdot s_{low}$, because $F_{crit}(7, 7, 0.05) \approx 3.8$. Should s_{high} only be $1.5 \cdot s_{low}$, then the experimental plan would call for $n_{high} = n_{low} \geq 18$.

Decisions   Because quadratic regressions are more difficult to handle and the individual coefficients of a three-parameter model are less well defined than those of a two-parameter one in the case of weak curvature, any gain from using a polynomial of higher order might well be lost through error propagation. The definite course of action was to accurately calibrate a part of the given concentration interval and to either dilute samples to fit the range, or then to use thicker (5 cm) cuvettes to gain sensitivity. In case this strategy should not work, it was decided to also calibrate the 0–10 μg/100 ml region (calibration points #28-33; for results, see Table 4.17. This regression line is indistinguishable from the other two as far as the coefficients are concerned, but the LOD and LOQ are much lower). The overall operating range thus covers 2 ppb (20 ml sample amount, 5 cm cuvette) to over 200 ppm (0.1 ml sample amount, 1 cm cuvette) CN^- in the sample, a factor of $1 : 10^5$, which is very large. In a screening run the sample is diluted according to a standard plan trimmed for speed and ease, and depending on this preliminary result, the sample is only then precisely diluted if there is impending danger of getting high cyanide levels that would require further treatment of the waste water. A simple linear regression is used for the approximately linear portion of the calibration function. Another course of action would be to improve the chemical work-up and the instrumental measurement procedures to obtain a linear calibration to higher concentrations. The profit/loss analysis of further method development *versus* occasional repetition of a dilution would have to be investigated. If a programmable sample carousel/diluter/UV-configuration were used, this repetition could be enacted automatically if an alarm limit is exceeded in the first measurement.

Since this situation does not impinge on pharmaceutical but on environmental quality, and perhaps worker safety, other guidelines apply (in the U.S. those of the EPA and/or the OSHA). It might come as a surprise to the scientist that government bureaucracies have the luxury of enacting disparate standards and technology, from sampling to specifications, for one and the same thing, as in the case of the FDA for innovator (brand name) and imitator (generic) products.

4.14 AMBIGUOUS AUTOMATION

A pharmaceutical specialty is produced in three dosage strengths (major component A); "A" and a second component "B" are controlled by HPLC for batch release purposes. It is decided to replace the manual injection of the sample solution by an automatic one. It is expected the means will remain the same but the standard deviations will be smaller for the automatic injection. Cross-validation of the methods is effected by running both methods on each of 10 samples. The mean and the standard deviation for each series of 10 measurements is given in Table 4.19.

Table 4.19. Cross-Validation Results[a]

	Manual Injection			Automatic Injection		
	Low	Med	High	Low	Med	High
Mean:						
A	497	750	992	493	753	1010
B	360	359	357	361	356	355
Standard Deviation (±):						
A	5.88	5.51	14.6	14.2	11.1	23.8
B	7.33	5.51	6.36	5.39	7.32	8.23
CV (±%):						
A	1.18	0.73	1.47	2.87	1.47	2.36
B	2.04	1.54	1.78	1.49	2.06	2.32
Residual Std. Dev. (of B):						
± Absolute	7.03	4.22	5.66	5.56	4.46	2.68
± Res. S.D./B [%]	1.95	1.17	1.59	1.54	1.25	0.75

[a]The best residual SD for the calibration measurements is an indicator of repeatability (i.e., ±0.75%); the rest of the overall spread of the results (e.g. ±2.32%) is due to manufacturing variability. Improvements due to automation are underscored, e.g., 1.59 → 0.75

Example 47: Observations and first interpretation:

1. It appears that automatic injection actually worsens precision (±5.88 → ±14.2, etc.). The relative standard deviation suffers, too: ±1.18 → ±2.87, etc.; the new variance component corresponds to $\sqrt{2.87^2 - 1.18^2} = \pm 2.6$ (2.6, 1.3, and 1.8 for component "A," 0, 1.4, and 1.5 for component "B").

2. The residual SD decreased markedly in three cases (7.03 → 5.56, 5.66 → 2.68, 1.59 → 0.75), remained about the same in the other cases; this shows that automation had the desired effect, only it was overshadowed by the earlier point 1. The average decrease in V_{res} is 42%.

Observation 1 is an illusion due to the fact that the above numbers measure the overall spread in one dimension (vertical or horizontal), and do not take into account the correlation between the results A and B that is very much in evidence in the right side of Fig. 4.22 (automatic injection), also see Fig. 1.23. The variability can be measured in five different ways (see Table 4.20):

- Calculations 1 and 3 are naïve because they ignore the correlation, cf. left side of Fig. 1.23.

- Calculations 2 and 4 each assume that one component's error completely dominates the other's (see Section 2.2.1), which is improbable given the fact that both A and B were measured under identical conditions and the signal ratios are in the range 1.4 ... 2.8. If some assumptions were made concerning the relative contributions of measurement errors s_x and s_y, the numbers could be sorted out (last paragraph in Section 2.2.10), but the burden of this additional assumption is not worth the trouble considering that the volume error measured under 5) is much bigger.

- Calculation 5 measures the SD of the Euclidean distance $\sqrt{(x_i - x_{mean})^2 + (y_i - y_{mean})^2}$, and is interpreted as the uncertainty associated with the injection volume and the proportionality constant $k \leq 1.0$ (bubbles in the injection loop can only decrease the effectively injected amounts A', B' relative to the nominal ones A, B) so that $A' = k \cdot A$ resp $B' = k \cdot B$, with k taking on a different value for each injection. An internal standard would serve to correct k to 1.00.

From the preceding discussion, it can be gathered that the automatic injection should eventually lead to more reproducible results (the residual standard deviation decreases by about 20%), but only if the spread along the

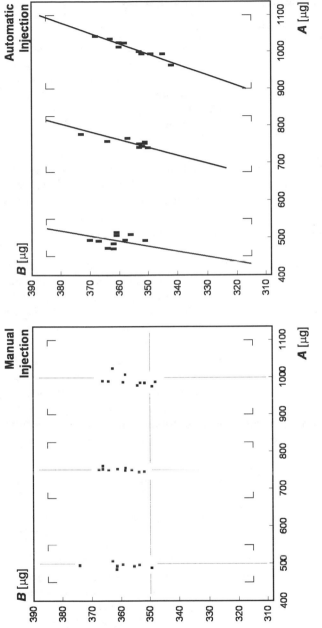

Figure 4.22. Correlation of assay values for components A and B, for three dosage levels of A, with 10 samples per group. The corner symbols indicate the ±10% specification limits for each component. For manual injection (left panel) only relative standard deviations of 1–2% are found, but no correlation. Automatic injection (right panel) has a lower intrinsic relative standard deviation, but the data are smeared out along the proportionality line because no internal standard was used to correct for variability of the injected volume. The proportionality line does not go through the corners of the specification box because component B is either somewhat overdosed (2.4%), analytical bias, or because an interference results in too high area readings for B. The nominal values are $A_l = 500$, $A_m = 750$, $A_u = 1000$, and $B = 350$, see dotted lines. The RSD is calculated according to Eq. (2.13).

Table 4.20. Measures of Variability

	Direction	Metric	Reference	Assumptions
1	vertically	$s_{y,\,total}$	y_{mean}	$s_y \gg s_x$, no correlation
2	vertically	$s_{y,\,res}$	regression line	$s_y \gg s_x$
3	horizontally	$s_{x,\,total}$	x_{mean}	$s_x \gg s_y$, no correlation
4	horizontally	$s_{x,\,res}$	regression line	$s_x \gg s_y$
5	along the diagonal	s_k	coordinate x_{mean}/y_{mean}	$s_x \approx s_y \ll s_{injection}$

regression line could be reduced. How repeatable could the results potentially be? The residual standard deviation is only $\pm0.75\%$ to $\pm1.54\%$ relative to the mean value B. The additional analytical variability is estimated as shown in Table 4.21.

This means that something like $\frac{1}{2}$ of the B variance is due to this lack of control; it might thus be possible to achieve a repeatability of $\pm1\%$ on both components. The fact that the potential precision (max $\pm 1.2\%$, with IS) for the automatic injection is hardly smaller than that achieved manually without the benefit of an internal standard (geom. average of $2.04 \ldots 1.78 \approx \pm 1.8\%$) shows that skillful work was being done. The question is now why in the case of the manual injection, which shows little or no correlation between A and B, repeatabilities of no less than $\pm1.5 - 1.8\%$ are observed. There is an explanation: The automatic injector is more reproducible in terms of the time necessary to turn the valve; this means that the injected volume is less smeared out in time and yields a better integrable peak form (cf. Section 3.3), a notion confirmed by the actual chromatograms. The interpretation for the medium and low dosage forms is essentially the same. Note that at high levels of component A the repeatability (standard deviation) for A sharply

Table 4.21. Potential Precision

Average (Manual + Automatic)	Range (Manual/Automatic)	
$\sqrt{1.95^2 - 1.54^2} = \pm1.2$	$(\pm1.0 \ldots \pm 1.4)$	Excess variance
$(1.2/1.95)^2 = 0.38$	$(0.38 \ldots 0.78)$	Contribution to variance
$\pm1.54 \cdot \sqrt{1 - 0.38} = \pm1.2$	$(\pm1.2 \ldots \pm0.4)$	Potential precision of assay for component B if the internal standard helps to eliminate the variance associated with the effectively injected volume

rises ($\pm11.1 \rightarrow \pm23.8$), but not so for B; the reason is related to saturation effects in the HPLC column. This could be avoided by injecting two separate sample dilutions, one optimized for the reproducibility of the A peak, the other for the B peak.

The moral of the story: The satisfaction of having all points inside the specification limits should not induce inaction. An internal standard was thereupon added to the procedure, and promptly, the dispersion along the diagonal in Fig. 4.22 (right) was eliminated.

4.15 MISTRUSTED METHOD

An in-process control (IPC) of a bulk chemical was augmented by a heavy-metals test because trace quantities of a catalyst were suspected to have a deleterious effect on the following synthesis step. Since the identity of the metal was known, a simple precipitation as the sulfide was deemed to give sufficiently accurate answers in a very short time (*proviso*: no other heavy metal present). A test along the lines of the official pharmacopoeial Heavy Metals Method (USP $\langle 231 \rangle$, Method I,[43]) is conducted wherein a reference solution containing 20 ppm of the metal chloride is treated in parallel to the sample, and the intensities of the coloration of the suspensions are compared 3–5 minutes after mixing (the finely divided suspension later coalesces and precipitates out of solution). Concentrations much higher than 20 ppm would be accommodated by further dilution of the sample. The relative confidence interval is judged to be around $\pm25\%$.

The four batches in question were found to contain about 20, 40, 20, respectively 90 ppm (cf. Fig. 4.23).

The Production Department was not amused, because lower values had been expected. Quality Control was blamed for using an insensitive, unselective, and imprecise test, and thereby unnecessarily frightening top management. This outcome had been anticipated, and a better method, namely polarography, was already being set up. The same samples were run, this time in duplicate, with much the same results. A relative confidence interval of $\pm25\%$ was assumed. Because of increased specificity, there were now less doubts as to the amounts of this particular heavy metal that were actually present. To rule out artifacts, the four samples were sent to outside laboratories to do repeat tests with different methods: X-ray fluorescence (XRF[180]) and inductively coupled plasma spectrometry (ICP). The confidence limits were determined to be $\pm10\%$ resp. $\pm3\%$. Figure 4.23 summarizes the results. Because each method has its own specificity pattern, and is subject to intrinsic artifacts, a direct statistical comparison cannot be performed without first correcting the "apparent concentrations" in order to obtain "presumably true

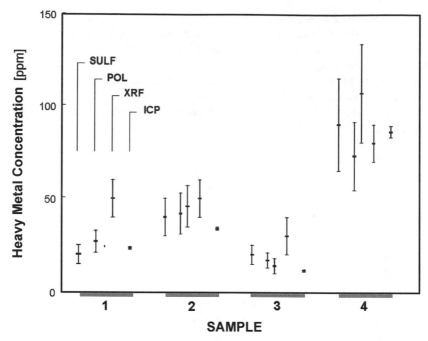

Figure 4.23. Comparison of results on four batches using four different methods. The results are grouped according to batch, and within a group, the methods are sulfide precipitation, polarography, X-ray fluorescence, and inductively coupled plasma absorption (left to right).

concentrations." Visually, it is quite evident, though, that all methods arrive at about the same concentrations of catalyst traces, roughly 20, 40, 20, and 90 ppm. For all practical purposes, the case could be closed. QC patted its own back, secretly acknowledged its streak of luck, and vowed never again to let itself be pressured into revealing sensitive results before a double-check had been run.

4.16 QUIRKS OF QUANTITATION

Situation Suppose a (monovalent) ionic species is to be measured in an aqueous matrix containing modifiers; direct calibration with pure solutions of the ion (say, as its chloride salt) are viewed with suspicion because modifier/ion complexation and modifier/electrode interactions are a definite possibility. The analyst therefore opts for a standard addition technique using an ion-selective electrode. He intends to run a simulation to get a feeling for the numbers and interactions to expect. The following assumptions are made:

Assumptions

- The electrode shows linear behavior in the immediate vicinity of the working point on the calibration curve EMF = E0 + $S \cdot \log_{10}(C)$.
- The sample has a concentration of about $C = 0.5$ mM.
- The term E0 remains constant over the necessary two measurements (a few minutes at most).
- The signal before digitization is sufficiently low-pass filtered so that noise is below 1 mV at the digital volt meter (DVM).
- 50.0 ml of the sample solution will be provided. Standard additions will be carried out using a 10 mM solution of the ion. The amount to be added was, by rule of thumb, set to roughly double the concentration, for ΔEMF $\approx$ 20 mV, a difference that can be accurately defined.

Note that a number of complicating factors have been left out for clarity: For instance, in the EMF equation, activities instead of concentrations should be used. Activities are related to concentrations by a multiplicative activity coefficient that itself is sensitive to the concentrations of all ions in the solution. The reference electrode necessary to close the circuit also generates a (diffusion) potential that is a complex function of activities and ion mobilities. Furthermore, the slope S of the electrode function is an experimentally determined parameter subject to error. The essential point, though, is that the DVM-clipped voltages appear in the exponent and that cheap equipment extracts a heavy price in terms of accuracy and precision (viz. quantization noise;[95] such an instrument typically displays the result in a "1 mV," "0.1 mV," "0.01 mV," or "0.001 mV" format; a two-decimal instrument clips a 345.678 ... mV result to "345.67 mV," that is it does not round up "... 78" to "... 8").

The questions to be answered are the following:

1. How much 10 mM solution must be added to get reliable results? What concentration difference must be achieved to get sufficient differences in signal and burette readings?
2. How accurately must this volume be added?
3. How accurately must this volume be read off the burette?
4. Must a volume correction be incorporated?
5. Is any other part of the instrumentation critical?

A simulation program is written that varies the amount added over a small interval around the nominal 2.5 ml and does each of the following:

1. Clips the simulated EMFs $E1$ and $E2$ to 1.0, 0.1, 0.01, resp. 0.001 mV resolution to emulate the digital voltmeter in the pH/Ion-meter.

2. Varies the last digits of $E0$ to evade artifacts (cf. Ref. 16).

3. Simulates an incorrect reading of the burette.

4. Allows for a volume correction.

First, a number of calculations are run without the above four features (*) in place (program lines 40, 110–130) to verify the rest of the program. Next, each of the features is introduced individually to capture effects, if any.

Note: In newer versions of BASIC, line numbers are no longer needed. Compare this BASIC code to Excel file ELECTRODE.xls to gain a feeling for the difference in approach necessary to obtain the same result on software platforms geared towards efficient and flexible programming, respectively user convenience.

Answers Concerning the questions posed above, the second one is easily answered by adding to or subtracting from $V2$ small volumetric errors in line 130. For the bias to remain below about 1%, the volume error must remain below 0.03 ml.

Table 4.22. BASIC Program for Evaluation of ISE Response

Program			Definition of Parameters
10	$V1 = 50$		Sample volume
20	$C1 = 5E - 4$		Estimated sample concentration
30	$C2 = 0.01$		Concentration of added stock soln.
40	$E0 = 300 + RND$	*	Intercept
50	$S = 59$		Slope in mV/decade
60	$R = 0.1$		DVM resolution in mV
70	FOR $V2 = 2.4$ TO 2.55 STEP 0.01		Added volume is varied
80	$C = (V1^*C1 + V2^*C2)/(V1 + V2)$		Dilution factor
90	$E1 = E0 + S^*LGT(C1)$		Calculation of reference EMF
100	$E2 = E0 + S^*LGT(C)$		Calculation of new EMF
110	$E3 = R^*INT(E1/R)$	*	Clipping to emulate DVM
120	$E4 = R^*INT(E2/R)$	*	Action on $E1$, $E2$
130	$V3 = V2 + 0.2$	*	Simulate incorrect reading of $V2$
140	$Q = 10^\wedge((E4 - E3)/S)$		Estimate concentration
150	$X = C2^*V3/((V1 + V3)^*Q - V1)$		
160	PRINT RESULT and PARAMETERS		A series of appropriate statements
170	NEXT $V2$		
180	END		

$^*V1$, $V2$, $V3$: volumes; $C1$, $C2$, C: concentrations; $E0$, $E1$, $E2$: EMFs (voltages).

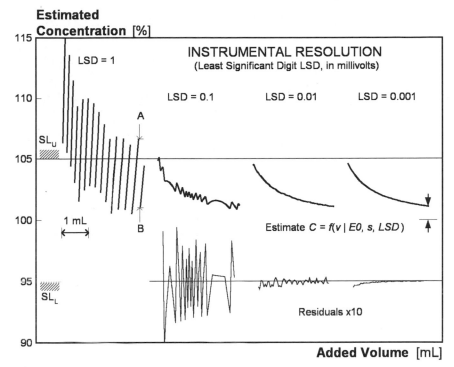

Figure 4.24. Estimated concentration of ion using the standard addition technique with an ion-selective electrode. The simulated signal traces are for DVM resolutions of 1, 0.1, 0.01, resp. 0.001 mV (left to right). For each resolution the added volume $V2$ is varied from 2.4 to 2.55 ml in increments of $V2 = 10\ \mu l$. The ordinate marks indicate the 95–105% SLs. The expanded traces for 0.1 ... 0.001 mV resolution are also given. The simulation was run for five different values of $E0 = 300 + RND$ [mV]. The vertical drops (e.g., A → B) occur at unpredictable values of $V2$: $\Delta V = 10\ \mu l$ would in this case entail an "inexplicable" $\Delta C/C_{nom}$ of nearly 8%! The traces do not reach the 100% level because a 50 μl error in reading off the dispenser or burette was assumed.

The first question is answered by noting that the exact volume $V2$ to be added is not critical as long as the DVM has "good" resolution (0.01 mV or better) and the volume is "correctly" read off (to 10 μl, or better). The volume $V2$ is thus set to 2 3 ml to retain sensitivity. Assume now that the instructions to the technician as far as instrumentation is concerned are ambiguous; that low-cost DVM in the corner and a plain graduated glass pipette are thought to do, and, upon repetition, some inconsistent results are obtained. Closer inspection using simulation reveals, however, that there is a systematic pattern (correlation) between volume added and estimate X (Fig. 4.24), largely because the least significant digit LSD of this low-resolution instrument cor-

Table 4.23. Simulation of ISE Response

Other Conditions	Numerical Results			
	$E0 = 300$ (mV)		$E0 = 300 + RND$ (mV)	
	Res(DVM) 0.001 mV	Res(DVM) 0.1 mV	Res(DVM) 0.001 mV	Res(DVM) 0.1 mV
$V3 = V2$	0.00 ± 0.00	0.08 ± 0.10	0.00 ± 0.00	-0.03 ± 0.25
$V3 = V2 + 0.2$	7.19 ± 0.00	7.27 ± 0.12	7.18 ± 0.00	7.21 ± 0.25
$V3 = V2 + 0.05$	1.81 ± 0.00	1.89 ± 0.00	1.81 ± 0.03	1.80 ± 0.23
$V3 = V2 + 0.01$	0.37 ± 0.19	0.44 ± 0.00	0.36 ± 0.17	0.36 ± 0.25
$V2 = 2 + RND$ and $V3 = V2$	0.00 ± 0.09	0.08 ± 0.24	0.00 ± 0.00	-0.04 ± 0.31

Mean and standard deviations, in % of the nominal concentration, found for simulations under various combinations of (a) random variation of $E0$, (b) volumetric (reading) error in $V2$, and (c) use of a pH/Ion-meter with a resolution of 0.1 or 0.001 mV. For the last line the exact volume $V2$ added was varied in the range $2 \ldots 3$ ml to simulate actual working conditions, and 100 repetitions were run.

responds to the 1 mV position. This is checked by varying $V2$ over a small interval, here from 2.40 to 2.55 ml in steps of 10 μl, a feat that is within the capabilities of a moderately-priced dosimeter. The effect is due to the fact that the volume correction monotonously changes, while the clipped EMFs $E3$ and $E4$ are step-functions; the interaction is commonly called a "quantization" effect or noise.[16] The improvement in resolution from a 0.1 to a 0.01 mV DVM is very striking in this case, which answers question 5.

A volume correction is necessary, as can be seen from a numerical experiment similar to the preceding one: If the increase in volume from $V1$ to $V1 + V2$ is ignored, a bias of about 7% is produced.

Finally, the calculations are repeated 100 times (with variation of $E0$ to simulate $E0$-jitter within the resolution window) to obtain statistically reliable means and standard deviations for the diverse combinations of factors (Table 4.23).

Excel file ELECTRODE.xls generates output similar to that of Fig. 4.24, with the following parameters to play with: $V1$, $C1$, $C2$, $E0$, S, R, step size $\Delta V2$, and volume bias $V3 - V2$; the randomization of $E0$, the volume correction $+V2$ in $(V1 + V2)$ in line 80, and the digitization of the EMFs in lines 110 and 120 can be activated.

Warning: One should realize that a dishonest analyst can willingly shift the result within a range of several percent of the true value, which would certainly suffice to make a slightly out-of-specification product suddenly "conform" to these limits. This could be accomplished with the following instrument configuration: A burette for delivering $V2$, a 0.1 or 1-mV-resolution of pH-meter, and a computer that immediately translates the actual $V2$, $E3$, $E4$, S, and $C2$ into X, and displays X. The analyst would simply have to stop the burette at the right moment to obtain highly "accurate" and "reproducible" results. If the method development had been properly done, however, then the work instructions ("SOPs" in GMP-parlance) would nail down the exact pieces of equipment to be used ("Model XYZ or equivalent"), and the method validation would have specifically tested for ruggedness against such operator-influenced variables as the titration volume.

Consequences

1. A pH/Ion-meter with a resolution of only 0.1 mV is not sufficient because the ensuing quantization noise introduces an apparent deviation of at least ±0.2%, and, more important in this particular case, these systematic effects lead to a bias that is strongly dependent on small shifts in $E0$. (See Fig. 4.24, left side.)

2. An electromechanical burette should be used that delivers volumes $V2$ with an accuracy of about 0.05 ml, or better.

The preceding simulation can be varied within a reasonable parameter space; the critical experimental conditions should be noted, and appropriate experiments made to confirm the model.

4.17   PURSUING PROPAGATING ERRORS

The salt of a carbonic acid A is contaminated by traces of water and a second organic acid B. The content of the three components is determined as in Table 4.24.

The number of replicate determinations and the typical relative standard deviations are noted, along with the average analytical response. Note that X is given in [%]! How pure is A? The answer is found as follows:

Table 4.24. Results of an Acid Analysis

	Component	Method	Amount	Repl	RSD
X	Water	Karl Fischer	0.85%	3	3.5%
Y	Acid B	Ion Chromatography	0.1946 mM/g	5	5.0%
Z	Sum $A + B$	Titration	7.522 mM/g	4	0.2%

1. Subtract B from the sum, and

2. Correct the difference for the water content of the sample, Eqs. (4.7)–(4.9).

3. Invoke error propagation by differentiating the expression for A with respect to X, Y, and Z. [See Eqs. (4.10)–(4.12).]

4. Sum over the squares of the products of the partial differentials and their respective typical errors. [See Eqs. (4.13)–(4.15).]

5. The typical error is here defined as a confidence limit. [See Eqs. (4.16)–(4.19).]

6. The confidence limits of this result are estimated to be ±0.028. [See Eqs. (4.21)–(4.26).]

$$A' = (Z - Y)/(1 - X/100) \tag{4.7}$$

$$= (7.522 - 0.1946)/(1 - 0.85/100) \tag{4.8}$$

$$= 7.3902 \text{ mM/g} \tag{4.9}$$

$$\delta A/\delta X = (Z - Y) \cdot (-1/100)/(1 - X/100)^2 \tag{4.10}$$

$$\delta A/\delta Y = (-1)/(1 - X/100) \tag{4.11}$$

$$\delta A/\delta Z = (1)/(1 - X/100) \tag{4.12}$$

$$\Delta A^2 = ((Z - Y) \cdot (1/100)/(1 - X/100)^2)^2 \cdot \Delta X^2 \tag{4.13}$$

$$+ 1/(1 - X/100)^2 \cdot \Delta Y^2 \tag{4.14}$$

$$+ 1/(1 - X/100)^2 \cdot \Delta Z^2 \tag{4.15}$$

$$\text{TE} = \pm\text{MEAN} \cdot \text{RSD} \cdot t\text{-factor}/\sqrt{n} \tag{4.16}$$

$$\Delta X = \pm 0.8500 \cdot 0.035 \cdot 4.3027/\sqrt{3} = \pm 0.0739\% \tag{4.17}$$

$$\Delta Y = \pm 0.1946 \cdot 0.050 \cdot 2.7764/\sqrt{5} = \pm 0.0121 \text{ mM/g} \tag{4.18}$$

$$\Delta Z = \pm 7.5220 \cdot 0.002 \cdot 3.1824/\sqrt{4} = \pm 0.0239 \text{ mM/g} \tag{4.19}$$

$$\Delta A^2 = 0.0739^2 \cdot ((7.522 - 0.1946) \cdot (1/100)/(1 - 0.85/100)^2)^2 \quad (4.20)$$

$$+ 0.0121^2/(1 - 0.85/100)^2 \quad (4.21)$$

$$+ 0.0239^2/(1 - 0.85/100)^2 \quad (4.22)$$

$$= 0.000030 + 0.000148 + 0.000583 \quad (4.23)$$

$$(\quad 4\% \quad + \quad 19\% \quad + \quad 77\% \quad) \quad (4.24)$$

$$= (\pm 0.028)^2 \quad (4.25)$$

$$\Delta A = \pm 0.028 \quad (4.26)$$

Example 48: The result is thus $CL(A) = 7.390 \pm 0.028$ mM/g, and should be either left as given or rounded to one significant digit in $\frac{1}{2}$CI: 7.39 ± 0.03: The %-variance contributions are given in parentheses (Eq. (4.24)). Note that the analytical method with the best precision (titrimetry), because of the particular numerical constellation, here gives rise to the largest contribution (77%).

4.18   CONTENT UNIFORMITY

Introduction   In order to assure constant tablet quality, the following requirements apply:

- Out of 20 tablets randomly pulled, one or two may deviate from the mean weight by more than 5%, and none may deviate more than 10%. The mean weight must be in the 95–105% range of nominal. Note: Because with today's equipment and procedures the drug is generally very well dispersed in the granulate, especially if the drug content is high, the weight can be used as indicator. Nevertheless, the content uniformity will be determined *via* assays, at least during the R&D phase, in order to validate the procedure.

- The assay is conducted on 10 randomly pulled tablets. Nine out of the ten assay results must be within 85–115% of the mean, and none may be outside the 75–125% range. (See Section ⟨905⟩ of the USP[43].) The mean content must be within the window given shortly.

- The coefficient of variation must be 6% or less. This figure includes both the sampling and the analytical variance.

Note on GMPs: The assays are conducted on individual dosage units (here: tablets) and not on composite samples. The CU test serves to limit the variability from one dosage unit to the next (the Dissolution Rate test is the other test that is commonly used). Under this premise, outlier tests would be scientific nonsense, because precisely these "outliers" contain information on the width of the distribution that one is looking for. The U.S. vs. Barr Laboratories Decision[55] makes it illegal to apply outlier tests in connection with CU and DR tests. This does not mean that the distribution and seemingly or truly atypical results should not be carefully investigated in order to improve the production process.

Situation

Example 49: A tablet weighing 340 mg, of which 50 mg are drug, is to be produced. It is known that the Content and Weight Uniformities, expressed as SDs, are ±2.25 mg, respectively ±6.15 mg. Since the weighing operation is very accurate and has an excellent repeatability in the 300 mg range (typically $\pm\frac{1}{2}$ LSD = ±0.05 mg or ±0.005 mg, that is a resolution of 1 : 6000 or even 1 : 60000), the variability of the tablet weight must be wholly due to processing. The HPLC analysis is found to be fairly precise (±0.5% or ±1.7 mg, double determination).

Question How much may the mean content and weight deviate from the nominal values and still comply with the requirements? Two approaches will be taken:

1. A purely statistical approach.
2. A Monte Carlo simulation.

Statistical Approach The minimal mean weight must be

- $w_{mean} \geq SL_L + t \cdot s_w = 0.95 \cdot 340 + 1.719 \cdot 6.15 = 333.6$ mg, which is 98.1% of the nominal weight; $t(p = 0.1, f = 19) = 1.719$. Obviously, the upper limit SL_u would be calculated analogously. The effective mean must then remain in the 333.3–346.4 mg window. The minimal mean drug content is similarly found to be
- $c_{mean} \geq SL_L + t \cdot s_c = 0.85 \cdot 50 + 1.833 \cdot 2.25 = 46.6$ mg, or 93.2% of nominal. From this it can be seen that there is some leeway in terms of required composition and homogeneity: The effective mean must remain in the 46.6–53.4 mg window; the sampling variance is approximately,
- $\sqrt{2.25^2 - 1.7^2} = \pm1.5$ mg. The true tablet composition could vary in the

approximate range drug content/total weight (47 : 346) to (53 : 334) and still comply. All the same, it is imprudent to purposely to stray from the nominal values 50 : 340.

Simulation Approach   The numerical simulations were carried out using program SIMGAUSS, see data file TABLET_C.dat for content uniformity, respectively TABLET_W.dat for weight uniformity. The mean weights and contents were varied over a range covering the nominal values.

In vectors (columns) 5 and 6 of file TABLET_C.dat, the results of $n = 10$ tablets subjected to individual assay are simulated for $\mu = 48$ mg and $\sigma = \pm2.25$ mg. The "observed" means are 47.53 and 48.87 mg (99, resp. 102% of nominal), and the corresponding "observed" standard deviations are ±1.71 and ±1.84 mg (76, resp. 82% of nominal). All four values are within the expected confidence limits $48 \pm 2.262 \cdot 2.25/\sqrt{10} = 46.4 \cdots 49.6$ mg respectively $\pm1.7 \ldots \pm3.3$ mg (use program MSD and a $n = 10$ data file, option ⟨Display Standard Deviation⟩. Here, only two simulations were carried out for each μ/σ combination; this is enough to make the point, but 20+ runs would have to be carried out to obtain a representative result. The analysis of data file TABLET_W.dat is analogous.

Example 50: Overall, using vectors 3 in the two data files TABLET_C.dat ($\mathbf{x}_{mean} = 47.5$, $s_x = 1.6$) and TABLET_W.dat ($\mathbf{x}_{mean} = 334.0$, $s_x = 5.6$), the following conclusions can be drawn (results rounded):

- The mean content of the drug is 2.5 mg (5.1%) below nominal.
- The CV is $\pm3.4\%$. *Note*: Had measurements similar to those in vector 4 been made, the CV of $\pm5.6\%$ would have come close to the $\pm6\%$ limit stipulated by the USP 1985, which would have necessitated a retest.
- There are no individual values outside the 85–115% range that would signal retest and perhaps rejection.
- The mean tablet weight is 334 mg, or 6 mg (1.8%) below nominal.
- The standard deviation (weight) is ±5.6 mg.
- There are no individual values farther than 12 mg (3.6%) from the mean weight.
- The (drug assay)/(tablet weight) ratio one should theoretically find is $(50.00 \pm 2.25)/(340.00 \pm 6.15) = 0.147 \pm 0.0071$. Effectively, one finds $(47.47 \pm 1.61)/(333.96 \pm 5.60) = 0.142 \pm 0.0054$ ($\pm3.8\%$). This is obtained by error propagation as $(1.61/333.96)^2 + (5.6 \cdot 47.47/333.96^2)^2 = 0.0054^2$. Since 80% of the variance is due to the first term, analytical and sampling errors dominate. The HPLC assay contributes ±1.7 mg, so that the sampling error comes to $\sqrt{1.61^2 - 1.7^2}$

$\approx \pm 0$ mg; mathematically, this is an impossible situation: A variance cannot be negative! The explanation is simple: First, the HPLC precision (± 1.7 mg) is an estimate based on experience with the analytical method, and can most probably be traced to the validation report; it is to be expected that the analyst who wrote that report erred on the conservative side by rounding up all precision figures, so as not to fall into the trap of promising precision that could not be upheld in later experiments. Secondly, the ± 1.61 mg figure originated from a different measurement run (a simulation in this case) with a much lower number of observations than went into the ± 1.7 mg figure. Under such circumstances it is easily possible that the relative size of the two numbers is reversed. As an aside, the confidence intervals $CI(\sigma)$ expected for $s_x = \pm 1.6/n = 10$ and, say, $s_x = \pm 1.7/n' = 50$, would be 1.1–2.9 resp. 1.4–2.1, which clearly overlap; (use program MSD, data file TABLET_C.dat with $n = 10$, option ⟨Display Standard Deviation⟩, to first obtain $CL_u/s_x = 5.26/2.88 = 1.83$ and then $1.83 \cdot 1.61 = 2.94$, and so on). More realistically, the sampling error is estimated at $\sqrt{2.25^2 - 1.7^2} = \pm 1.5$ mg. The two quotients 0.147 and 0.142 are indistinguishable; any significant difference would have implied inhomogeneity, or, in practical terms, segregation of drug from the matrix during processing. *Note*: When powders of unequal size-distribution, particulate shape, and/or density are well mixed, all it takes is machine vibration and the ever-present gravity to (partially) de-mix the components! This phenomenon is used in industrial processes to concentrate the commercially intersting component from available stock, but is a nuisance in a tableting operation.

4.19   HOW FULL IS FULL?

Applicable Regulations   Assume that a cream is to be filled into a tube that has "20 g" printed on it. The lot size is 3000 units. The filling equipment's repeatability is known to be $\approx \pm 0.75$ g ($\pm 3.75\%$). Two somewhat simplified regulations will be investigated that epitomize the statistical and the minimal individual fill weight approaches:

Example 51

1. (European Union EU: test $n = 50$ units

1a. The mean fill weight must not be less that 20.0 g. The effectively found means must meet the requirement $x_{mean} \geq 20.0 - 0.379 \cdot s_x$.

The factor 0.379 corresponds to $t/\sqrt{n}$ for $p = 0.005$ (one-sided), $n = 50$, and $f = 49$. (See Eq. 1.12a).)

1b. No more than two units may be below 91% of nominal. In the case a retest is necessary (three or four units failing this criterion), no more than six out of the cumulated 100 units may fail.

1c. If five or more units ($\geq$ seven units in the retest case) fail the requirement, the product may not carry the coveted "e" logo that signals "European Quality Standard."

2. (CH: Switzerland; no minimal sample number n)

2a. The mean fill weight must not be less that 20.0 g. Up to 5% of the units filled may contain between 87.5 and 95% of the nominal fill weight. Units containing less than 87.5% of the nominal fill weight may not be marketed. Systematic bending of procedures to profit from these margins is forbidden.

2b. Overfilling according to given equations becomes necessary if the experimental relative standard deviation (CV) exceeds ±3% resp. ±4.5%. Since a filling error of $s_x \approx \pm0.75$ g (±3.75%) is associated with the equipment in question, the regulations require a minimal mean fillweight of $20.00 + 1.645 \cdot s_x - 0.05 \cdot 20.00 = 20.23$ g; 20.35 g was chosen so that a margin of error remains for the line-operators when they adjust the volumetric controls.

2c. The "no marketing" *proviso* for seriously underfilled units forces the filler to either systematically overfill as foreseen in the regulations or to install check-balances or other devices to actively control a high percentage (ideally 100%) of the containers and either discard or recycle under-filled ones.

Data file FILLTUBE.xls.dat contains a set of 20 in-process controls (IPC) of $n = 50$ simulated weighings each. The first 10 vectors are for EU conditions ($\mu = 20.02$ g), the others for Swiss regulations ($\mu = 20.35$ g); $\sigma = \pm0.75$ g. The default settings can be changed. Pressing [F9] initiates a new simulation. The results can be captured and incorporated into a .dat file, see program DATA, option ⟨Import Data from Excel⟩. For one specific simulation, the results were as follows.

EU case: The following means were found: 20.11, 19.96, 19.82, 20.05, 19.97, 19.98, 20.04, 20.03, 20.14, and 19.94. In each of 4 IPC-runs, one tube was found with a fill weight below 91% of nominal. In all 10 IPC-runs, the calculated mean was above the $20.00 - 0.379 \cdot s_x$ criterion. Thus the batch(es) conform(s) to regulations and can carry the "e" logo (as a matter of fact, one IPC-run of $n = 50$ would have sufficed, for a sampling rate of 1.7%). If the mean (target) fill weight μ were reduced to below 19.8 g, the probability of not meeting the above requirements would increase to virtual certainty. This can be tested by setting up a series of alternate hypotheses $H_{1...n}$ for $\mu_{1...n} = 20.1 \cdots 19.0$ and determining β. (See Table 4.3.) Alternatively, change the default "mu" in FILLTUBE.xls until most the "OK" in the "RELEASE" row are turned off.

Swiss case: The following means were found: 20.32, 20.43, 20.34, 20.60, 20.35, 20.36, 20.45, 20.40, 20.30, and 20.31. The number of tubes with fill weights below the -5% limit was 4, 1, 0, 0, 4, 1, 3, 3, 3, and 3, for a total of 22, and none below the -12.5% one. Twenty-two tubes out of 500 tested correspond to 4.4%. Since the limit is 5% failures, or 2.5 per 50, fully six out of 10 IPC inspection runs at $n = 50$ each did not comply. At a total batch size of 3000 units, eventually 1/6 of all packages were tested. Evidently, unless the filling overage is further increased, a sampling rate of well above 10% is necessary to exclude these stochastic effects, and so the 10 inspections were combined into one test of $n = 500$.

When investigating filling records, one occasionally stumbles across values that seem to be way out of line. Does such a value represent "normal operation," or has some other mechanism taken over, such as the blocking of a filling nozzle, poor synchronization in a cutting operation, or delivery of improper material? In order to rectify the situation and avoid it in the future, it is important that the probable cause can be assigned.

Many times, there is physical evidence to either bolster a hypothesis or to dismiss it. If this should not be so, then the case might be decided by determining the probability of a given value belonging to the "normal" population (P_0). Essentially, as part of the failure investigation one conducts an outlier test for the value in question *versus* a set of values known to belong to this population, or better yet, *versus* the values acquired immediately before and after the questionable event. If there is reasonable certainty about what effect could have occurred, and data to match, an alternate hypothesis is set up and the probability of the value in question belonging to this population (P_1) can be assessed. If the populations P_0 and P_1 are well represented by $ND(x_1, s_1^2)$ and $ND(x_2, s_2^2)$, t-values are calculated according to Eq. (1.13). The population for which t is smaller is considered to better explain the observation. Should the t-values be relatively similar, the decision might swing the other way with small changes in the size of the data sets; in such a case it is better to rely on experience and common sense rather than on theory.

An eye should also be kept on the absolute size of the standard deviation before and after a proposed elimination. If the elimination of a questionable point results in a standard deviation that is markedly smaller than what is common experience for the test at hand, the F-test cannot be used for confirmation, unless s_2 is replaced by either the s_{method} obtained during the validation, or its lower confidence limit.

Example 52: Assume that a method typically yields $s_x = 0.3$; for a given set of analytical samples the values $x() = 99.5, 99.3, 99.4, 99.3, 99.6, 99.4, 99.5, 99.3, 99.4,$ and 100.5 are found: $s_x = 0.105$ for the first nine points and $s_x = 0.358$ for $n = 10$; this yields $F = 11.56$, which makes the 10th point look like an outlier. With $s_2 = 0.3$, F is reduced to 1.43, which is completely unspectacular. The lower CL(0.3) is about 0.2 ($CL_L/s_x \approx 0.68$ for $n = 9$), but this does not push F beyond 2.9, which is still insignificant at the $p = 0.05$ level. Thus, the unusually low s_x, which might be due to chance or to the operation of quantization effects, does not justify the elimination of x_{10}.

Example 53: If the standard deviation before elimination of the purported outlier is not much higher than the upper $CL(s_{method})$, as in the case ($s_x = 0.358 < CL_U(0.3) \approx 0.57$; factor $CL_U/s_x \approx 1.9$ for $n = 9$, see program MSD), an outlier test should not even be considered; both for avoiding fruitless discussions and reducing the risk of chance decisions, the hurdle should be set even higher, say at $p \leq 0.01$, so that $CL_U/s_x > 2.5$.

With small data sets or if there is reason to suspect deviations from the Gaussian distribution, a robust outlier test should be used.

The individual values should also be examined, e.g., by using program HUBER: Column 12 of file FILLTUBE10.dat is characterized by a relatively tight cluster of values, and one value (19.88 g) is somewhat farther removed from the mean, namely at $(x_i - \mu)/s_x = t = -2.39$. The probability of such a deviation is assessed by using the single-sided Student's t-table for $f = 9$: $p = 0.1/t_c = 1.383$, $p = 0.05/t_c = 1.833$, $p = 0.025/t_c = 2.262$. The deviation was accorded significance (Huber's $k = 5.37$). If this "observation" were eliminated as an outlier, the changes in the median (+0.05 g), the mean (+0.05 g, $p \approx 0.25$), and the standard deviation (-0.079, $F = 3.0$) are nowhere near significance. (Use program TTEST.) This means that the presence of this purported outlier only marginally influences the summary indicators x_{mean} and s_x. The deviation is at 99.4% of nominal and at 97.8% of the "observed" mean; thus there was no reason to discard this point particularly. Again, outlier tests are acceptable for research and failure investigation purposes, but not in connection with release testing under GMPs.

The point that needs to be made is that with sample size as small as it is here ($n = 10$), the distribution can strongly vary in appearance from one sample to the next, much more so than with $n = 100$ as in Fig. 1.10; for example, vectors C_46 (column 1) and C_55 (column 20) of file TABLET_C.dat are the extremes, with standard deviations of ±2.88 and ±1.21. The corresponding Huber's k-values for the largest residual in each vector are 6.56 (this looks very much like an outlier, $k_c = 3.5$) and 2.11 (far from being an outlier). The biggest k-value is found for vector C_49 (column 8) at 7.49; Fig. 4.25 shows the results for this vector as they are presented by program HUBER.

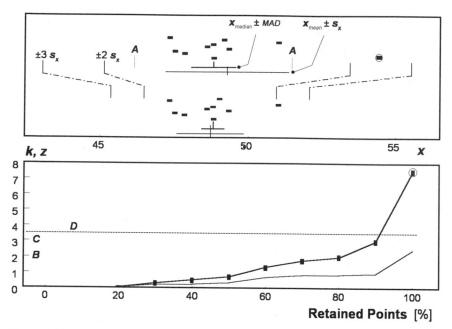

Figure 4.25. Detection of an outlier by Huber's method. Upper panel: Abscissa: Content [mg]; dark points: Measurements; (A) |: Huber's cut-off at $k = 3.5$. The median is marked with an "up" tic; x_{mean} is marked with an "up, down" tic. The circled point is the one that is proposed for elimination. Lower panel: Abscissa: Data ordered according to $|x_i - x_m|$; (B, C) ——...——: Classical $\mu \pm 3\sigma$ and $\mu \pm 2\sigma$ cut-offs. (D) M: Huber's k-value.

In the upper half of the upper panel, the original data are plotted on the content [mg] axis. Since one point is far outside the acceptance range (at about $+2.46 \cdot s_x$), eliminating this one yields the modified distribution pictured in the lower half of this panel. Note that the $\pm 2 \cdot \sigma$ and $\pm 3 \cdot \sigma$ values are now much closer to the mean than before because $s'_x \ll s_x$ (the appropriate cut-off values are connected by dotted lines). The $x'_m \pm 3.5 \cdot$ MAD' cut-off values moved closer together also; one of the remaining $n' = 9$ points is just outside this new acceptance range. Program HUBER does not automatically repeat the elimination procedure to avoid cascades of possibly unwarranted eliminations. This example clearly shows that after the first elimination there is no justification for identifying further outliers: The long-term experience is that a standard deviation of ± 2.25 mg is typical for this production process, and at $n' = 9$ we already have $s'_x = \pm 1.1 \ll \pm 2.25$. Also, note that x_{mean} and x_m moved much closer together from $n = 10$ to $n' = 9$.

The lower panel gives the points ordered according to (absolute) deviations: The abscissa is ordinal, the value with the smallest deviation $|x_i - x_m|$

being plotted at $x = 100/N\%$, the one with the highest deviation at $x = 100\%$; since the first point to be eliminated would be situated at the right-hand end, the abscissa is labeled "% points retained." The ordinate gives Huber's k-factor (solid line "D") and the corresponding Student's t-factor (lines) using s_x. Huber's critical k is by default set to 3.5 (Ref. 21, dashed horizontal), but can be changed. If a $\pm 2 \cdot \sigma$ or a $\pm 3 \cdot \sigma$ cut-off rule is to be applied, marks at $y = 2$ resp. $y = 3$ on the frame indicate where to draw the line. Evidently, in the example given by vector 8, nine points are closely clustered and one is far removed; the choice of Huber's k-factor is not critical.

Example 54: One point is eliminated for $2.97 \leq k_c \leq 7.48$. If no point is eliminated ($k_c > 7.49$), we have $s'_x = s_x$, so only one dotted cut-off curve is plotted.

4.20 WARRANTY OR WASTE

Introduction Pharmaceutical preparations are exposed to all kinds of insults during their lifetime, chief among them being temperature excursions, high humidity, light, oxygen, and packaging components. Shelf-lives can be as high as 5 years, and because the consumer must be protected from undue dangers, the health authorities have issued guidelines[178] as to how a product must be tested; an excerpt:

- Storage at controlled room temperature (the earth is divided into four climate zones for this purpose), with sampling times in intervals of 3, 6, or 12 months.

- Storage under stress conditions according to ICH Guidelines,[132] i.e. 25°C/60% RH, 30°/65% RH, or 40°C/75% RH (Conditions I, II, resp. III; temperature controlled to within ± 2°C, relative humidity to within ± 5%), with sampling intervals of 2 weeks or 1 month.

- Storage in the same primary container as will be used on the market, e.g., blister strips for tablets or ovules, tubes for creams, or vials and ampules for injectables.

- During the development phase a series of laboratory or pilot-scale batches will be subjected to this stability program. As soon as the process is scaled up to production-size batches, the first few, and at least one per year thereafter will also go on stability. Submission is only possible if the product completes a minimal combination of tests, e.g., one full-size batch for 12 months and two reduced-size batches for 6 months at Condition I. Despite much harmonization, the health authorities of the U.S., the EU, and Japan have still not agreed on a minimal combination of tests for registration.

Since full analyses are carried out, a lot of data are generated. Every parameter is reviewed for trends that signal product aging or outright decomposition of the active principle; this can be as cosmetic in nature as discoloration or as potentially hazardous as buildup of toxic derivatives. If the drug substance is an ester, for example, hydrolysis, particularly if moisture penetrates the primary packaging material, will decompose the compound into its acid and alcohol components. From a pharmaceutical or medical viewpoint, even if there is no toxicity issue involved, this will result in a loss of bioavailability. Even this is to be avoided because subpotency introduces therapeutic uncertainty and can go as far as lethal undertreatment.

Situation   A cream that contains two active compounds was investigated over 24 months (incomplete program if today's ICH standards are applied, which require testing at 0, 3, 6, 9, 12, 18, and 24 months). The assays resulted in the data given in file CREAM.dat. Program SHELFLIFE performs a linear regression on the data and plots the (lower) 90% confidence limit for the regression line. For each full time unit, here months, it is determined whether this CL drops below levels of $y = 90\%$ resp. $y = 95\%$ of nominal. Health authorities today require adherence to the 90% standard for the end-of-shelf-life test, but it is to be expected that at least for some products the 95% standard will be introduced.

Interpretation   Active component A is so stable that a shelf-life in excess of 60 months could be assigned (it is unusual for a pharmaceutical to be approved for more than 5 years). Component B, however, undergoes hydrolysis (this fact has to be independently established, i.e., by GC/MS techniques, or equivalents). (See Fig. 4.26.) The data points cover an incomplete 24-month stability program ($T = 0, 3, 7, 24$). The intercept is at 104.3%, an indication for over-dosing, and the slope is ≈ -0.49 [%/month]. The linear regression line is extrapolated until the lower 90%-confidence limit (two-sided) for $Y_B = a + b \cdot x$ intersects the SL, here at 26 months (arrow), the integer value of the real intersection point; this limit is equivalent to the 95% one-sided CL. In regulatory practice, this would translate to an officially approved shelf-life of 2 years. If the specification were raised to SL = 95%, a shelf-life of 18 months might be granted because real-time data is available that goes even beyond this time, but only if the authorities do not object to overdosing. Without systematic overdosing (cf. Fig. 4.46), the shelf-lives drop to 8 resp. 18 months (at SL = 95, resp 90%), which is no longer

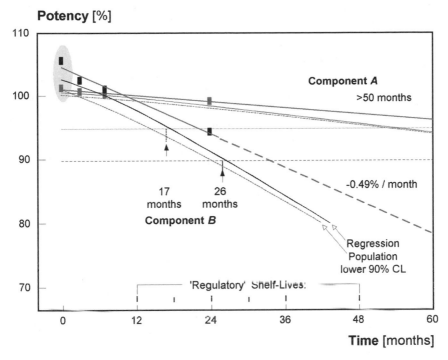

Figure 4.26. Shelf-life calculation for active components A and B in a cream; see data file CREAM.dat. The horizontals are at the $y = 90$ (specification limit at $t = $ shelflife) resp. $y = 95\%$ (release limit) levels. The linear regression line is extrapolated until the lower 90%-confidence limit for $Y_B = a + b \cdot x$ intersects the SLs; the integer value of the real intersection point is used. The intercept is at 104.3%.

interesting from the commercial point of view. Any new data point outside the population-CL [Eq. (2.25)] would raise suspicions; the regression-CL delineates the quality of the extrapolation [Eq. (2.16)].

The statistical interpretations are "there is a 5% chance that the extrapolation is below 90% at $t = 26$" and "there is a 5% chance that a further measurement at $t = 26$ months will yield a result below $y \approx 89\%$ of nominal." Every batch in the stability program is subjected to this procedure; the batch that yields the shortest shelf-life sets the expiration date.[178] Possible solutions are as follows:

1. If the authorities request an overage-free product, for an SL of 90% an expiration date of 18 months results (this would be a pain for the Logistics and Marketing Departments because

    a. most companies consider it unethical to ship any goods to the wholesaler with less than 12 months of shelf-life left;

    b. quality control testing and release already take 2–4 weeks, so only 5 months remain to sell the product;

    c. reducing batch size so inventory is sold within 5 months increases production costs and overhead, more frequently exposes the manufacturer to "out-of-stock" situations, and strains the goodwill of a workforce whose production schedules are already being modified the moment they are distributed.

2. Refer the project back to the R&D department for an improvement of the formula or the packaging (this can easily cost huge amounts of money and delay market introduction by years, especially if alternatives had not been thought of or had been prematurely deleted from the stability program).

This decision-making guideline involving the lower 90% CL is now accepted worldwide. In setting up the rule, science and bureaucracy was involved. A more complex and scientifically more logical rule could have been made, but the lawyer's penchant for hard and simple-to-enforce rules won out.

So-called accelerated testing is used to get a feeling for the long-term trend without having to wait for years[58] (see the next section).

It is common to have data for several batches available, and pooling that would lead to higher degrees of freedom and improved prediction capabilities. The full procedure is explained in Refs. 58, 178; this very conservative, purely statistical approach can create the absurd situation that high-quality data is excluded from and low-quality data is included in the pooling.[181] The use of the concept of constant power instead of arbitrary criteria is discussed in Ref. 181. If data for a large number of batches are available[182]—usually late in the life-cycle of a product, just before the next formulation or packaging improvement is due to be introduced—data sets can be randomly picked for pooling.[183]

Because the individual measurements are carried out over a period of several years, it is likely that the working- or even the secondary standard that is used to calibrate the analytical method is used up or has degraded, and thus has to be replaced; for the relationship amongst such standards. (See Fig. 4.44.) Laboratory errors occur now and then that are small enough to go undetected, particularly if the opportunities for scientific plausibility checks are squandered by overly bureaucratic practices (see Fig. 4.49); such errors can be modelled by invoking a large-SD Gaussian error in a few percent of the Monte Carlo trials. (See file ASSAY_AB.xls.) File DEGRAD_STABIL.xls combines the effects and allows the results to be viewed in the standard format (assay *vs.* time on stability), which masks systematic errors such as a miscalibration on a given day, and in the calendar format, where these things show up.

4.21   ARRHENIUS-ABIDING AGING

Introduction   During the early stages of development of a dosage form or a drug substance one rarely has more than a notion of which temperatures the material will be processed or stored at once all has been said, done, and submitted. Also, the decomposition of a component might be subject to more than one mechanism, often with one reaction pathway dominating at lower, and another at higher temperatures. For these reasons, stress tests are conducted at a number of different temperatures that, first, are thought to bracket those that will eventually be quoted on the box ("store at 25°C or below"), and second, are as high as the dosage form will accommodate without gross failure (popping the vial's stopper because the water boiled) and still deliver results in a reasonably short time (a few weeks). Plots using program SHELFLIFE would in such a case reveal increasingly negative slopes the higher the storage temperature. Program ARRHENIUS allows for all available information to be put into one file, and then calculates the slope for every temperature condition. Given that the basic assumptions underlying Arrhenius' theory hold (zero-order kinetics, i.e., straight lines in the assay-*vs.*-time graph, activation energy independent of temperature, Ref. 58), the resulting slopes are used to construct the so-called Arrhenius-diagram, i.e., a plot of ln(slope)-*vs.*-$1/T$, where T is the storage temperature in degrees Kelvin. If a straight line results (see Fig. 4.27, left panel), interpolations (but never extrapolations!) can be carried out for any temperature of interest to estimate the slope of the corresponding assay-*vs.*-time plot, and then therefrom the probable shelf-life. The Arrhenius analysis is valuable for three reasons: (1) detection of deviations from Arrhenius' theory (this would indicate complex kinetics and would lead to a more thorough investigation, but often requires the degradation to be followed to low assay levels before it can be diagnosed[58]), (2) planning future experiments, and (3) supporting evidence for the registration dossier.

Situation and Interpretation   A series of peptides was assessed for stability in aqueous solutions. The data in files ARRHENX.dat ($X = 1$, 2, or 3) was found in a doctoral thesis.[184] Figure 4.27 shows one case where a temperature range of 50°C was covered. (See also Table 4.25.)

Since the Arrhenius diagram is linear and the collision parameter A is constant over the whole temperature range, the activation energy can be cal-

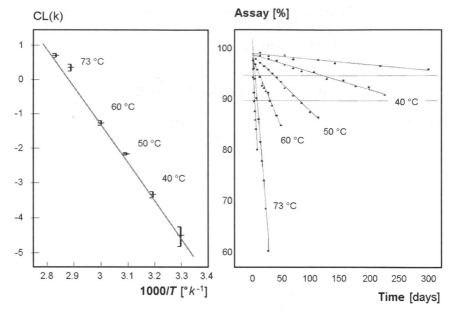

Figure 4.27. Arrhenius analysis: The right-hand panel shows the assay-vs.-time data for an aqueous solution of a peptide. The regression lines are for storage temperatures of 80°, 73°, 60°, 50°, 40°, and 30°C. The left-hand panel gives the ln(-slope)-vs.-1/T Arrhenius plot.

culated to be $E \approx 94.8 \pm 5.6$ [KJ/mol] and it can be safely assumed that this temperature range is ruled by fairly simple kinetics.

Example 55: An interpolation for 35°C on the assumption that the same sampling plan would apply as for the 80°C results yields an implausibly short shelf-life of 50 days (90% level); the reason is that both x_{mean} and S_{xx} are

Table 4.25. Results of Arrhenius Analysis

Item	set_1	set_2	set_3	set_4	set_5	set_6	DIM
Temp.	80	73	60	50	40	30	[°C]
1000/T	2.831	2.888	3.001	3.094	3.193	3.299	[1000/°K]
Slope	−2.014	−1.423	−0.281	−0.115	−0.035	−0.011	[%/day]
ln(−s)	0.70	0.35	−1.27	−2.16	−3.34	−4.50	[−]
Shelf-life							
90%	4	6	28	76	240	715	[days]
ln(A)	33.0	33.3	33.0	33.1	33.1	33.1	[−]

very different for the two conditions. In this case the second option given in program ARRHENIUS must be used, that is, the individual time-points appropriate to a test at 35°C are entered (0, 180, 360, 540, 720, and 900 days): The estimated shelf-life is now 410 days (90% level; 170 days for 95% level). This example demonstrates that both a statistical evaluation and the program must offer a level of sophistication that is up to the job. Not adding the second option would have made the programmer's job easier, and perhaps the too short shelf-life would have just been accepted by the users as given because nobody pointed out to them the statistical details surrounding S_{xx}.

4.22 FACTS OR ARTIFACTS?

In file JUNGLE2.dat, a simulated data set is presented that obeys these rules:

- All results are clipped to a specific number of digits to simulate the operation of digital read-outs of analytical equipment; p is the precision, d is dimension, and σ is the superimposed $ND(0, \sigma^2)$ noise in Table 4.26.
- Impurities **A**, **B**, and **C** are in the ppm-range; this would be typical of residual solvents or volatile by-products of a chemical synthesis. **B** and **C** are coupled to **A**. Impurity **C** has its noise reduced by $\frac{1}{2}$ in the range above 695 ppm, an effect that can occur when a detector is switched to a lower sensitivity in anticipation of a large signal.
- Compound **D** is in the low %-range, and extremely variable, as is often the case with impurities that are not specifically targeted for control; under these circumstances, less is known about the reaction path-

Table 4.26. **Simulation Parameters**

Item	Equation for μ		σ	d	p
A	$A - 125$		30	ppm	1
B	$B = 15.2 + (A - 125)/5$		2	ppm	0.1
C	$C = 606 + (A - 125)$:	$C < 695$	40	ppm	1
		$C \geq 695$	20		
D	$D = 0.3$	$D \geq 0.01$	0.5	%	0.01
pH	$pH = 6.3 + (C - 600)/500$		0.2	–	0.1
Color	$Color = 23$		7	mAU	1
HPLC	$HPLC = 99 - A - B - C - D$		0.3	%	0.01
Titr.	$Titr. = 98.3 + (HPLC - 99)$		0.4	%	0.05

Table 4.27. Probability Categories

Fraction of Results	Probability-of-Error Category	Interpretation
$\approx 1/12$	$0.00 \le p < 0.04$	Significant to highly significant
$\approx 1/12$	$0.04 \le p < 0.08$	Marginally significant
$\approx 1/6$	$0.08 \le p < 0.16$	Insignificant
$\approx 2/3$	$0.16 \le p < 0.50$	Essentially random

ways that generate or scavenge these compounds than about high-impact impurities. Also, there are generally so many target variables in the optimization of a reaction that compromises are a way of life, and some lesser evils go uncontrolled.

- The **pH** is relatively well-controlled, but coupled to **C**.
- **Color** is the result of an absorption measurement, commonly carried out at $\lambda \approx 410$ nm, to assess the tinge that is often found in crystallization liquors that impact an off-white to yellowish aspect to the crystalline product.
- The **HPLC** assay is fully coupled to the impurities **A–D** on the assumption that there is a direct competition between the major component and some impurity-producing reaction pathways. The basis-value 99 was introduced to simulate other concentration losses that were not accounted for by impurities **A–D**.
- The **Titr**ation result is just a bit lower than the **HPLC** result and is strongly coupled to it.

This example was set up for a number of reasons:

- To allow exploration using programs CORREL, LINREG, HUBER
- To demonstrate the effect of the size of vectors on correlation
- To smuggle in typical artifacts

The effect of the size of the compared vectors is shown by taking JUN-GLE2.dat, and using program DATA, option ⟨DEL row⟩ to cut down the number of rows to, for example, 24, 12, or 6. The probability levels change in both directions, e.g., **A/C** from the category "large square: $p < 0.01$" to the category "small dot: $p > 0.16$," that is from highly significant to no significance whatsoever. Whenever there is a large number of measurements available, even very tenuous links turn out to be highly significant, and spurious correlations are to be expected. (See Fig. 2.2.) WARNING: Data explo-

Legend (single-tailed test, p = alpha / 2)

$0.00 \leq p < 0.01$ ■□	$0.02 \leq p < 0.04$ ■ □	$0.08 \leq p < 0.16$ ■ □
$0.01 \leq p < 0.02$ ● ○	$0.04 \leq p < 0.08$ ◆ ◇	$0.16 \leq p < 0.50$ . .

	IMP 2	IMP 3	IMP 4	IMP 5	IMP 6	IMP 7	IMP 8	solv	REAG	PRODUCT
IMP 1	◆	□	.	■	.	○	◆	□	.	◆
IMP 2		■	■	.	.	■	.	■	■	□
IMP 3			●	.	.	■	.	■	■	□
IMP 4				■	.	■	.	■	■	□
IMP 5					.	.	.	.	.	.
IMP 6						.	.	.	.	□
IMP 7							.	■	●	□
IMP 8								.	.	.
solv									■	□
REAG										○

Figure 4.28. Correlation graph for file PROFILE.dat. The facts that (a) 23 out of 55 combinations yield probabilities of error below $p = 0.04$ (42%; expected due to chance alone: $\approx 8\%$) and (b) that they fall into a clear pattern makes it highly probable that the peak areas [%] of the corresponding chromatograms follow a hidden set of rules. This is borne out by plotting the vectors two by two. Because a single-sided test is used, p cannot exceed 0.5.

ration without graphical support (program CORREL, option ⟨Graph⟩) and a very critical mindset can easily lead to nonsensical interpretations! Running files TABLET_C.dat, TABLET_W.dat, and ND_160.dat through program CORREL proves the point: These are lists of fully independent, normally distributed numbers, and correlations turn up all the same! Independent of n, the distribution in Table 4.27 is found.

Thus, one must expect about 5–10% apparently significant correlations; fortunately, these false positives appear in a random arrangement, so that when a really significant connection turns up, the human visual system perceives a clearly recognizable pattern, for example for file PROFILE.dat. (See Fig. 4.28.)

A further incentive for supplying file JUNGLE2.dat is the possibility of smuggling in some numerical artifacts of the type that often crop up even in one's own fine, though just a bit hastily concocted, compilations (see JUNGLE3.dat):

- Rows 1–7, item **Titr**: 0.5% subtracted; typical of temporary change in the production process, calibration procedure, or analytical method, especially if values 1–7 were obtained in one production campaign or measurement run.
- Rows 9 + 10, items **A**, **B**, and **C**: Factor of 2 introduced to simulate the operation of absentmindedness in sample and hardware preparation, or computation: An extra dilution step, a sensitivity setting, or transcription errors.
- Row 14, item **B**: Deleted decimal point.

- Row 38, item **Titr**: An assay value of 101.1 might be indicative of a calibration or transcriptional error if the analytical procedure does not admit values above 100.0 (the value can be a realistic result, though, if an assay value of close to 100% is possible and a huge analytical uncertainty is superimposed).

- Row 45, item **pH**: A value of 4.8 in a series of values around 6.2 is highly suspicious if the calibration of the electrode involves the use of an acetic acid/sodium acetate buffer (pK $\approx$ 4.6)

- When JUNGLE2.dat and JUNGLE3.dat are analyzed with program CORREL, a plausible pattern turns up: Impurities **A**, **B**, and **C** and the **pH** are mutually correlated; impurity **D** is correlated with **Titr** and **HPLC**. The artifacts in JUNGLE3.dat cause a reduction in the strengths of correlation in four bins (three of them over two to four classes), and an increase in three bins over one class each.

4.23   PROVING PROFICIENCY

In today's regulatory climate, nothing is taken for granted. An analytical laboratory, whether in-house or in the form of a contractor, no longer gets away with the benefit of the doubt or the self-assured "we can do that for you," but has to demonstrate proficiency. Under the Good Manufacturing Practices (GMPs) a series of do's and don'ts have become established that are interpreted as minimal expectations. Since one never knows today which results will be declared "crucial" when the Regulatory Department collates the submission file a few years down the road, there is tremendous pressure to treat all but the most preliminary experiments as worthy of careful documentation under GMP. Besides, some manifestations of modern life—job-hopping and corporate restructuring are examples—tend to impair a company's memory to the point that whatever happened more than 6 months ago is lost forever, unless it is in writing. For hints on how to proceed, see Section 3.2, "Method Validation."

Setting   An established analytical method consisting of the extraction of a drug and its major metabolite from blood plasma and the subsequent HPLC quantitation was precisely described in a R&D report, and was to be transferred to three new labs across international boundaries. (Cf. Section 4.32.) The originator supplied a small amount of drug standard and a number of vials containing frozen blood plasma with the two components in a fixed ratio, at concentrations termed "lo," "mid," and "hi." The report provided for evaluations both in the untransformed (**lin**ear/**lin**ear depiction)

and the doubly logarithmized (**log**arithmic/**log**arithmic) format. The three
files VALIDX.dat were selected from among the large volume of data that
was returned with the three validation reports to show the variety of problems and formats that are encountered when a so-called "easily transferable"
method is torn from its cradle and embedded in other "company climates."
The alternatives would have been to

- Run all analyses in the one and established central laboratory; the logistical nightmares that this approach convokes in the context of a global
  R&D effort demands strong nerves and plenty of luck. (Murphy's law
  reigns supreme!) Regional laboratories make sense if the expected volume of samples is so large as to overwhelm the existing laboratory, and
  if the technology is available on all continents.
- Run the analyses in regional labs, but impose a military-style supply-and-control regime; this exposes the "commander" to cultural cross-currents that can scuttle his project.

Problems

- In one new laboratory the instrument configuration was reproduced as
  faithfully as possible: The instrument was similar but of a different
  make.
- The HPLC-column with the required type and grade of filling material (so-called stationary phase) was not available locally in the 2.1 mm
  diameter size set down in the report, so the next large diameter (4.0
  mm) was chosen, in full cognizance that this would raise the detection
  limit by an estimated factor of $(4.0/2.1)^2 \approx 3$–4. For the particular use,
  this was of no concern, however. The problem was encountered because
  the R&D people who developed the method had used a top-of-the-line
  research instrument, and had not taken the fact into account that many
  routine laboratories use robust but less fancy instruments, want to write
  them off before replacing them, and do not always have a laboratory
  supply house around the corner that is willing to deliver nonstandard
  items at reasonable prices and within a few days (some manufacturers will not, cannot, or are prohibited from delivering to all countries).
  As a sign of our times, some borders are very tight when it comes to
  smuggling X-ray-opaque metal HPLC columns across; to the customs
  officers, the white powder in them looks suspicious.
- An inadvertent, and at first sight trivial shift in conditions raised
  the extraction efficiency from the stated $\approx 85\%$ to nearly 100%: This
  together with the different model of detector (optical path!) caused

the calibration line slope to become much higher. There was concern that saturation phenomena (peak-broadening, shifting of retention times) would set in at lower concentrations.

- Because two laboratories were involved, there was a certain risk that this foreign contract laboratory in the crucial calibration run would, despite cross-validation,[26] come up with results for the QC standards that were at odds with those in the primary laboratory; this would necessitate explanations.

- In a multi-year, multi-laboratory situation it is unlikely that the amount of primary standard (PS) will suffice to cover all requests. The next best thing is to calibrate a larger amount of lower-quality secondary standard (SS) against the PS, and to repeat the calibration at specified intervals until both the PS and the SS have been consumed. In this way, consistency can be upheld until a new lot of PS has been prepared and cross-validated against the previous one. In practice, a working standard will be locally calibrated against the SS and be used for the daily method calibration runs. (See Section 4.32.)

Results   The raw data consisted of peak height ratios of signal : internal standard, see data files VALID1.dat (primary validation; $m = 10$ repeats at every concentration), VALID2.dat (between-day variability), and VALID3. dat (combination of a single-day calibration with several repeats at 35 and 350 [ng/ml] in preparation of placing QC-sample concentration near these values). Fig. 4.29 shows the results of the back-calculation for all three files, for both the lin/lin and the log/log evaluations. Fig. 4.30 shows the pooled data from file VALID2.dat.

The data in VALID1.dat show something that is characteristic of many analytical methods: The standard deviation steadily increases, particularly above 250 ng/ml, from low concentrations, where it is close to zero, to ±0.14 at 500 ng/ml while the CV drops, from >10% to about 5%. The residuals at $x \approx 250$ ng/ml do not appear to lie on a straight line; this notion is strongly reinforced when the data are viewed after double-logarithmic transformation (program VALIDLL; see later). Note that the back-calculated values for the two lowest concentrations are far above 100%. The same observation in two other laboratories where the same method was run confirmed that this non-linearity is real. A cause could be assigned after the conventional three-step liquid-liquid extraction (aqueous-to-organic, clean-up of organic phase with aqueous medium, pH-change and back-extraction into water) was replaced with a single-step procedure involving selective extraction cartridges, which brought perfect linearity (the higher cost of the cartridges is more than jus-tified by the solvent and manpower savings). Obviously the cartridges elim-

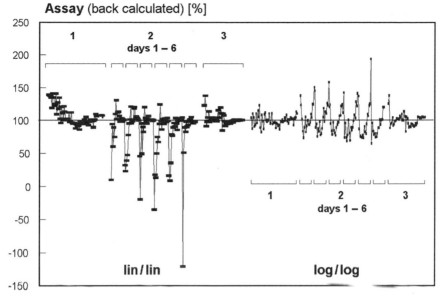

Figure 4.29. Back-calculated results for files VALIDX.dat. The data are presented sequentially from left to right. The ordinate is in % of the nominal concentration. Numbers $X = 1, 2$, and 3 indicate the data file. Each bracket indicates a day's worth of results (sorted by concentration). The log/log format tends to produce positive deviations at low concentrations, while the lin/lin format does the opposite, to the point of suggesting negative concentrations! The reason is that the low concentration values are tightly clustered at the left end of the lin/lin depiction whereas the values are evenly spread in the log/log depiction, with commensurate effects on the position of x_{mean}, the sum S_{xx}, and the influence each coordinate has on the slope. The calibration design was optimized for the log/log format.

inated a component that interfered at low analyte concentrations. The question came up whether the method should be changed. Since this technological improvement came after a lot of work had already been done, and this particular study late in the product development cycle did not require full exploitation of the available concentration range and precision, it was decided to leave things as they were and not go into the trouble of validation of the new, cross-validation of the new against the old, and registration of the new method with the health authorities in all the involved countries. Despite the obvious curvature, the coefficient of determination r^2 was larger than 0.9977 throughout.

Back-calculation is achieved by equating the individual calibration signal $y(i)$ with y^*, using Eq. (2.19), and calculating $100 \cdot X(y^*)/X_{nominal}$ [%]. The estimated standard deviation on $X(y^*)$, s_x, is transformed to a coefficient of variation by calculation of either $CV = 100 \cdot s_x/X_{nominal}$ or $CV =$

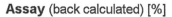

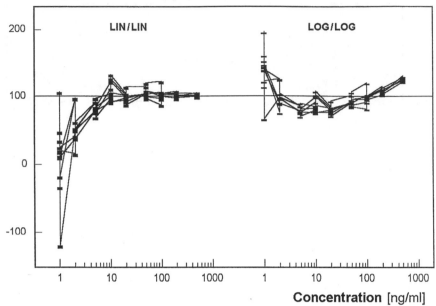

Figure 4.30. Back-calculated results for file VALID2.dat. The data from the left half of Fig. 4.29 are superimposed to show that the day-to-day variability most heavily influences the results at the lower concentrations. The lin/lin format is perceived to be best suited to the upper half of the concentration range, and nearly useless below 5 ng/ml. The log/log format is fairly safe to use over a wide concentration range, but a very obvious trend suggests the possibility of improvements: (a) nonlinear regression, and (b) elimination of the lowest concentrations. Option (b) was tried, but to no avail: While the curvature disappeared, the reduction in n, $\log(x)$ range, and S_{xx} made for a larger V_{res} and, thus, larger interpolation errors.

$100 \cdot s_x/X(y^*)$; the distinction is negligible because any difference between $X_{nominal}$ and $X(y^*)$ that shows up here would have caused alarm above.

The options ⟨Table⟩ and ⟨Results⟩ repeat the ordinate and abscissa values and (see summary in Table 4.28) provide:

1. The absolute and the relative residuals in terms of concentration X.
2. The back-calculated values (also in x-coordinates) as absolute and relative values.
3. Program VALID adds the symmetric CV, while program VALIDLL gives the (asymmetric) low and the high values.
4. The slope and the intercept with the appropriate relative 95% CLs, the residual standard deviation, and r^2.

Table 4.28. Overview of Key Statistical Indicators Obtained from Three Different Laboratories that Validated the Transfer of the Analytical Method

Slope [-]	CL(b) [%]	Intercept [-]	CL(a) [%]	s_{res} [-]	r^2 [-]	LOD [ng/ml]	LOQ [ng/ml]	X15% [ng/ml]	FDA [ng/ml]	Transform
0.00527	1.8	-0.0198	106.0	0.0616	0.9955	4.0	7.9	16	10	NN [1]
0.00525	0.9	0.01790	49.5	0.0143	0.9997	1.7	3.4	1.3	2	NN [2]
0.00528	1.9	0.02368	77.8	0.0311	0.9987	3.5	7.0	6.2	1	NN
0.00521	0.6	0.01583	37.4	0.0100	0.9999	1.1	2.3	0.2	5	NN
0.00529	1.5	0.00690	207.8	0.0424	0.9992	2.7	5.4	6.8	10	NN
0.00526	1.5	0.00477	310.8	0.0228	0.9994	2.8	5.6	7.3	1	NN
0.00509	0.9	0.01374	59.8	0.0133	0.9998	1.6	3.2	1.9	5	NN
0.00523	**0.6**	**0.01400**	**40.1**	**0.0241**	**0.9991**	**1.1**	**2.1**	**0.8**	**2**	**NN**
0.00751	0.4	0.00209	295.7	0.0164	0.9998	0.3	1.6	2.5	1	NN [3]
1.01	1.5	**	**	0.0347	0.9966	**	**	**	**	LL [1]
0.861	4.6	**	**	0.0635	0.9930	**	**	**	**	LL [2]
0.861	5.3	**	**	0.0789	0.9901	**	**	**	**	LL
0.849	5.0	**	**	0.0740	0.9910	**	**	**	**	LL
0.874	5.3	**	**	0.0798	0.9902	**	**	**	**	LL
0.859	6.3	**	**	0.0884	0.9880	**	**	**	**	LL
0.870	6.6	**	**	0.0943	0.9859	**	**	**	**	LL
0.862	**2.0**	**	**	**0.0778**	**0.9893**	**	**	**	**	**LL**
0.961	1.3	**	**	0.0378	0.9983	**	**	**	**	LL [3]

LEGEND: "NN": linear regression without data transformation; "LL": idem, using logarithmically transformed axes; "**": not interpretable "X15%": concentration x at which σ_Y/Y is 0.15; "FDA": lowest calibration concentration for which $s_Y/Y < 0.15$. The two lines of bold numbers pertain to the pooled data. [X = 1, 2, or 3] identifies the appropriate data file VALID_X.dat.

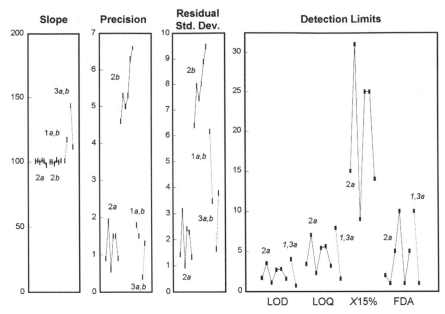

Figure 4.31. Key statistical indicators for validation experiments. The individual data files are marked in the first panels with the numbers 1, 2, and 3, and are in the same sequence for all groups. The lin/lin respectively log/log evaluation formats are indicated by the letters a and b. Limits of detection/quantitation cannot be calculated for the log/log format. The slopes, in percent of the average, are very similar for all three laboratories. The precision of the slopes is given as $\pm 100 \cdot t \cdot CV(b)/b$ in [%]. The residual standard deviation follows a similar pattern as does the precision of the slope b. The LOD conforms nicely with the evaluation as required by the FDA. The calibration-design sensitive LOQ puts an upper bound on the estimates. The $X15\%$ analysis can be high, particularly if the intercept should be negative.

5. The LOD and the LOQ (in the case of program VALIDLL, these values cannot be calculated)

6. For each group of repeatedly determined signals ($m_j \geq 2$) the basic statistics are given.

Fig. 4.31 gives the key indices in a graphical format.

Observations

a. The slopes cluster tightly around their respective means, even though they are less well defined, see CL(b) in the log/log depiction; The slope found for VALID3.dat is definitely higher and very well defined ($0.007511 \pm 0.4\% > \approx 0.0052 \pm 1.3\%$). The fact that the log/log slopes for VALID2.dat ($\approx 0.862 \pm 5.6\% < 1.00$) do not include $\mu = 1$ points to curvature in the non-transformed depiction.

b. Some intercepts in the nontransformed depiction do not include $\mu = 0$; this is due to the curvature at low concentrations which is traceable to a lack of selectivity in the extraction process. As was to be expected, the coefficient of determination r^2 is not very informative. The reduction of the heteroscedacity brought about by the log/log transformation is not reflected by an improvement in r^2, which decreases from, e.g., 0.9991 to 0.9893 (s_{res} cannot be compared because of the units involved).

c. The calibration-design-sensitive concept of calculating limits of detection yields LOD-values in the range 1.1–3.5 ng/ml and LOQ-values in the range 2.3–7 ng/ml. The dotted curve in the option ⟨Display Valid Graph⟩ (see point (5) in the list that follows) suggests a range $X15\%$ ≈ 9–31 ng/ml, three to four times the LOQ. The LOD according to the FDA concept depends on the data set one chooses, values of 1–10 ng/ml being found (if the standard deviation for DAY_4 and $x = 10$ ng/ml, see file VALID2.dat, had been only 5% smaller, x_{LOD} would have been 2 instead of 10 ng/ml)! The very tight definition of slope b in VALID3.dat pushes the LOD to below 1 ng/ml.

d. The analysis of the variances is inconclusive: The group SDs for the individual days are to all intents and purposes identical to the SDs of the pooled data; the same is true for the residual SDs. If there is a between-days effect, it must be very small.

Option ⟨Display Valid Graph⟩ assembles all of the important facts into one graph (see Fig. 4.32).

1. The abcissa is the same as is used for the calibration line graph. The ordinate is $\pm 30\%$ around $Y = a + b \cdot x$

2. At each concentration, the relative residuals $100 \cdot (y(i) - Y(x))/Y(x)$ are plotted.

3. At each concentration x_j the CV found for the group of m_j repeat determinations is plotted and connected (dotted lines).

4. At each concentration x_j the relative confidence limits $\pm 100 \cdot t \cdot s_y/y_{mean,j}$ found for the group of m_j repeat determinations is plotted and connected (dashed lines). Because file VALID2.dat contains only duplicate determinations at each concentration, $n = 2$ and $f = 1$, thus $t(f, p = 0.05) = 12.7$, the relative CL are mostly outside the $\pm 30\%$ shown. By pooling the data for all 6 days it can be demonstrated that this laboratory has the method under control.

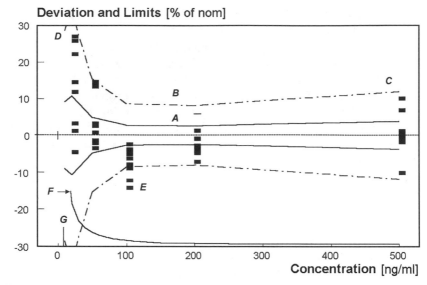

Figure 4.32. Graphical summary of validation indicators for file VALID1.dat. In this depiction, the typical "trumpet" form (A, B) of the c.o.v./CL curves is seen; the fact that there is a spread toward the right-hand edge (C) suggests that the measurement errors grow in a slightly over-proportional fashion with the concentration. The narrowing down of the "trumpet" at the lowest concentration (D) is a sign that due to the proximity to the LOD, the measurement distribution is not truly Gaussian but has the low side clipped. The data points for the lowest concentration (10 ng/ml) are off scale ($+31.2 - +65.9$), while the fourth-smallest concentration (E) yields negatively biased results (effct of background interference at the lowest concentrations due to the extraction's lack of selectivity, ensuing curvature). If the same plot were made for one day's data from VALID2.dat ($m = 2$ at each concentration), the dashed CL-curves would be off scale because of the large Student's t at $f = 1$. The curve (offset by -30) that starts at the lower right corner stops at the point x where the c.o.v. is 15% of the estimate Y (F). The LOQ calculated according to Fig. 2.14 appears in the lower left corner (G). The curves were obtained as $A = \pm t \cdot \text{CV}$, $B = \pm 100 \cdot t \cdot \sqrt{(V_Y)/Y}$, and $F = 100 \cdot \sqrt{(V_Y)/Y} - 30$, and the points as $y = 100 \cdot r_i/Y_i$.

5. Beginning at the lower right corner, a dotted curve is plotted that gives the CV for the estimated signal Y and $m = 1$, namely, the lower edge of the graph corresponds to 0%; the value $y = 60\%$ would be reached near the top left corner, but the curve is not plotted beyond the point where it reaches $y = 15\%$; the corresponding x-value, $X15\%$, is given. This result is sensitive to the calibration design, just as is the LOD/LOQ concept presented in Section 2.2.7, and is achieved through extrapolation to below the lowest calibration concentration. The two equations (4.27) and (4.28) give the general and the explicit formulation. In case the intercept should be negative, $X15\%$ can reach very high values.

$$y = 100 \cdot \sqrt{V_Y}/Y(x) \text{ in } [\%] \tag{4.27}$$

$$y = 100 \cdot \sqrt{V_{\text{res}} \cdot \left(1/N + \frac{(Y - Y_{\text{mean}})^2}{S_{xx}}\right)} \cdot \frac{1}{A + B \cdot X} \tag{4.28}$$

In contrast to this, the FDA requests a limit of acceptance defined by $100 \cdot s_Y/Y \leq 15\%$ for the lowest calibration standard $x \geq x_{\text{LOD}}$. This, of course, is insensitive to the calibration design and depends exclusively on the precision of the measurements performed at this low concentration; x_{min} should be chosen so that signals in the lower half of the size distribution are not overwhelmingly determined by base-line noise. GMPs require that further calibration concentrations in this region may only be added with a full six-day revalidation, even if it should turn out that the lowest x was initially chosen too high or too low. The analyst, therefore, is torn between (a) sacrificing valuable sensitivity (this can be disastrous for a pharmacokinetic study), (b) redoing the whole validation program with an adjusted lowest concentration, or (c) adding a series of concentrations designed to cover the suspected region of the LOD from the beginning. The danger is that the signal-distribution is no longer symmetric and residuals tend to become positive. Fig. 4.31 shows how closely the estimates match.

4.24 GOTTA GO GAMBLING

If an individual measurement is out of specifications (OOS), the interpretation should logically include at least the following items: The total number of repeat measurements, the distance between this measurement and the rest, the complexity of the analytical protocol, the plausibility of such a deviation, and the (medical) risk associated with analytical noise. In the pharmaceutical industry the debate about what to do with OOS results has raged on and off for many years, for example see Ref. 185. The Barr ruling, passed by Judge Wolin, and what the FDA interpreted into it[55] has sent shivers down many a spine. It is clear that a balanced solution should be found that customers, manufacturers, scientists, and regulators can agree to; perhaps a recently-issued draft guidance[178] is an omen for the better. At the moment, the regulators have the upper hand, and they seem to know only two shades of gray: black and white. Managers squirm when they think of the difference a blip of instrumental noise[186] can make: being a free man, or a jailbird accused of fraud. What is the issue, what can be done about it?

Shades of Gray   When repeat determinations are carried out on a sample and one or more of the individual results are outside the specs, the following criteria could apply:

- The sample complies if *the mean is within or just touches the SL*, e.g., $x_{mean} \leq 105\%$ for a $\pm 5\%$ specification range (all observers agree on this). In this most liberal interpretation, individual measurements may be outside the specification limits, but the amount of dispersion would be capped by the understanding that the standard deviation as found for the samples may not be larger than the standard deviations resulting from the system suitability test and/or the method validation. The clause "not larger than" could be defined so as to most easily satisfy legal thinking and/or numerical comparisons (cf. The USP XXIII Content Uniformity test which stipulates $n = 10$ and $CV \leq 6\%$, and has two sets of limits[43]), or could be defined statistically (F-test, predefined number of repeat determinations $n \geq n_{min}$ in order to prevent misuse with very small n. (See Figs. 1.27 and 4.48).) One should of course not employ imaginative rounding to gain an advantage, that is effectively "widen" the available specification interval, from, say, $\langle 95.0 - 105.0 \rangle$ to $\langle 94.56 - 105.44 \rangle$ by willfully changing the number of decimal places to suit the circumstances. It is understood that the right procedure is to round to a pre-determined number of digits after all of the calculations are performed with a sufficiently large number of significant digits. (See Section 1.1.5.)

- The *means resulting from several (k) repeat determinations on every of (j) samples* and/or sample preparations must comply; several such sample averages $x_{mean,j}$ would go into the over-all average $x_{mean, total}$. In the simplest case of $j = 2/k = 2$ (duplicate determinations on each of two sample work-ups), both sample averages x_{j1} and x_{j2} would have to comply. It could actually come to pass that repeat measurements, if foreseen in the SOP, would in this way not count as individual results[178]; this would cut the contribution of the measurement's SD towards the repeatability by $\sqrt{2}$, $\sqrt{3}$, etc. For true values μ that are less than 2–3 σ from the limit, the OOS-risk would be reduced.

- In the strictest interpretation, *each and every measurement* would have to comply.

Freshman and Industrial Science Settings

- *The simple situation*: The analytical method uses an instrument that provides an immediate readout of the definitive result, for example a pH-meter. If the operator notices an OOS result, this is only seconds removed from the preparatory steps leading up to the actual measurement. Any links to observations of potentially error-causing details (insufficient rinsing, dirty spatula, or glassware) are still fresh in mind. A plausible case for *operator error* can be made, even if the observation had not yet been written up.

- *The complex situation*: The analysis involves some instruments (e.g., dissolution apparatus, HPLC, balance), dozens of samples, and several analysts, and could take days to perform. If nobody notices anything particular about the individual snippets of work or any of the results, no observations of probable operator error are documented. When the supervisor then inserts number after unspectacular number into the formula, all results turn out to be within specification limits, with the exception of one or two that are barely outside. But, no documentation, no operator error, just inexplicable OOS, full investigation!

The simple situation is what the lawyers had in mind when drafting the rule, science Hollywood style! The other situation is illustrated with files OOS_RISK_X.xls. (See Example 56.)

The Strictest Interpretation   In the following, the language of the Barr decision (see notes at end of this section) is interpreted in the most restrictive sense, namely that each and every measurement has to comply with the rule "no values outside the limits."

The Math   As is immediately apparent, a mean cannot coincide with a SL if all measurements that go into it must also conform (cf. Fig. 2.13, distribution for $p = 0.5$), unless $s_x = 0$. Any attempt to limit the individual measurements to the specification interval will result in a narrowing of the available margin for error in x_{mean}, be it manufacturing bias or inhomogeneity. This may be acceptable as long as one has the luxury of 90–110% release limits, but becomes impracticable if these are reduced to 95–105%.

The Solution   Increasing the number of determinations sounds like a good way of reducing the manufacturer's risk, after all, Equation (1.12) and Fig. 1.16 show that increasing n reduces the uncertainty of the result. However, as conveyed by Fig. 1.2, doing more determinations increases the probability of a large deviation from the true mean. The two trains of thought need to be combined to define the best strategy available to the manufacturer, even if this implies very tight inhouse limits.

Assumptions

- European specification limits (assay: 95–105%).
- Zero sampling error: Eq. (1.6) reduces to $V_{reprod} = V_{repeat}$.
- Independent individual samples/measurements in the sense of Fig. 1.5.
- A result that is nearer to the SL_U and an accepted risk of 5% (alpha/2 = 0.05 for the single-sided test; use the alpha = 0.1 column in the t-table).
- A CV for the population, $100 \cdot s_x/x_{mean}$, of exactly 1%.

Questions

- What is the largest true mean μ that complies for $n = 4$ (double sampling, two determinations for each sample)?
- How does the situation change if additional samples are taken?
- What happens if intermediate results are rounded?

Example 56: Using file OOS_RISK_5.xls, a typical calculation in a photometric assay might look like this: 262.3 mg = 0.765 AU · 348.34 mg · 0.98433 AU^{-1}, where the three numbers after the equals sign are the actual absorbance and tablet weight and a constant that combines all dilution factors and the corresponding measurements for the perfect tablet (0.747 AU instrument reading, 340 mg tablet weight, of which 250 mg are drug substance). The registered specifications are 250 mg ± 5%; note that the exact tablet weight is not of regulatory interest as long as the assay is within limits, but is useful to check for deviations from perfect mixing, and a convenient in-process measurement. Five extracts are assayed with the raw data given in Table 4.29; the analyst did not note any suspicious events.

The individual results are 262.3 mg/tablet, etc., for a mean of 260.46 and a standard deviation of ±1.90 (CV = 0.73%). The CL for the mean are 258.65 and 262.27. Because the upper specification limit is nearer, the relevant z-value is $(250 \cdot 1.05 - 260.46)/1.90 = 1.073$, and therefore CP = −0.8582. The probability of one or more of the five measurements being above 262.5 is $100 \cdot (1 - (0.8582)^5) = 53.4\%$. The probability that the mean is OOS is calculated as follows: $z_{mean} = \sqrt{5} \cdot (262.5 - 260.46)/1.90 = 2.40$; → CP = 0.9918; → p = $100 \cdot (1 - 0.9918) = 0.82\%$. This product complies with the SL_U of 262.5, and the reproducibility is about as good as one can expect for an extraction followed by UV-photometry. All the same, the product might have to be destroyed because the fourth value is OOS and an obvious laboratory error cannot be demonstrated.

In the foregoing discussion, all numbers were carried along with the maxi-

Table 4.29. Frequency of OOS Results

Absorbance Reading [AU]	Tablet Weight [mg]	Assay [mg] result in raw form:	Assay [mg] result rounded to 1 decimal place:	Calibration Factor	Type of Result
0,747	340.00	250,00	250,0	0,98433	← nominal
				Interpretation:	
0.765	348.34	262.3044	262.3	Ok	← assay
0.761	347.27	260.1313	260.1	Ok	← assay
0.760	346.12	258.9292	258.9	Ok	← assay
0.765	348.66	262.5453	262.5	**OOS**	← assay
0.759	345.87	258.4017	258.4	Ok	← assay
0.7620	347.252	260.4624	260.44	mean	
0.0028	1.261	1.899885	1.89	Std. Dev.	
0.3712	0.363	0.72943	0.73	CV	

SL_U (+5%) = 262.5

		z-value:	CP:	% Probab.	Calculate
$z(SL_U - x_{mean})$:	rounded:	2.431790	0.992488	0.7	Risk that average is OOS
	full prec.:	2.398184	0.991762	0.8	
$z(SL_U - x_i)$:	rounded:	1.087529	0.861599	52.5	Risk that one result is OOS
	full prec.:	1.072501	0.858252	53.4	

mum number of significant digits, which resulted in one OOS result. Now, if the relevant SOP had stated that the individual result had to be rounded to four significant digits before being compared against the SL_U, the deviant "262.55" (actually 262.545) would have become the acceptable "262.5," and no OOS-investigation would have had to be started. There is even some statistical merit in restricting the result to three of four significant digits because the least accurate measurement does not offer more precision (0.765 AU; the balance might read two more: 348.66 mg). OOS_RISK_X.xls allows setting of the number of decimal places.

Example 57: The three files can be used to assess the risk structure for a given set of parameters and either four, five, or six repeat measurements that go into the mean. At the bottom, there is an indicator that shows whether the 95% confidence limits on the mean are both within the set limits ("YES") or not ("NO"). Now, for an uncertainty in the drug/weight ratio of 1%, a weight variability of 2%, a measurement uncertainty of 0.4%, and μ 3.5% from the nearest specification limit, the ratio of OOS measurements associated with "YES" as opposed to those associated with "NO" was found to be 0 : 50 ($n = 4$), 11 : 39 ($n = 5$), respectively 24 : 26 ($n = 6$). This nicely illustrates that it is possible for a mean to be definitely inside some limit and to have individual measurements outside the same limit purely by chance. In a simulation on the basis of 1000 sets of $n = 4$ numbers $\in$ ND(0, 1), the x_{mean}, s_x, and CL(x_{mean}) were calculated, and the results were categorized according to the following criteria:

1. "Both CL(x_{mean}) in the SI" (Yes, No), respectively
2. "No OOS value involved" (True, False), e.g.:
   - Case YT (496 events) is accepted by everyone.
   - Case NT (257 events): The x_{mean} and one of its confidence limits is inside the SL, and no OOS results were detected. This is not mentioned in the guideline, and appears to be accepted by the FDA, but poses a risk for the manufacturer if the SD is large in that the probability of a retest (e.g., at a customer's lab) with a mean outside the SI is larger than 5%. The largest s_x that would still be acceptable is 11.5 (x: 90, 90, 110, 110; $p(x \notin SI) \geq 0.08$), about an order of magnitude larger than the typical repeatability!
   - Case YF (five events) is the one that is here proposed for acceptance because the mean is better controlled than in case NT if the SD is reasonably small; then the OOS value is a chance result within the ND(0, 1) population.
   - Case NF (242 events) comprises all situations combining an OOS

result with a x_{mean} that has at least one of its CL outside the SI, whether x_{mean} itself is inside or outside the SL; the latter case would lead to a rejection anyway. The former case, an OOS value and a $CI(x_{mean})$ straddling the SL, is a risky situation, but does not, statistically speaking, imply that the product is defective.

This demonstrates that, even for specification limits at $\pm 2 \cdot \sigma$, the probability is close to 1% that both $CL(x_{mean})$ are contained in the SI and s_x is reasonable (say, smaller than 1.5), but one of the four values is just outside a SL. These simulations were expanded to $n = 5$ and $n = 6$, and the number of data sets was increased to 350 000 for confirmation. At $n = 6$ and SL = ± 2.5, the frequency of such "OOS" occurrences of type YF is on the order of several percent, since for $\mu = 0$ the expectation is $CP(2.5) = 0.99379$, the probability of observation is $6 \cdot (1 - CP) = 0.0373$ for the single-tailed situation (one SL), resp. 0.0745 for the two-tailed one (symmetrical SL).

Winners and Losers To whose benefit is such a strict interpretation of the Barr ruling?

- For manufacturers that try to cut corners, the writing on the wall has become clearer, and the customer will be thankful.
- For those who act responsibly, the legal and regulatory compliance departments and the health authorities benefit because complex scientific judgment is reduced to the mechanics of wielding checklists.
- The manufacturing and quality control departments face higher costs because they have to eliminate process and measurement variability, even if they are already operating at the technological limit. They will have to add people to their staffs to run all of the investigations and handle the additional paperwork (because malicious intent is suspected, peers and supervisors have to sign off at every step to confirm that each SOP was strictly adhered to; whether the SOPs made sense, scientifically speaking, or were installed to satisfy formalistic requirements is of no interest here).
- Marketing will be able to advertise that their product complies with tighter limits. (See Fig. 4.35.)

The customer might now be led to believe that the products are "safer," because biological variability (huge, by comparison), medical practise (it is left to the doctor's discretion to adjust the dose), or compliance (dismal) are outside this discussion.

Will Science Succumb to Legal Correctness?   The strict interpretation is unscientific because it distinguishes between natural variability (a statistical property of every measurement) that is "legally acceptable" as long as every measurement is inside the limits, and natural variability that is called "lack of control" because a value is outside some limit. In human terms, this purported difference between "skillful work" and "sloppyness" is not apparent, and nature does not split such hairs either. A scientist might not recognize an OOS occurrence without doing a calculation, but takes the blame if he failed to document "laboratory error"; his scientific training, his experience with a product and a method, and his honed skills of judgment tell him that there is no laboratory error and the product is good, but these are all declared void.

Since it does not appear that a lower limit was set as to what constitutes "OOS," a similarly legalistic situation holds as when the infamous Delaney clause* was on the books: What today still passes as "≤ 105%" (105.005 ... 105.049, assuming an instrumental resolution equivalent to 0.1), will become OOS if instrumental resolution improves by a factor of 10.

Is the Judge's Back Exit a Trapdoor?   Continue testing: The Judge left the door ajar by adding that retesting is appropriate when analyst error is documented or the review of analyst's work is "inconclusive." There are three problems with this: First, retesting without *a priori* definition of the circumstances is forbidden, second, if the "inconclusive" label pops up too often this looks suspicious, and third, retesting could be interpreted by some as coming awfully close to the rightly banned practice of "testing a product into compliance." So, careful rephrasing of SOPs so that retesting is allowed under certain circumstances is imperative.

Document till you drop: While it looks as if a good explanation would help overcome individual OOS results, daily practice demonstrates that lab technicians, not being lawyers, never document, date, and sign every move or observation they make, largely because this would fill volumes with everrepeating trivia and keep them from doing productive work. Thus, managers often are at a loss to explain some scientifically not so terribly exciting number, and start to write "this combination of ordinary noise, base-line drift, and signal asymmetries tells me that ... but could be stretched to mean

---

*The Delaney Clause was introduced to protect the American consumer from dangerous substances. By the "if proven toxic, X is forbidden" logic that did not include any reference to medical risk, the inexorable progress of instrumental analysis naturally lead to the situation where, in a few years, apple pie would have been declared *off limits* had the clause not been struck from the books, because it is certain to contain a one part per trillion of some X that is toxic at the kg-per-day level.

...." Well, that will not necessarily convince the FDA that expects crisp and unequivocal answers[**].

An investigation would already be called for if both x_{mean} and $CL(x_{mean})$ were inside the limits and a single measurement were outside by just "1" in the least significant digit. In FDA coinage, this constitutes "product failure"; Judge Wolin used the analytically more appropriate "out-of-specification result"[***]. Mincing words does not help in this case, though, because an OOS result must be classified through investigation as either (1) a laboratory, (2) a non-process operator, or (3) a process-related manufacturing error. Notice that the only available categories are human errors that result in bias: Design and execution of a process; it apparently did not cross anybody's mind that measurement noise is a stochastic component that can only marginally be influenced by man. (In this connection, see file PEDIGREE.xls.) In other words, the operator is culpable even if Mother Nature throws dice. Since an investigation cannot explain away stochastic variability as laboratory error ("Laboratory errors occur when analysts make mistakes in following the method of analysis, use incorrect standards, and/or simply miscalculate the data."), the result stands. The same is true for the occasional instrumental hiccup, unless there is a smoking gun in the form of a blown seal, a thunderstorm-induced brown-out, or some other glaring deficiency.

Maybe lawyers will come to recognize the stochastic nature of measurements and pass regulations that take this into account. (It does not help that in English the word *error* is used for both the human and the random form; in German, for example, the distinction is *Ausführungsfehler/Zufallsfehler*, with only the former carrying a connotation of guilt). The probability of this happening is remote, though.

Otherwise, one is left with the possibility to accept that an individual OOS result constitutes "failure," and must be avoided. This can be done by reducing the probability of occurrence of one OOS result out of N measurements to some predefined level, e.g., $p \leq 0.05$. The acceptable level of risk is management's decision, because they will have to face the press (Wall Street, FDA, etc.) if scandal errupts, and approve the budget overruns incurred by increased testing and wasted batches.

[**]This is a quirk of American customs that could become the undoing of natives of other countries. Even a knowledgeable and honest person is expected to fire back "yes Sir, no Sir"-type answers and the slightest wavering is interpreted as a sign of having something to hide. In many countries, though, politeness demands that one does not act assertively, and that one make a distinction between government agents and drill sergeants.

[***]FDA-speak: Until recently the term *product failure* was used, even if for a perfectly good product the deviation was traceable to a mistake in adding up some numbers in the lab. This is as preposterously strict as not graduating any student who ever forgot to switch off the dormitory light.

Probability density

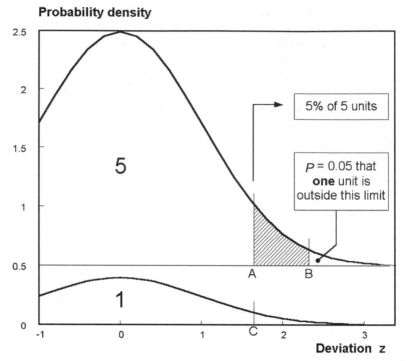

Figure 4.33. Probability density for one, respectively five measurements. The areas to the right of points B and C are the same; the chances of observing <u>one</u> event this far from the mean are 5%. The probability of observing five events to the right of point A are 5% (equivalent to a 25% chance of observing one event). The upper curve is offset by +0.5.

In the following example, the practical consequences of choosing the second way are explored.

Example 58: Fig. 4.33 explains the theoretical basis and Fig. 4.34 depicts increasingly sophisticated ways of reducing the risk of having a single measurement beyond the specification limit. Between two and 10 measurements are performed and the mean is calculated in order to decide whether a product that is subject to the specification limits ⟨ 95 ... 105% of nominal⟩ can or cannot be released. A normal distribution $ND(\mu = $ variable, $s_x^2 = 1.00)$ is assumed; the case $x_{mean} \approx 105$ is explained; the same arguments apply for $x_{mean} \approx 95$, of course. Tables 4.30–4.32 give the key figures.

1. The most obvious way is to restrict x_{mean} to ≤105, as was traditionally done. It is evident that for $x_{mean} = 105$ the probability of each mea-

Assay [%]

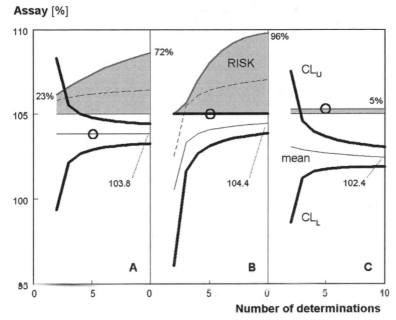

Number of determinations

Figure 4.34. The confidence limits of the mean of 2 to 10 repeat determinations are given for three forms of risk management. In panel **A** the difference between the true mean (103.8, circle!) and the limit L is such that for $n = 4$ the upper confidence limit (CL_U, thick line) is exactly on the upper specification limit (105); the compound risk that at least one of the repeat measurements $y_i \geq 105$ rises from 23 ($n = 2$) to 72% ($n - 10$). In panel **B** the mean is far enough from the SL_U so that the CL_U (circle) coincides with it over the whole range of n. In panel **C** the mean is chosen so that the risk of at least one repeat measurement being above the SL_U is never higher than 0.05 (circle, corresponds to the dashed lines in panels A and B).

surement x_i being above the limit is precisely 50%. This case is not depicted.

2. A reasonable procedure for a QC lab using HPLC is to take two samples, work up each one, and inject two aliquots of each solution; a

Table 4.30. The True Mean μ has to be at Least this Far from the Nearest SL for CV - 1%

n	2	3	4	5	6	7	8	9	10
A	1.18	1.18	1.18	1.18	1.18	1.18	1.18	1.18	1.18
B	4.46	1.69	1.18	0.95	0.82	0.73	0.67	0.62	0.58
C	1.95	2.12	2.23	2.32	2.39	2.44	2.49	2.53	2.57

Table 4.31. The Maximal OOS-Risk [%] This Implies for any Single Measurement as Created by the Assumptions in Table 4.30

n	2	3	4	5	6	7	8	9	10
A	12.0	12.0	12.0	12.0	12.0	12.0	12.0	12.0	12.0
B	0.0	4.6	12.0	17.0	20.5	23.1	25.1	26.8	28.1
C	2.5	1.7	1.3	1.0	0.85	0.73	0.64	0.57	0.51

limit $\mu \leq 103.82$ (circled line in panel A) then reduces the probability of the average x_{mean} being larger than 105 to $p \leq 0.05$ (use Student's t-table for $p = 0.1$ because a one-sided test is being done, $t = 2.3534$ for 3 degrees of freedom, $\sigma \cdot t/\sqrt{4} = 1 \cdot 2.35/2 = 1.18$; $105 - 1.18 = 103.82$; cf. Eq. (1.12a). The probability $p(x > 105)$ is found by first determining the z-value as $z \approx |105 - 103.82|/1.00 = 1.18$, and then looking up the cumulative probability (integral over ND(0, 1) over the range $-\infty$ to 1.18), which is CP = 0.881, and converting to percent $(1.000 - 0.881) \cdot 100 = 11.9\%$. For $n = 1$ this translates to roughly one measurement out of 9, for $n = 5$ it is every other one! The gray background area, which increases from 23% at $n = 2$ to 72% at $n = 10$ (scale compressed by a factor of 20 relative to the 95–110% assay scale), shows that the risk of an OOS result is 39.8% at $n = 4$ and increases for $n = 2 \rightarrow 10$ despite the tightening of the confidence interval (bold lines). The laudable intention to reduce the risk of overdosing the patient by increasing the number of repeat determinations actually punishes the manufacturer by boosting the risk of a "product failure because of higher than allowed potency." The calculations were done for $\mu = 103.82$.

3. If the risk of OOS results is disregarded, a refined strategy would be to use a flexible limit, namely $\mu \leq L$ for $L = 100.56 \ldots 104.42$, which would make the upper confidence limit CL_U (x_{mean}) coincide with the 105% specification limit (circled bold line in panel B). The OOS risk in panel B is larger than that in A for $n > 4$.

Table 4.32. The Joint OOS-Risk [%] Associated with n Repeat Measurements

n	2	3	4	5	6	7	8	9	10
A	23	32	40	47	54	59	64	68	72
B	0	13	40	61	75	84	90	94	96
C	5	5	5	5	5	5	5	5	5

True Mean μ [% nom]

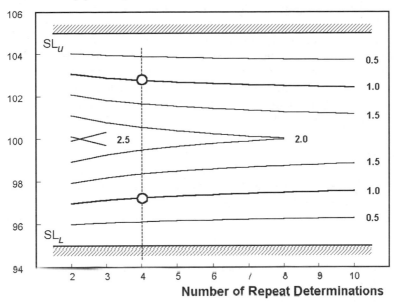

Figure 4.35. The range available for the true mean μ as a function of the number of repeat measurements and the CV. The case discussed in the text is indicated by thick lines and circles. The SL are assumed to be 95 and 105%. For a CV = 2% the OOS risk is above 5% for $n > 8$, and for CV = 2.5%, n is restricted to 2. For SL = 90 ... 110%, the figure must be split in the middle and the upper part shifted by +5%, the lower part by −5%.

1. The best bet a manufacturer can make is to set an upper limit on the OOS risk he is willing to take, e.g., 5%, and combine that with a risk-reduction strategy for the release/reject decision by increasing the number of repeat determinations to $n \geq 4$, see panel C. The penalty he pays is that the margin for manufacturing error (high/low dosing and/or content uniformity) $|\mu_{max} - \mu_{min}|$ that still fits within the specification interval SI = $|SL_{U} - SL_{L}|$ is reduced. Cf. Figs. 2.12 and 2.13.) The relevant equation is $p_{total} = (1 - (1 - p_{individ})^n)$, where $(1 - p_{individ})$ is identified with the CP for an individual measurement being inside the limit. Figure 4.35 expands the concept to $s_x = 0.5 \ldots 2.5$. The consequences are absolutely clear: If the analytical method does not achieve a reproducibility of 1.5% or better, the standard European assay limits for pharmaceutical products of 95–105% cannot be met. An American manufacturer still has more room to breathe because the corresponding limits are typically 90–110%.

Without good "laboratory error" explanations to ward off REJECTED labels, batch failure rates are likely to increase under the strict interpretation. The health authorities are waiting: The rejection rate is a criterion they use to judge the trustworthiness of the companies they inspect and license.

The FDA did not include outlier tests in the USP for chemical assays, but allowed the practice for biological tests. The reason for this could be that because of the high precision, n is usually small in chemical testing; with $n < 3$, outlier tests cannot be conducted. It appears that Judge Wolin followed this recommendation when deliberating his decision.[185]

Conclusion   It should be possible to design a system of rules[187] that allows natural variability and all-too-human mistakes to be detected and distinguished from fraud, risky products, and defective processes. Precision indices taken from validation reports and recent calibration runs should allow the risk of product failure to be assessed, even if a single value is outside the limits; outlier tests might have a role in this; nobody should have to feel responsible for Mother Nature's poor habits. As long as controls and corrective mechanisms are built in, an operator should be able to acknowledge a mistake and properly fix it without feeling the heavy hand of the law on his shoulder.**** Laboratory errors are part of life, and by criminalizing the participants the frequency of errors goes up, not down, because of the paralysis and fear that sets in.

JUDGE WOLIN'S INTERPRETATIONS OF GMP ISSUES CONTAINED IN THE COURT'S RULING IN USA vs. BARR LABORATORIES, 2-4-93; available from FDA's homepage.[55] *"AVERAGING RESULTS OF ANALYSIS. Averaging can be a rational and valid approach, but as a general rule this practice should be avoided because averages hide the variability among individual test results. This phenomenon is particularly troubling if testing generates both OOS and passing individual results which when averaged are within specification. Here, relying on the average figure without examining and explaining the individual OOS result is highly misleading and unacceptable.*

"PRODUCT RELEASE ... the USP standards are absolute and cannot be stretched. For example, a limit of 90 to 110 percent of declared active ingredient, and test results of 89, 90, 91, or two 89s and two 92s all should be followed by more testing."

****Fear of Fraud: It is illuminating that there are companies that lock the input to a computer system (specifically a so-called LIMS, a Laboratory Information Management System) in such a manner that it takes the supervisor's password and signature to correct a missed keystroke, even if this is noticed before the ⟨Enter⟩ key is pressed. If the unwitnessed use of the ⟨Backspace⟩ key betrays "malicious intent to falsify raw data," then every scientist should be in jail.

[Author's comment] Because a general rendition of the "Scientific Method" cannot be cast in legally watertight wording, all possible outcomes of a series of measurements and pursuant actions must be in writing before the experiments are started. This includes but is not limited to the number of additional samples and measurements, and prescriptions on how to calculate and present final results. Off-the-cuff interpretations and decisions after the fact are viewed with suspicion.

4.25   DOES MORE SENSITIVITY MAKE SENSE?

A quality control laboratory had a certain model of HPLC in operation. One of the products that was routinely run on the instrument contained two compounds, A and B, that were quantitated in one run at the same detector wavelength setting. At an injection volume of 20 μL, both compounds showed linear response. The relatively low absorption for compound B resulted in an uncertainty that was just tolerable, but an improvement was sought.

The problem could have been resolved by running two injections, either with different wavelength settings and/or with different dilutions injected, but this would have appreciably increased the workload of the technicians and the utilization factor of the instrument. Without an additional instrument, the laboratory would have lost much of its flexibility to schedule additional analyses at short notice. Another solution would have been to utilize a programmable detector that switches wavelengths between peaks, but that only works if the weakly detected component strongly absorbs at some other wavelength.

On checking through instrument specification sheets, the lab's director realized that another model of HPLC had a sensitivity (ΔSignal/ΔConcentration, usually determined as the first derivative of the calibration function near the origin) that was about twice as high as that of the first HPLC. By switching the product from one instrument to another, he would have the problem off of his desk. He ordered a comparison using the method as written, and for good measure added some injections at 10 μL. The results can be seen in Fig. 4.36. For both compounds, the 10 μL injections seemed to fill the bill: The signals had increased to about twice the height. While this was more or less true over the 50–100 range for compound B, compound A's calibration curve (here approximated by $Y = c \cdot \sqrt{x}$) indicated that a linear interpolation, particularly if one contemplated one-point calibrations (dotted line!), would be strongly biased. A multipoint nonlinear calibration approach, on the other hand, while theoretically feasible, would consume an inordinate proportion of the available resources before the reliability of the result came anywhere near what was already available on the model 1 instrument.

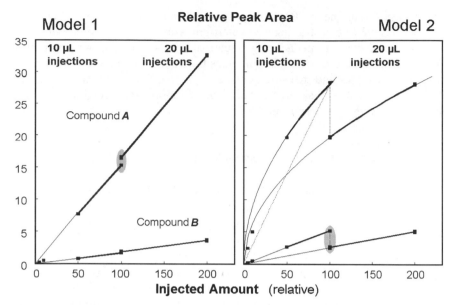

Figure 4.36. Cross validation between two HPLCs: A stock solution containing two compounds in a fixed ratio was diluted to three different concentrations (1 : 10 : 20) and injected using both the 10 and the 20 μl loop on both instruments. The steps observed at Amount = 100 (gray ellipses) can be explained with effective loop volumes of 9.3 and 20 μl (model 1) and 14.3 and 20 μl (model 2) instead of nominally 10 and 20 μl. This is irrelevant as both a sample and the calibration solution will be run using the same equipment configuration. The curved portion of the model 2 calibration function was fitted using $Y = A \cdot \sqrt{x}$; this demonstrates the nonlinearity of the response at these high concentrations. The angle between the full and the dotted line indicates the bias that would obtain if a one-point calibration scheme were used.

When the results for the 20-μL injections are included in the analysis, it is quite clear that the situation was much more complex than the comparison of sensitivities would suggest:

- The chosen concentrations overload the detector of the model 2 instrument (curvature in the 0–5–10–100 and 0–100–200 range).
- It would be impossible to determine both compounds in the same run.
- The selection of an instrument is not just a question of comparing specification sheets. Furthermore, since vendors usually include legal wording in their brochures that allows them to change components (detector geometry, plumbing details, electronics, etc.) without notification, even two instruments of the same make, model, and nominal performance

may be sufficiently different to require revalidation and/or re-optimization. This is all the more true if the method requires the instrument's characteristics to be fully utilized.

4.26 PULL THE BRAKES!

Three test batches of a chemical were manufactured with the intention of validating the process and having a new product to offer on the market. Samples were put on stability under the accepted ambient (25°C, 60% relative humidity) and accelerated (= stress; 40°C, 75% rh) conditions cf. Section 4.20. One of the specification points related to the yellowish tinge imparted by a decomposition product, and an upper limit of 0.2 AU was imposed for the absorption of the mother liquor (the solvent mixture from which the crystalline product is precipitated) at a wavelength near 400 nm.

At $t = 0$, the only thing that could have been a reason for concern was the fact that two batches started with a much higher amount of the impurity than the third, see Fig. 4.37, right side. At $t = 1$ and 2, the stressed samples suggested a slowing down of the degradation. These points are underlaid by gray ellipses to alert to the fact that the common trend, high at 1, low at 2 months for all three samples, could be due to a calibration error. At $t = 3$ it became obvious that all samples had suffered. A quick graphical extrapolation pointed to a probable shelf life of only about 6 to 9 months, too little to make the product commercially interesting (cf. Section 4.20 for calculation of shelf-life). Still, some harbored hope that while the stressed

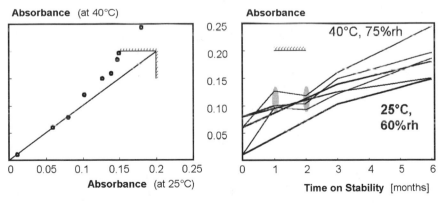

Figure 4.37. Build-up of a decomposition product over time. Two climatic conditions were evaluated: The higher the temperature, the more rapid the process.

samples might not make it, those stored at ambient would survive to at least 12 months. Unfortunately, the 6-month samples confirmed the trend, and the project died. When the data were routinely put through a shelflife analysis (program SHELFLIFE or file SHELFLIFE.xls, data set BUILD_UP.dat), the result came out as "4 months" at 40°C and just a month more at 25°C.

Conclusions   Regulators, having learned their chemistry, expect either simple mechanisms (e.g., zero-order hydrolysis kinetics under controlled standard conditions if the packaging limits the amount of water diffusing through to the product), or a good justification for a more complicated degradation model. Hoping for some unspecified chemical wonder to avert an unfavorable trend borders on self-delusion, and if it does occur in the end, then one is often hard-pressed for an acceptable model or sufficient data to write up a plausible explanation. The Marketing Department—unaware that "statistics" would add a safety margin to any conclusions drawn from the little data available—asked for a re-analysis in the belief that there was a way to make the numbers pronounce "stable for 12 months"; they had to be politely rebuffed.

What was the influence of the storage temperature? In the right panel, the stressed samples degrade faster, at least in hindsight, but the trend was not all that clear before the 6-month results became available. A correlation between the two storage conditions (left panel) leaves no doubt: A temperature-dependence is there, and there might even be some form of curvature, but there is hardly enough data to confirm any specific theoretical model. The samples do not all start the stability program with the same amount of the degradation product (difference = 0.07 at $t = 0$, about half of what accrues over six months at 25°C), but the points fall on the same curve. This makes it unlikely that the degradation process involved during manufacturing is different from that encountered during storage, but it proceeds faster. If the product had survived, a test could have been run at the intermediate condition, 30°C/−65% rh, to increase the range of acceptable storage conditions.

4.27   THE LIMITS OF NONLINEARITIES

Many biologically interesting molecules, for instance hormones, can be determined using any of a number of analytical methods, such as GC, GC-MS, and RIA. In blood serum and similarly complex matrices, the more traditional methods (colorimetry, titration, TLC) suffer from interference and/or lack of sensitivity.

The selectivity issue might be manageable for a single subject, but if dozens of humans or animals are to be monitored, differences in genetic makeup and metabolic status could well call for individual adjustments of extraction and/or chromatographic conditions, which would drive up costs and make the scheme impossible to validate. Immuno-assays, on the other hand, while being less precise and perhaps even less specific, offer the distinct advantages of working with body fluid from a variety of individuals and being fast. The boundary conditions imposed by the problem setting will allow a selection to be made on the basis of required accuracy and precision, number of samples and allowable turn-around time, available skill base and instrumentation, cost limits, etc.

This example assumes that RIA was chosen. The principle behind RIA is the competition between the analyte A and a radioactively tagged control $^{*}C$ (e.g., a ^{125}J-marked ester of the species in question) for the binding site of an antibody specifically induced and harvested for this purpose. The calibration function takes on the shape of a logistic curve that extends over about three orders of magnitude. (Cf. Fig. 4.38a.) The limit of detection is near the $B/B_0 = 1$ point (arrow!) in the upper left corner, where the antibody's binding sites are fully sequestered by $^{*}C$; the nearly linear center portion is preferably used for quantitation.

Here the results of 16 calibration runs are presented: For each run, the concentration standards 0, 5, 10, 20, 50, 150, and 500 [ng/ml] were done in triplicate, and the four parameters of the Rodbard[132] equation were determined using the 7 $(B/B_0)_{mean}$ values; each (B/B_0) was then run through the interpolation as if it were an unknown's signal. The results of these back-calculations are depicted in Figs. 4.38c,d; the obviously linear function shows that the calibration and the interpolation are under control. The limit of detection, expected at a concentration of 0.1 before the experiment, was determined for each calibration run and is seen to vary between about 0.026 and 0.65, see file RIA_Calib.xls; the "broom" shape is not an artifact of the method. Figure 4.38d shows the same data as Fig. 4.38c but in a format more likely to appear in a method validation report. In Fig. 4.38b, the precision information is summarized: For the three concentrations 5, 20, and 150 ng/ml, the %CV for each triplicate determination is given as a dot, and the %CV for the back-calculated concentrations are shown as a line. Because of the logarithmic scale, the average CV is not found in the middle of a range, but closer to the top (horizontal lines).

The difference, e.g., 5.0 − 1.4 in the column marked "20 ng/ml," must be attributed to the "interpolation error," which in this case is due to the uncertainties associated with the four Rodbard parameters. For this type of analysis, the FDA-accepted quantitation limit is given by the lowest calibration concentration for which CV ≤ 15%, in this case 5 ng/ml; the cross indicates

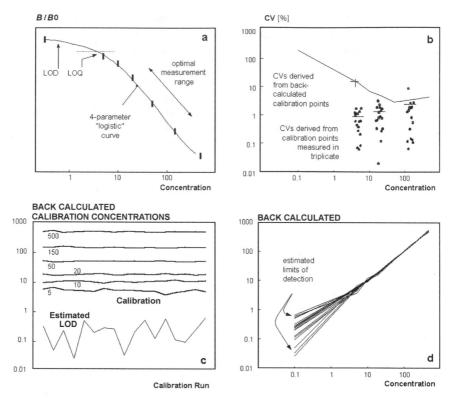

Figure 4.38. Validation data for a RIA kit. (**a**) The average calibration curve is shown with the LOD and the LOQ; if possible, the nearly linear portion is used which offers high sensitivity. (**b**) Estimate of the attained CVs; the CV for the concentrations is tendentially higher than that obtained from QC-sample triplicates because the back transformation adds noise. Compare the CV-vs.-concentration function with the data in Fig. 4.6! (**c**) Presents the same data as (d), but on a run-by-run basis. (**d**) The 16 sets of calibration data were used to estimate the concentrations ("back-calculation"); the large variability at 0.1 pg/ml is due to the assumption of LOD = 0.1.

the intercept of the CV-line with the 15% level, which is at 4 ng/ml. The dotted horizontal in Fig. 4.38a points to the corresponding B/B_0 level for 4 ng/ml.

When reporting a result above this concentration level (e.g. 73.2 ± 4.2), one should append a precision statement ("73 ± 4," resp. "73 (±6%)").

A result below the quantitation limit would be given by "<LOQ." If the data are used in further calculations, a number of either zero or, in this example, five, would be substituted, the rule being that option is chosen which makes it harder to prove a hypothesis or lowers the risk of a false statement.

Table 4.33. Estimation of CV.

	5 ng/ml	20 ng/ml	150 ng/ml
Average CV for triplicate calibration measurements	0.91	1.4	2.5
Average CV for back-calculated concentrations	13.0	5.0	3.4
Estimated CV due to interpolation, e.g. $\sqrt{13^2 - 0.91^2}$	13.0	4.8	2.3

4.28   THE ZEALOUS STATISTICAL APPRENTICE

An existing HPLC method was modified to assay the active component in a line of products that were to comprise, amongst other dosage forms, a gel and an emulsion, both containing 2% of the drug. Because there were no complicating factors involved, such as excipients that could co-elute with the drug substance or give rise to a signal at the chosen wavelength, the method was thought to be applicable to both forms. A validation protocol was drawn up that involved, along with other tests, ten repeat injections of each sample solution. The sample solutions were obtained by thoroughly shaking an accurately weighed quantity of gel or emulsion in a given volume of a water/alcohol mixture that would easily dissolve the drug and would allow the excipients to form a second phase. Vials containing either sample or calibration solution were placed on the HPLC's carousel along with routine samples in a predetermined order so that the evaluation program would aquire all of the chromatograms, establish the calibration line's parameters, automatically interpolate the individual sample's drug content in percent of the nominal concentration, and depict the results are tables and in the form of an old-fashioned line-printer "plot." The next morning, the technician immediately sat down to evaluate the chromatograms: The peak shapes looked fine and there was no reason to improve on the base-lines that the signal integrator's algorithm suggested. The numbers from the summary table, ordered by size, are given in column 2 of Table 4.34.

The results appear to cluster around the values 101 with the first two and the last gel samples falling out of line. The technician reported this to his supervisor and the two thereupon decided to cast out "bad results"; by recalculating the standard deviation for the seven remaining gel results, they easily demonstrated a big improvement in s_x (from 0.47 to 0.115) and "confirmed" the bad results as being really bad: The outliers no longer hovered around $t = 1.4 \ldots 2$, but stuck out at $t' = 6 \ldots 8$.

**Table 4.34. Content Uniformity of Dosage Form; Results
After Elimination of Three "Outliers" Are in Italics**

Sample	GEL [% of nominal]	
1	100.3	
2	100.4	
3	100.9	100.9
4	101.0	101.0
5	101.1	101.1
6	101.1	101.1
7	101.2	101.2
8	101.2	101.2
9	101.2	101.2
10	102.0	
Mean	101.04	*101.10*
Std. Dev.	0.47	*0.115*
CV	0.43	*0.11*
N	10	7
F test	$H1$; $F = 16.7$ ($F_{crit, 0.05} = 4.1$)	
t test	$H0$; $t = 0.39$ ($t_{crit} \approx 2.6$; $p \approx 0.71$)	

Example 59: $(102.0–101.04)/0.47 = 2.04$, $(102.0–101.10)/0.115 = 7.8$). Thus, the interpretation could have been that a fantastic reproducibility had been achieved, and that three of the 10 results could not be relied upon because of operator error (pipetting, weighing, whatever ...).

The pair set out to write up the first section of the validation report but their triumph was short-lived. A dispassionate colleague pointed out one fallacy and two GMP violations:

- A sample of only 10 repeat determinations is "statistically small," and far-reaching conclusions are hard to draw, particularly as far as the distribution of values is concerned. Claiming an extraordinarily high precision in the face of the alternative explanation—an average and very plausible precision—should raise eyebrows.

- Since the U.S. vs. Barr decision in 1993 (relevant to pharmaceuticals and related fields, rules applied by the Federal Food & Drug Administration, FDA), outlier tests may no longer be applied to physicochemical tests, under the assumption that such test methods, having been optimized and validated for the particular set of circumstances, rarely produce outliers. These tests may not be applied to CU results at all.

- Good manufacturing practices mandate that operators work according to pre-set procedures and write down any observed irregularities as they occur; the "laboratory error" clause cannot be invoked without good reason, particularly if decisive experiments are under way.

In retrospect, the zeal came about by the will to deliver "perfect results" and the failure to stand back and see the big picture before plunging into the details:

- The calibration is as good as can be expected.
- The repeatability patterns look perfectly acceptable.

A number of lessons were learned:

- Use graphics instead of or as adjunct to tables.
- Know the method and what it can deliver.
- Do not try to wring anything out of the data beyond what can be reasonably expected under routine, or at most R&D conditions.
- Adhere to protocols and accept the results.

Pure logic would forbid any attempt to further reduce standard deviations if the results already demonstrate excellent proficiency: A CV of around 0.5% is about as good as one can hope for in an HPLC method; a CV of 0.1% can then only come about as a chance result or because the instrument's display should offer a digit more. In a poorly designed method, however (e.g., an intrinsic method variability of ±6 mAU in connection with an instrumental resolution of 10 mAU), this would be expected and would be a reason for an investment decision. The quotient of observed-to-expected variability should have been looked at more closely. A graphical depiction showing the specification limits and the individual results would have helped. (See the summary sheet shown in Table 4.35). As experience with the manufacturing process and the analytical method accumulate, one glance at the printout will tell whether there is a problem or not.

If one lives by the regulations and does not cast out "outliers," then a comparison between the gel and the emulsion results shows that the standard deviations are comparable ($F = 1.57$), but the means are significantly different; action is now called her to determine the cause for this discrepancy: Manufacturing error, incomplete extraction, or interference by excipients?

Summary Sheet for HPLC   Since all 10 calibration results are very close to 100% and no trend is apparent, the HPLC system is regarded as being "in control."

The routine gel results indicate either a manufacturing bias (overdosing)

Table 4.35. Printer Graphics as a Simple Means to Check Quality of Data

```
5    6    7    8    9    0    1    2    3    4    5 Scale: 95-105%
02468024680246802468024680|2468024680246802468024680 of label claim
-----------------------------------------------------------------
|                          **                     | 99.8, 100.0 Calibr
|                          *  *                   | 100.0, 100.2
-----------------------------------------------------------------
|                                R                | 1   101.2 GEL
|                                R                | 2   101.2
|                                   R             | 3   102.0
|                             R                   | 4   101.0
|                             R                   | 5   101.1
|                          R                      | 6   100.4
|                             R                   | 7   101.2
|                             R                   | 8   101.1
|                          R                      | 9   100.9
|                          R                      | 10  100.3
|                        -==A==-                  | MEAN 101.0 +/- 0.5
-----------------------------------------------------------------
|                          *  *                   | 99.9, 100.2 Calibr
-----------------------------------------------------------------
|    R                                            | 1    96.8 EMULSION
|         R                                       | 2    97.3
|        R                                        | 3    96.9
|      R                                          | 4    96.0
|       R                                         | 5    96.2
|           R                                     | 6    97.3
|          R                                      | 7    96.9
|          R                                      | 8    97.0
|             R                                   | 9    98.1
|        R                                        | 10   97.2
|      ===A===                                    | MEAN  97.0 +/- 0.6
-----------------------------------------------------------------
|                          H                      | 100.1, 100.2 Calibr
|                          **                     | 99.9, 100.1
-----------------------------------------------------------------
02468024680246802468024680|2468024680246802468024680 105
5    6    7    8    9    0    1    2    3    4    5
```

```
Symbols used: (*) calibration result, (H) duplicate calibration
results in same 0.2% class, (1,2) duplicate measurements,
(X) duplicate measurements in same 0.2% class, single results in
(U) Content Uniformity or (R) 10x Repeatability test
(#) out-of-specification measasurement,
(A) mean with standard deviation in 0.2% (=) or 0.1% (-) increments
```

of about 1%, (A), a calibration error (B), or a superimposed extraneous (non-specific) signal (C). Since placebos (dosage form without drug) run under the same conditions give a base-line result (peak area = 0), the conclusion must be that hypotheses (B, C) do not apply and that most probably a manufacturing error (A) is the cause.

The emulsion results can be interpreted either as manufacturing error (about 3% underdosing (D)), incomplete extraction (E), or inhomogeneity (F). Possibly, there is a trend towards an increasingly positive signal (G).

Possible Explanations/Courses of Action

- Interpretations A and D can easily be checked by taking additional samples from later production runs. If the means seen for many production lots tend to concentrate near 100%, then the emulsion results seen here would just be chance result on the low side; if the means concentrate near some other number, then a formulation error, a manufacturing loss, or sample work-up could be to blame.

- Incomplete extraction (E) would always result in low values.

- Product inhomogeneity (F) for replicate samples pulled from the same production lot would show up in sample means that scatter much more than the method repeatability (=A=).

- Sample solution instability or incomplete extraction/separation would show up if several aliquots from the same sample work-up were put in a series of vials that would be run in sequence that would cover at least the duration of the longest sequence that could be accommodated on the autosample/instrument configuration. For example, if an individual chromatogram is acquired for 5.5 minutes, postrun reequilibration and injection take another 2.75 minutes, and 10 repeat injections are performed for each sample vial in the autosampler, then at least $15 \cdot 60/(5.5 + 2.75)/10 = 11$ vials would have to be prepared for a 5 P.M. to 8 A.M. (= 15 hour) overnight run. If there is any appreciable trend, then the method will have to be modified or the allowable standing time limited.

Conclusions Concerning an Investigation   The effective standing time of a given sample, its position in the sequence of samples, and many other details are normally not reported in the manufacturing batch records and have to be dug out of the analytical raw data. Because it is the thick signed-off reports that the data analyst in the statistics department tends to see first—he might need help in understanding the method- and instrument-specific notebook entries, if he ever learns of them—many investigations come to improper conclusions. It is absolutely essential that the investigators are thoroughly familiar with the laboratory, method, and product particulars.

The raw data should be accessible in summary sheets; the semi-graphical format is a big help because it exposes the problematic cases and provides a pointer to the notebook entries.

4.29   NOT PERFECT, BUT WORKABLE

A pharmaceutical product contains two active principles, A and B, in three fixed combinations. In order to validate the HPLC method, five calibration solutions are made up that contain compounds A and B. The specific concentrations for A are 0.006, 0.012, 0.018, 0.024, and 0.03 mg/ml (LO, respectively the 10- and the 100-fold thereof (MID, HI), and those for B are 0.05, 0.1, 0.15, 0.2, and 0.25 mg/ml.

Calibration   Each of the solutions is injected once and a linear regression is calculated for the five equidistant points, yielding, for example, $Y = -0.00064 + 1.004 \cdot X$, $r^2 = 0.9999$. Under the assumption that the software did not truncate the result, an r^2 of this size implies a residual standard deviation of better than 0.0001 ($\approx 0.5\%$ CV in the middle of the LO range; use program SIMCAL to confirm this statement!); the calibration results are not shown in Fig. 4.39.

Repeatability   Fiveteen placebo tablets are dissolved in water and spiked with the appropriate amount of a stock solution LO and/or HI so as to obtain the same concentrations of A and B as for the calibration solutions. Aliquots of each of these solutions are injected three times, for a total of 45 results of A and 45 for B.

Content Uniformity   Ten tablets per formulation are worked up one by one and aliquots of the so obtained solutions are injected.

An internal standard is used throughout.

In Fig. 4.39, results for spiked placebo and for the verum tablets are given for compound A (bold lines) and B; all horizontal bars should be at 100%, and the vertical lines should be centered at the same height. The gray trendlines, particularly for the LO- and HI-range A-values indicate a systematic difference in response between the calibration solutions and the spiked placebo tablets (extraction efficiency, interference, etc.). For same ranges, the verum-tablets assays either underestimate the content of A by 4–5%, or A is underdosed. For compound A the repeatability figures are as follows (%-of-nominal, see file Fig4_39.dat), see Table 4.36.

Concentration c_{bc} [%-nom]

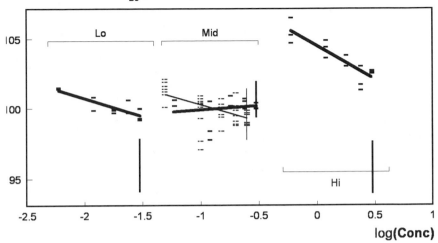

Figure 4.39. Variability of back calculated concentrations $Conc_{bc}$. For each concentration range five calibration points were measured, over which a separate regression was run (not shown). Placebo tablets were spiked to the same concentrations and measured in triplicate (short horizontal lines; gray trend lines in background). Ten repeat determinations of actual product (vertical bars = Mean ± SD) were done. The bold lines pertain to compound A in all concentration ranges, the thin lines to compound B (middle concentration range only).

When the group means (e.g., 101.4, column 3) are subtracted, the residuals in column 4 result, the standard deviation over which is 0.33 ($f = 14$). The standard deviation within the individual groups of "LO" measurements is 0.52 or lower (number in parentheses); when the squares of these numbers are multiplied by 2 ($f = 2$ each) and summed, an overall variance of 1.52 is found, the same as when 0.33 is squared and multiplied by 14. The difference between the numbers 0.33 and 0.76 is due to the variability of the group means. When the same treatment is applied to the MID and the HI groups, the standard deviations in the next two lines are found. That the difference for the HI group is particularly large is no surprise: The means drop over the concentration range. The regularity of the pattern is an indication that a systematic effect is at work. Because the corresponding pattern for compound B falls out of line, too, an explanation involving a calibration error is close at hand. The common stock solution or one of the dilution steps performed before the spiking of the placebo matrix is a conceivable explanation, but since the patterns for compounds A (down) and B (up or level) point in different directions, that theory falls flat. Another explanation is that Compound A in the formulation HI is at a concentration that is 10–100

Table 4.36. Excerpt of Triplicate Determinations of Spiked Placebo

Concentration	Assay	Mean (Std.Dev.)	Within-Group Residuals
0.006	101.4	101.4	−0.03
	101.5	(0.06)	0.07
	101.4		−0.03
0.012	99.9	100.2	−0.30
	99.9	(0.52)	−0.30
	100.8		0.60
0.018	99.7	99.9	−0.17
	99.9	(0.15)	0.03
	100.0		0.13
0.024	100.5	100.0	0.60
	99.7	(0.52)	−0.30
	99.7		−0.30
0.030	99.2	99.5	−0.30
	99.3	(0.44)	−0.20
	100.0		−0.50
Std. Dev. LO		(0.4)	0.33
Std. Dev. MED		(0.7)	0.60
Std. Dev. HI		(0.7)	0.57
Std. Dev. A		(1.93)	0.50

times higher than in the other two formulations, and so could contribute to a competitive interaction of A and B with the matrix, i.e., in adsorption. The question could only be answered through further experimental work, but this was not done because the observed irregularities are far within the required specification limits.

The average within-group repeatability of ±0.5% most likely describes detector noise that leads to misassignments of the integration endpoints and peak-area variability, and can be considered acceptable for low-dose products.

Example 60: If compound samples that were actually composed of five individual tablets had been analyzed instead of the spiked matrix, the CV would be expected to be larger than ±0.5% on account of the additional manufacturing error, but by a factor $\sqrt{5} = 2.2$ lower than the content uniformity CV. (Cf. Eq. (1.5).) Since the average CV for CU was found to be ≈1.76% (±1.97, 1.28, resp. 1.95%), this would have to be in the region of about $1.76/2.2 ≈ ±0.8$, which is still well within the range of accepted instrumental noise.

The following conclusions are drawn:

- The repeatability is limited by instrumental noise.
- The content uniformity is within the accepted range (1.93% is less than 6% CV; maximum one out of ten values is outside the 85–115% range, but within the 75–125% range; no value below 75 or above 125% of label claim.
- As long as the specification limits remain at 90–110%, there is no reason to improve the method at high concentrations of A; this could be done with a view towards a future tightening to 95–105%, but would not at the moment impact the regulatory position or improve the medical risk, however.

Note that this case study was calculated on the basis of an old report in which all assay values were rounded to the 0.1% position; if the raw data had still been accessible, the conclusions would probably have remained the same, but some specific numbers could have changed. This situation is very common if data trends over several years are investigated. It is not unusual that raw data from routine production QC release tests are destroyed a year or two after the expiration date of the product because local laws do not require longer retention.

4.30   COMPLACENT CONTROL

A tablet containing two drug compounds, A and B, is being scaled up from kilogram to half-ton batches in preparation for a regulatory submission. The applicable specifications and sample work-up methods are

- Assay: 90–110% of label claim (pick 20 tablets at random; grind and mix; weigh an amount of powder corresponding to five tablets; dissolve compound sample and centrifuge excipients; run a HPLC analysis on an aliquot of the supernatant), and
- Content uniformity: Nine out of 10 randomly picked and individually analyzed tablets must yield drug contents between 85 and 115% of label claim; all 10 results must be within 75–125% of label claim.

It is to be expected that the compound sample gives a well-defined mean drug content due to the large sample size that averages out tablet-to-tablet inhomogeneities and also because the large amount of powder taken through the work-up reduces any effects that losses might have.

CU ASSAYS

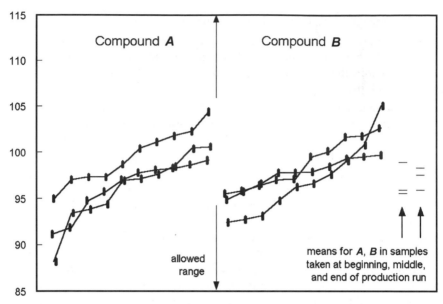

Figure 4.40. Content uniformity samples (10 each) taken from the beginning, middle, and end of a production run and each assayed for the compounds *A* and *B*. The results were sorted according to size; both the slopes and the averages are very similar and well within the allowed range.

The content uniformity results for compounds *A* and *B* are shown in Fig. 4.40: the three data sets from the beginning, middle, and end of the manufacturing run were individually ordered by size and plotted. The results are as follows (use program MULTI and data file CU_Assay1.dat; see Tables 4.37 and 4.38):

Example 61: The raw data, given as %-of-nominal values with one decimal place, are found in Table 4.37: For each group of 10 values the mean and the standard deviation were calculated. Using these, the *t*-values for the differences $|L - \text{mean}|$, with $L = 75, 85, 115$, resp. 125% were determined; they are all above 2.9, indicating low risk. The corresponding CP-values were calculated; the differences $\Delta CP_{75\text{-}85}$ and $\Delta CP_{115\text{-}125}$ were added and multiplied by 100 to obtain the approximate risk, in %, of finding a result between the inner and the outer limits. For a content uniformity test with $n = 10$ tablets, a risk of 0.003872% translates into a deviant result once every 20–25 trials, or, with six CU runs per batch, every third or fourth batch.

Table 4.37. Content Uniformity of Tablets and Determination of Risk of Noncompliance

Item	Compound A			Compound B			Comp. A	Comp. B
$x(i)$	91.1 91.8 94.7 95.7 96.9 97.7 98.0 98.2 98.6 99.1	94.9 96.9 97.2 97.2 98.6 100.3 101.0 101.7 102.2 104.3	88.1 93.3 93.7 94.3 96.8 97.0 97.5 98.4 100.3 100.5	95.5 95.8 96.4 96.9 97.0 99.5 100.0 101.6 101.7 102.6	94.8 95.7 96.6 97.7 97.7 97.8 98.4 99.3 99.5 99.7	92.4 92.7 93.1 94.7 96.2 96.6 97.5 99.0 100.7 105.0		
mean	96.18	99.43	95.99	98.70	97.72	96.79	97.20	97.74
SD	2.83	2.93	3.75	2.69	1.63	3.98	3.48	2.93
n	10	10	10	10	10	10	30	30
$t(75)$	7.487	8.335	5.604	8.818	13.938	5.477	6.387	7.750
$t(85)$	3.952	4.923	2.934	5.097	7.803	2.964	3.510	4.341
$t(115)$	6.653	5.312	5.075	6.064	10.600	4.577	5.121	5.884
$t(125)$	10.188	8.724	7.745	9.785	16.735	7.091	7.998	9.292
$CP(75)$	1.000000	1.000000	1.000000	1.000000	1.000000	1.000000	1.000000	1.000000
$CP(85)$	0.999961	1.000000	0.998327	1.000000	1.000000	0.998479	0.999776	0.999993
$CP(115)$	1.000000	1.000000	1.000000	1.000000	1.000000	0.999998	1.000000	1.000000
$CP(125)$	1.000000	1.000000	1.000000	1.000000	1.000000	1.000000	1.000000	1.000000
ΔCP	0.000039	0.000000	0.001673	0.000000	0.000000	0.001521	0.000224	0.000007
ΔCP	0.000000	0.000000	0.000000	0.000000	0.000000	0.000002	0.000000	0.000000
Risk [%]	0.003872	0.000048	0.167299	0.000017	0.000000	0.152320	0.022437	0.000709
Compound assay	95.91	98.95	95.56	98.31	97.54	95.91		

Table 4.38. Effect of Raw Data Rounding on Bartlett and ANOVA Tests

Compound A			Compound B				Raw Data
96.23	99.50	96.07	98.75	97.76	96.83	Mean	CU, full precision
2.83	2.93	3.75	2.69	1.62	3.99	SD	
96.18	99.43	95.99	98.70	97.72	96.79	Mean	CU, rounded to 1
2.83	2.93	3.75	2.69	1.63	3.98	SD	decimal
96.24	99.50	96.08	98.74	97.76	96.82	Mean	Compound assay,
2.83	2.92	3.74	2.69	1.61	3.99	SD	full precision

The standard deviations are not distinguishable (Bartlett test). Conclusions are valid for all three data sets. All means belong to the same population (ANOVA test). Overall result: 97.5 ± 3.2 (compound assay).

The six data sets do not differ in variance (Bartlett test) or in means (ANOVA), so there is no way to group them using the multiple range test. This being so, the data were pooled for compounds *A* and *B*, yielding the two columns at right (data in CU-Assay2.dat).

The risk of noncompliance is seen to be negligible. The last row gives the results for the compound assays for comparison.

Table 4.38 repeats the Mean/SD rows of the previous table and compares these results to those for full-precision content uniformity and the corresponding compound sample assays; because the raw data were normalized to %-of-nominal, many more decimal places were available than are quoted in Table 4.37; this shows that the rounding procedure must be carefully considered, and a reduction of significant digits does exact a price. Fortunately, the conclusions that are drawn do not differ. This problem of slight differences in the last digit has a very practical aspect, though. Under GMPs, a basic requirement is that the data going into submission documents must be peer-reviewed for transcription and calculational errors. This is usually interpreted to mean a peer witnesses the data acquisition and the supervisor reviews the calculations (standard practice in pharmaceutical labs), and on top of this a QA person checks everything in the submission documents against the raw data on a *dotted Is & crossed Ts* level. The willingness to bear responsibility sometimes drops to zero and a report quoting some number as "98.7%" is rejected on the grounds that the laboratory note book says "98.696%." Because not all software packages, calculators, or instruments allow the number of decimal places to be freely chosen, changing one component in the equipment train could mean all relevant SOPs have to be adapted if one does not want to risk having to come up with long justifications for mathematical trivialities.

4.31 SPRING CLEANING

GMP regulations call for an annual product review; essentially, the intention is to detect irregularities and emerging trends so the process can be fixed before the product drops out of specifications. This is a common-sense approach considering the large number of things that can go wrong with multistep processes involving dozens of raw materials, elaborate equipment trains, and a multitude of operators, supervisors, and managers. Such a complicated manufacturing process is robust and delivers a product with a reproducible quality if and when each element conforms to a given set of criteria such as impurity profile, crystal size distribution, stabilizers, colorants (chemicals), machine settings, wear, vibration, replacement parts, temperature, humidity (equipment, infrastructure), calibration status (analytical instruments), and skills (operators).

Over the life cycle of a product, many changes take place, some of which are known and controllable (e.g., upgrading an analytical method, installing a new mixing vessel), and some that can only be suspected (e.g., a supplier changes his process or modifies the design: The small print says something along the lines "the manufacturer retains the right to improve the product without notification"). In both cases, the problem boils down to complexity that is only incompletely described by a specification sheet. A rough calculation illustrates the point.

Example 62: If a manufacturing process involves two raw materials, each defined by three sets of specification limits, and four pieces of equipment with one control knob, then a complete validation protocol would ask for (Three settings: Lo, Target, Hi)^(2 materials · 3 specs + 4 machines · 1 control) = $3^{10} \approx 59\ 000$ experiments, even without repetitions. This product would never reach the market if one did not employ experience and scientific rationale to simplify development by testing only the presumed critical issues, say a total of three specification points for $3^3 = 27$ experiments.

Since both the manufacturer and his suppliers are forced to work by the same strategy, each company's raw materials QC and finished product QC laboratories test only for the most obvious characteristics. An undetected change in a raw material's properties might change a product characteristic that could also go unnoticed until much later.

The gradual accumulation of many such changes is likely to show up in the quality of the product if enough numbers are compared over a sufficiently large number of production runs. Anyone undertaking such a review must keep an eye out for the effects of human frailty and "mission creep." Accepting a list of numbers at face value can be dangerous. Feeding such a list through some fancy statistics packages without doing a lot of filtering beforehand is careless at best.

The Excel-file PRODUCT_BFG.xls contains an excerpt from a large spreadsheet put together over the course of 2 years by the chemists responsible for a particular synthesis step. A small portion is reproduced in Tables 4.39 and 4.40. This took care of the first step of the review process, namely finding the relevant items in the manufacturing batch records and writing them into the spread sheet. Incidently, the trend towards a paperless shop floor will not necessarily help: Transcription errors will be down, but the items that are acquired on-line, such as temperature-, pressure-, and r.p.m.-readings will be so much more easily available that the sheer quantity of these apparently relevant and high-quality data will keep people from doing what they are best at: Casting a critical eye and seeing connections.

PRODUCT_BFG.xls consists of the chemist's table, a reduced table obtained by casting out data that is not ammenable to statistical analysis such as subjective assessments ("off-white color, characteristic crystal form, pungent smell"; STEP#1), and the final table that was freed of all inconsistencies by going back to the original data (STEP#2). Various interesting items are highlighted:

- The items and the appropriate dimensions and specifications are given in the first three columns.
- A dimension "txt" indicates a subjective description, something that obviously cannot be analyzed by statistical means, particularly if two individuals use differing terminology (batches 12 and 13). Enforcing the use of pre-defined terminology, e.g., by way of pick-lists at the data-entry terminal, would have prevented this particular mistake from happening but would potentially have provoked a loss of information right up front: "If it's red/blue/green they want, let them have that; something could be wrong because I see pink, but I'll put that in the 'R'-box."
- Identity tests have binary character and are completely irrelevant from the chemometric point of view: Only gross failure would yield a product that would differ from the reference.
- In the two rows for TLC, the legends "total impurities" and "single impurity" are reversed.
- The TLC method is good enough to in all cases determine that the impurity in question is below 0.5%, the criterion for product release. However, the data cannot be used in a quantitative statistical analysis because several levels of LOQ are involved: from "<0.5%" down to "0.05%." All entries would have to be re-coded as "<.5%," with the total loss of information, because otherwise distinctions would be introduced where

Table 4.39. Excerpt from Trend Analysis Table for Product BFG^a

		Specification	BFG.8	BFG.9	BFG.10	BFG.11	BFG.12
Lot no.			BFG.8	BFG.9	BFG.10	BFG.11	BFG.12
Batch no.			18	19	20	21	22
Manufact. date			Okt 96	Okt 96	Dez 96	Dez 96	Dez 96
Input in kg			1226,0	305,8	1204,3	1462,5	305,2
Amount in kg			1224,2	308	1200,0	1466,0	301,4
Yield in %			99,9%	99,4%	99,6%	100,2%	98,8%
Remarks							
Appearance		White to almost white, cryst. substance	Almost white, cryst. substance	Almost white, cryst. substance	Almost white, cryst. substance	idem	Almost white cryst. substance
Odor		Odorless or almost odorless	Almost odorless	Almost odorless	Almost odorless	idem	Almost odorless
Identity	by UV	Conforms to ref. spec.	Conforms	Conforms	OK	Conforms	Conforms to spec.
	Basic org.	Positive	Positive	Positive	Positive	Positive	Positive
	Sulfate	Positive	Positive	Positive	Positive	Positive	Positive
Loss on drying		0.00–2.00%	0,05	0,08	0,01	0,01	0,01
TLC purity	Single impurity	NMT .5%	<1.0	<1.0	<1.0	<1.0	<1.0
	Total impurities	NMT 1%	<0.5	<0.5	<0.5	<0.5	<0.5
Assay		98.0–100.5%	99,7	99,6	100,1	100,2	100,0
Sieve analysis	>250 μm	For information	—	—	—	—	—
	>150 μm	NMT 10%	1,5	0,6	2,5	0,8	3,9
	<106 μm	For information	—	—	—	—	—
	<75 μm	For information	26,6	12,4	25,2	18,5	18,8
	>45 μm	For information	95,6	91,1	97,1	96,2	88,4
Tapped volume		0.600–1.100 g/mL	0,717	0,784	0,755	0,748	0,815

Table 4.39. (*Continued*)

		BFG_8	BFG_9	BFG_10	BFG_11	BFG_12
Lot no.						
Batch no.		18	19	20	21	22
Manufact. date		Okt 96	Okt 96	Dez 96	Dez 96	Dez 96
Input in kg		1226,0	309,8	1204,3	1462,5	305,2
Amount in kg		1224,2	308	1200,0	1466,0	301,4
Yield in %		99,9%	99,4%	99,6%	100,2%	98,8%
Remarks						
pH (1%, aqueous)	3.5–5.5	4,4	4,5	4,4	4,5	4,4
Clearness and color of soln.	Not more than slightly turbid and not more than slightly yellow sol.	Clear and slightly yellow	Clear and slightly yellow	Clear and slightly yellow	Clear and slightly yellow	Clear and almost colorless
Heavy metals	NMT 20 ppm	<20	<20	<20	<20	<20
Residue on ignition	0.00–0.20%	0,01	0,02	0,09	0,02	0,04
Chloride in ppm	NMT 350 ppm	<350	<350	<350	<350	<350
HPLC purity [area-%] Main comp.	NLT 99.0%	99,61	99,81	99,8	99,80	99,81
Single impurity	NMT 0.5%	$\le$ 0.23	$\le$ 0.11	$\le$ 0.11	$\le$ 0.10	$\le$ 0.10
Total impurities	NMT 0.1%	n.d.	n.d.	n.d.	n.d.	n.d.
DSC melting point	No peak max. $\le$240°C	No peak maximum	No peak maximum	No peak maximum	None	No peak maximum
Residual solvents Methanol	NMT 0.2%	0,10	0,08	0,09	0,10	0,09
Ethanol	NMT 0.5%	0,02	0,02	0,02	0,02	0,03
Acetone	NMT 0.2%	<0.02	<0.02	<0.02	<0.02	<0.02
Toluene	NMT 0.2%	—	—	<0.01	<0.01	<0.01
Ethyl acetate	NMT 0.2%	—	—	<0.01	<0.01	<0.01

[a]NMT: not more than; NLT: not less than.

Table 4.40. Excerpt from Cleaned-up Trend Analysis Table for Product BFG

Lot No.				BFG_8 1	BFG_9 2	BFG_10 3	BFG_11 4
Input in kg	Size			1226.0	309.8	1204.3	1462.5
Yield	Yield			0.999	0.994	0.996	1.002
Loss on drying	LOD		0.00–2.00%	0.05	0.08	0.01	0.01
TLC purity	TLC1	Single impurity	NMT .5%	1.0	1.0	1.0	1.0
	TLC2	Total impurities	NMT 1%	0.5	0.5	0.5	0.5
Assay	ASSAY		98.0–100.5%	99.7	99.6	100.1	100.2
Sieve analysis	Sieve_150	>150 μm	NMT 10%	1.5	0.6	2.5	0.8
	Sieve_75	<75 μm	For information	26.6	12.4	25.2	18.5
Tapped volume	T-Vol		0.600–1.100 g/mL	0.717	0.784	0.755	0.748
pH (1%, aqueous)	pH		3.5–5.5	4.4	4.5	4.4	4.5
Residue on ignition	ROI		0.00–0.20%	0.01	0.02	0.09	0.02
HPLC purity	HPLC Pur	Main comp.	NLT 99.0%	99.61	99.81	99.80	99.80
[area-%]	HPLC1	Single impurity	NMT 0.5%	0.23	0.11	0.11	0.10
	HPLCtot	Total impurities	NMT 0.1%	0.01	0.01	0.01	0.01
Residual solvents	MeOH	Methanol	NMT 0.2%	0.10	0.08	0.09	0.10
	EtOH	Ethanol	NMT 0.5%	0.02	0.02	0.02	0.02

none existed (does "<0.5%" mean "0.1 ... 0.5%," or "<0.1%" next to an entry "<0.1%"?).

- There are errors due to Excel's automatic time-format: The "4. Jul" and the "35889" (Julian date) in batches #11 and #18 can be traced to the use of both the Continental ("4,7", sieve analysis) and the corresponding Anglo-American notation ("4.7") in the same table (many people get so used to decimal points that they switch back and forth between commas and points). Nothing happens as long as the entries remain in text mode; when conversions to specific formats are necessary in order to do calculations, auto-formatting takes over and in this case decided to interpret one notation as a number, the other as a date. In international companies, where a lot of data gets pushed around, one way to circumvent renewed data-entry is to export the text-version to a word-processor program, edit commas to points, or vice versa, and reimport. If column-separation is achieved by inserting commas or semicolons ("3.45, 6.87, ..." vs. "3,45; 6,87; ..."; e.g. *.csv files), one has to be careful not to completely garble the message.

- Three different batch sizes were used, namely ≈400, ≈800, respectively ≈1200 kg; in the chemical industry it is standard practice to validate a process for the available kettles and then adjust the particular batch size (minimum filling so the stirrer bar remains covered, maximum filling so no overpressure occurs and vents stay free) to achieve the requested tonnage; often, campaigns of batches are run before the equipment is cleaned and refitted for the next product, see batch numbers #19 ... 23, followed by #1, ...

- The input to batch #23 contains an unwarranted decimal point, which causes the yield-figure to go way beyond 100%.

- The sieve analysis notation changed at some point in time, see batches #1, #23 and #24 in columns N–P, lines #19 and 21: The ">45 μm" value shifts from ≈92 to ≈7%. The customer had requested an additional specification "<106 μm", which explains the row of dashes in line 19 up to column N; an appropriate footnote should have been inserted. In connection with this, the specification in line 21 was changed from ">45" to "<45"; correspondingly, all values had to be converted by hand, e.g., 93 = 100 − 7. (See table Step #1.)

- In table Step #1 it becomes apparent that 8 rows of data contain no information or information that cannot be interpreted (TLC, heavy metals, chloride, residual solvents C, D, and E) and 4 where further clarification is needed (Sieve analysis (as above), HPLC impurities (changing LOQ?)).

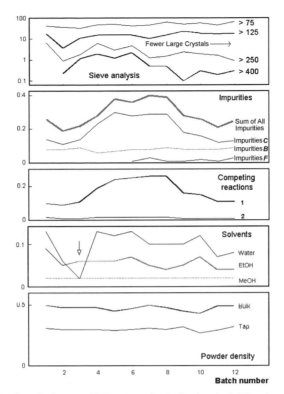

Figure 4.41. Trend analysis over 12 batches of a bulk chemical. The sieve analysis shows that over time crystals larger than 250 μm were reduced from a weight contribution in the range of a few percent of the total to about 1% in favor of smaller sizes. Impurity *C* appears to follow the trend given by the lead compound for the competing side reaction #1. The very low moisture found for sample #3 could be due to a laboratory error because during drying one would expect ethanol to be driven off before water. Methanol is always below the detection limit.

- The resulting table Step #2 retains less than half of the original information.

Fig. 4.41 (*n* = 12 batches) and 4.42 (*n* = 46 batches) depict what can and what cannot be gleaned from a detailed study of such cleaned-up tables: Unless a connection is fairly obvious, such as between impurities *B* and *C* in Fig. 4.42, comparisons can resemble small-talk about the weather, in that there is always a subset of data to prove a given pet notion, and another to disprove it. Why? Over the course of a few months a series of insignificant "changes" in raw materials (a new batch), equipment (higher-capacity steam

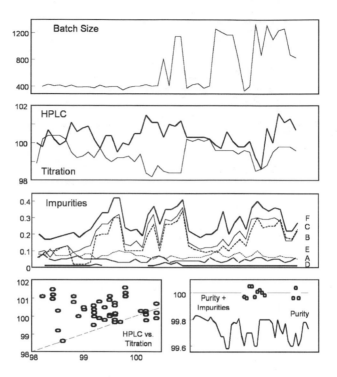

Figure 4.42. Trend analysis over 46 batches of a bulk chemical produced according to the same manufacturing procedure: Small and scaled-up batch size [kg], HPLC and Titration assays [%], resp. individual HPLC impurity levels [%], *versus* batch number. The lack of full correlation between assays indicates that the titration is insensitive to some impurities detected by HPLC. The mass balance, where available, suggests that all relevant impurities are quantified. Impurities B and C, for instance, are highly correlated ($r^2 = 0.884$, $p = 0.0002$).

line), process (slightly faster temperature ramp), personnel (switch from early to late shift), or analytical method (new HPLC column oven) can sufficiently modify the situation so that batches can only be compared within groups ("campaigns"). The more elements that a system (all of the above and more) encompasses, the harder it becomes to keep conditions truly constant over any length of time. (See also Section 4.38.)

The legalistic notion that only validated processes are to be used assumes that the chain of events from raw materials to analysis of the final material can be validated *in globo*, something that is patently impossible with the given number of adjustable parameters, not to mention unforeseen glitches. Doing the validations in bits and pieces (modules "process," "sampling," "analysis," "data evaluation," etc.) certainly helps, but does not cover the

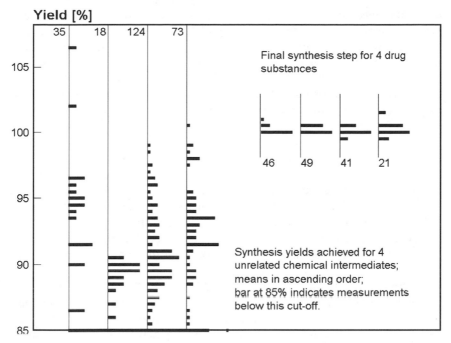

Figure 4.43. Yields determined for a large number of batches of four bulk chemicals and four drug substances. During the early synthesis steps the yields are variable; 10 batches run during one campain of the first compound clump near 95%. The result at about 107% is due to analytical variability and/or calibration bias. The final synthesis step of a drug compound, for which the aggregate value of a kilogram of material is much higher, has been trimmed for maximal yield.

interaction between the modules. The validation philosophies practiced today (FDA: Run three batches at target conditions; European health authorities: Also consider Hi/Lo variation of the most important parameters) more or less cover the ideal process, but by necessity require a reassessment if anything changes. The distinction between "minor" and "major changes" allows for some scientific reasoning about the probable outcome without subjecting a company to million-$ revalidation campaigns every time a detail changes. By necessity, this implies surprises.

Because of the relatively small number of experiments done on commercial-scale equipment before submission, and the often very narrow factor ranges (Hi/Lo might differ by only 5–10%), if conditions are not truly under control, high-level models (multi-variate regressions, principal components analysis, etc.) will pick up spurious signals due to "noise" and unrecognized drift. For example, Fig. 4.43 summarizes the yields achieved for cheap intermediates at the beginning of (unrelated) complex syntheses, and very expensive final products. The variability in the former reflects yield as such (except in the two results above 100%), whereas in the latter it is mainly a question of analytical precision. Effort to improve a chain of synthesis steps is channeled into the areas that add the most to the overall yield, waste management, or costs; a comparatively cheap step with a 90% yield may have to wait until resources become available for improvements.

4.32   IT'S ALL A QUESTION OF PEDIGREE

In all of the above discussions the existence of a calibration standard was taken for granted. It was assumed that such a calibration standard is extremely pure and is available in such quantities that quality and/or supplies could not lead to a bottleneck. The analyst could do his experiments in a reasonably short time and mostly use one and the same calibration solution throughout. The exceptions to this rosy situation were barely hinted at in Sections 1.1.4, 1.8.4, 4.7, 4.20, and 4.23, namely that in batch release, in stability, and in validation work individual results would have to be compared over a long time frame or across large distances. (The time factor can take its toll within minutes or hours if instrumental drift is a problem[188]; flame photometers as used for sodium or potassium determination are a real pain in this respect.) For all relative methods (titration, chromatography, non-MS spectroscopy, etc.), this implies comparison against a reference. In this section, the device the pharmaceutical industry resorts to, the calibration-standard hierarchy, is described. In contrast to some high-quality physical standards (e.g., the atomic clock for frequency, time, and (short) distances, or single-isotopic materials for mass), chemical standards, especially of the organic variety, are not so stable as to be beyond doubt. Some of their more unpleasant characteristics are as follows:

- Affinity for water or other solvents, leading to a change in weight.
- Partitioning between matrix compartiments and/or container walls.[7,10,11]
- Recrystallization to a thermodynamically more stable form.
- Susceptibility to hydrolysis and/or oxidation.[9–11]
- An impurity profile that depends on precise experimental conditions.
- Impurities that are hard to detect and/or separate from the main compound, etc.

If the chemical moiety in question is a commodity or has been on the market for a long time, it is likely that there is a source of high-quality and

Table 4.41. Calibration Hierarchy[a]

Quality Level	Highly Purified PRIMARY STANDARD	Purified SECONDARY STANDARD	Purified WORKING STANDARD
Price	$15000/g	$2000/g	$100/g
Available quantities	mg-to-grams	grams	kilograms
Shelf-Life	5-years @ $-20°C$	12-months @ 5°C	1-month @ RT
Purpose	WWLT calibration	Within-lab calibration	analysis

[a]RT: ambient; WWLT: worldwide or laboratory-to-laboratory long-term standard. Requalification of standards: against predefined criteria using validated methods at stated intervals.

fully documented material, such as the USP-standards. For all proprietary chemicals, such as drug substances under development, the owner will have to come up with his own standards. These will be at least of two types, namely pure compounds (main compound and all known process and degradation impurities, and possibly metabolic derivatives) and "limit" samples (main compound spiked with such amounts of controlled impurities that all impurity specifications are just barely met).

A hierarchy of standards will be established, if only for economical reasons, see Table 4.41; the quoted prices are for ordinary organic chemicals with a molecular weight of up to a few hundred and are intended as a rough guide only. The multi-stage cleaning process (e.g., recrystallization + prep-scale HPLC) needed to obtain samples that are so clean that full characterization and documentation is justified is so expensive and works only for relatively small quantities that it is usually out of the question to supply such material to every involved laboratory in amounts sufficient for day-to-day calibration. The situation depicted in Fig. 4.44 is that where three regional chemical plants produce a bulk chemical according to one and the same process, and each purifies a sample of its own material, and submits it to the central laboratory for cross-validation against the top-level standard. These secondary standards are a lot less expensive and available in much larger amounts than the primary standard. The reason for using regional samples is that despite the nominal identity of the manufacturing processes, it is quite normal that the impurity "fingerprint" betrays local particulars. The secondary standards are then used to create local working standards that are actually issued to the routine control labs. The assigned expiry dates are derived from stability studies run on standard on one level against standard at the next higher level. During an actual analysis of a production control

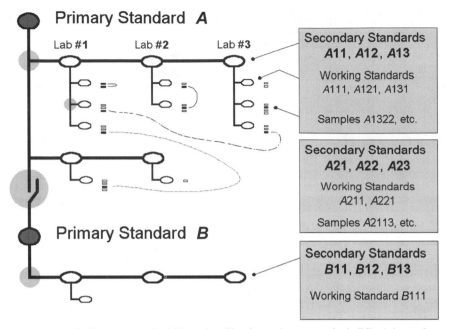

Figure 4.44. Calibration standard hierarchy. The first primary standard (PS) *A* is used to calibrate local secondary standards (SS), which themselves serve to calibrate working standards (WS). The three levels of standards and the analysis runs are linked through a code that includes the PS-generation (letters *A*, *B*, etc.), the SS-generation (first number), the site (second number), the WS-generation (third number), and the analysis run (fourth number); each analysis run includes a calibration and up to a few dozen analyzed samples. Some of the links that are established through cross validation are indicated by light gray disks. Four of the six possible levels of relationship are indicated by arcs: (1, not shown) within calibration, (2) within WS, (3) WS to WS, (4, dashed) SS to SS within the same generation, (5, dotted) SS to SS between generations, and (6, not shown) PS to PS generation.

sample, the lot number of the working standard will be noted, and its quality is traceable to the primary standard. After a certain time, even the primary standard will have expired or have been used up. The next lot of primary standard needs to be ready for cross-validation against the previous one. The provision of standards has to be planned so that the lineage covers several sites for the lifetime of the product (20 or even more years). Even so, quality assurance will have to perform direct comparisons between the working standards of two or more sites should any discrepancies turn up, such as a release at 100.8% of label claim in Factory B followed a few months later by a complaint from the local affiliate in country Z to the effect that the shipment would have to be returned because 106.3% exceeded both the local limit (105%) and the allowed discrepancy relative to the certificate of

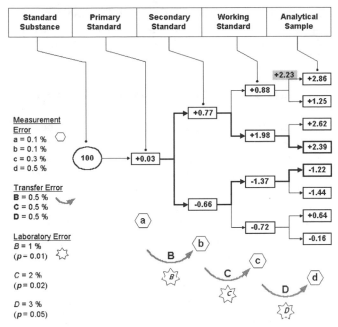

Figure 4.45. Divergence of results between related calibration runs. Simulation program PEDIGREE.xls simulates the effect of cumulating multiple within-run and between-laboratory analytical uncertainties and the occasional systematic laboratory error. The reference laboratory is expected to work with a precision of 0.1%; the cross-calibration from primary to secondary standard suffers from a transfer error (between-run or between-instrument error) of 0.5% and, on top of that, a measurement error of 0.2%. The further one moves to the right, the larger the errors; a laboratory error of typically 3% could occur with a probability of p = 0.05 in the last step and is observed in one case: The bias is +2.23%. The numbers in the right column indicate what would be found if the primary standard were tested twice against each of the four working standards. The two highlighted results are coupled through six links.

analysis (say, 1%). Most likely, the reason for the artifact is a degraded standard in affiliate Z's lab, lack of instrument maintenance, or poor work practices.

File PEDIGREE.xls illustrates how a number of determinations on a sample, all tied together by a hierarchy of working-, secondary-, and primary standards, diverge if some simple assumptions about achieved within- and between-group precision are made. For good measure, it is possible to assign the risk of making a laboratory error and the typical size of it. A sample output is given in Fig. 4.45.

Overall, the story shows that a manufacturer must go to extreme lengths to establish a satisfactory QA-network; without the effort, statistical evaluations would be worthless and futile, because one would be comparing "apples and oranges" despite the same trade name on the packages. Only with a fully operational QA-system in place can discrepancies be resolved.

4.33 NEW TECHNOLOGY RATTLES OLD DREAMS

A minor constituent in an old product had traditionally been determined using UV-spectroscopy after dilution or extraction. While the compound (drug substance, DS) was known to be susceptible to hydrolysis, the wide specification limits (90–110% of label claim, partially justified by the large analytical uncertainty at the low concentration) assuaged all worries, even in the face of occasional discrepancies between what was found upon analysis and what had gone into the product. By a fairly common sleight-of-hand that sailed under the "manufacturing losses" claim, namely adding a 3, 5, or even 10% overage to the manufacturing formula, Quality Control would find values above 100% at release ($t = 0$; see arrow in right panel in Fig. 4.46), and below 100% towards the end of the official shelf-life ($t = 36$ months). Formally, all regulations have to be followed if technically feasible, though medically the loss of active drug was not significant because in everyday life a cream is not dosed very accurately. The hydrolysis (= degradation) products H, H′, and H″, fortunately, did not pose a toxicological risk:

$$DS \quad \rightarrow \quad H \quad \rightarrow \quad H' \quad \rightarrow \quad H'' \quad \text{Degradation pathway}$$
$$+ \qquad\quad (+) \qquad (-) \qquad\quad (-) \quad \text{UV-activity}$$

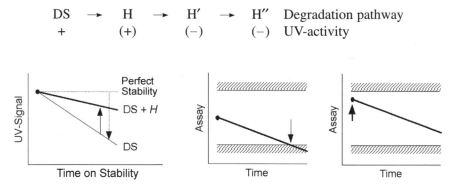

Figure 4.46. Insufficient specificity in UV measurements and the concept of manufacturing overage. The left panel shows what happens when a UV-active compound DS degrades over time to compound *H*, which has a slightly lower molar extinction coefficient ϵ. Instead of observing the decrease to the DS line (down arrow ↓), *H* contributes (up arrow ↑) sufficiently to the total signal to make the product appear relatively stable (line DS + *H*). When the rapid decomposition of DS is recognized, one option used to be to tacitly add an overage to the manufacturing formula (↑ in right panel) to shift the intersection of the trend line with the SL_L (↓ in middle panel) towards higher shelf-life.

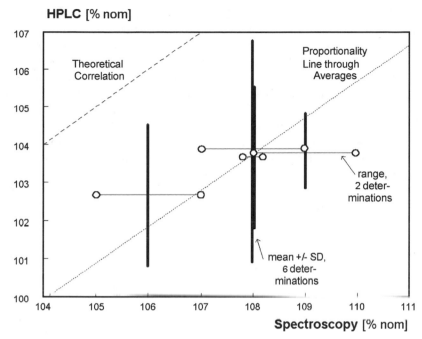

Figure 4.47. Drug assay using HPLC respectively UV Spectroscopy. Correlation of HPLC and UV results obtained on four batches of a cream. The vertical error bars each give the mean ± standard deviation of 6 HPLC determinations; because the Student's t-factor for five d.f. is nearly equal to $\sqrt{6}$ (see Section 1.3.2), the bars can also be interpreted as 95% confidence limits. The circles connected by a line indicate the corresponding duplicate UV determinations. The proportionality line passes through the origin and the center of mass for the four coordinates. The drug is slightly overdosed ($\approx$ 103–104%; the traditional UV assay apparently is not as selective as it should be; an interference adds about 4% to the result.

After HPLC had become so established that no excuse remained not to convert this product from UV to HPLC, the method was applied and results for four batches of product were obtained. (See Fig. 4.47.) The concentration range covered by these first samples is too small to allow any statement to be made concerning the proportionality of UV and HPLC results. Since a manufacturing overage of 10% was built into the formula, a cluster of points near the 110/110 coordinate would be expected. The (higher) UV-results indicate that on the order of 1–4% of the drug substance must have been lost during manufacturing (the physical amount of the high-potency material actually brought into the vat is so small that any losses through adsorption to metal surfaces have a noticeable effect). The fact that the theoretical correlation line $y = x$ does not intersect the vertical error bars marks the $\approx$4% difference between the two lines as statistically significant. The chromatograms confirmed what had been suspected all along: The primary degradation product H (after the hydrolysis step), while cleanly separated from the main peak by HPLC, had in fact been lumped together with the drug substance on account of the essentially identical UV-spectrum, see the left panel of Fig. 4.46. A thorough investigation of aged samples showed a similar pattern: The small loss of activity observed with TLC was due to (1) the extinction coefficient of H being somewhat smaller than that of DS so that $\Sigma(DS + H) < 100\%$, and (2) the further degradation of the hydrolysis product to compounds H' and H'' that do not exhibit UV activity at the chosen wavelength. The shelflife of the product had to be reduced to 24 months to bring it in line with the actual content of DS at that age. File DECOMPOSITION.xls allows the extinction coefficient, the degradation rate, and the analytical uncertainty to be played with.

The GMP guidelines now require "stability-indicating" methods to be used when laying claim to a specific shelf-life. This implies both theoretical and practical degradation chemistry (combinations of light, pH, temperature, moisture, etc.), extensive high-resolution chromatography under a variety of separation conditions, and the use of pure degradation products for peak-identification/calibration. In this way, self-delusion (poor science at the bench or judgment at managerial levels, e.g., Section 4.7) and deceit (e.g., Section 4.8) would be obvious.

This example nicely demonstrates that straight statistics could (1) unsuspectingly lead to wrong conclusions ("DS is quite stable"), or (2) be used on purpose to cover up something that was already known or suspected ("powerful statistical analysis, conclusion must be true").

4.34 SYSTEMS SUITABILITY

Under GMP rules, a system suitability test must be carried out before the instrument is used. Relatively frequently, the corresponding SOP states something like "three times inject solution X, and determine the CV for the peak area; if it is larger than 2%, the instrument is out of order." Figure 4.48 depicts one result obtained with file SYS_SUITAB.xls: The thick gray line gives the calculated SD $vs.$ the number of repeat measurements, and the thin lines indicate the corresponding upper confidence limits for $p = 0.2 \ldots$ 0.025 (one-sided test, since s_x must be tested against an upper limit). For $n = 3$ the CL_U is above 4σ, which means the CV would have to be <0.5% if the imposed limit of 2% is not to be violated too often (by changing the contents of cells B3 and E3, the presumed s_x and the SL_U can be directly defined; cell C3 allows the instrument's digitizer resolution to be set). The gist is that

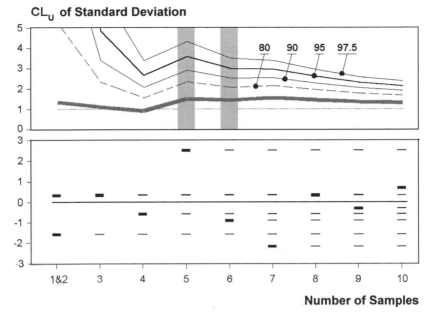

Figure 4.48. The calculated standard deviation and its upper CL. A series of 10 measurements was simulated, bottom panel), with the newest addition at each step given in bold. The corresponding SD is given by the thick line in the top panel, and the 80 ... 97.5% CL_U by thin lines. Notice that point 5, which is high, drives the SD up from ≈ 0.9 to ≈ 1.5 ($E(\sigma) = 1$); the 95% CL is at $2.38 \cdot \sigma$, respectively 3.6. The ordinates are both scaled in units of σ. This depiction, for just one level of p, is part of the display of program CONVERGE.

n must be at least 5 or 6 before such a limit becomes meaningful and the result does not fluctuate too much.

Example 63: Assuming the commonly-seen requirement [$n = 3$] combined either with [$s_x < 2$] or [$R(n) < 2$] and measurements $x_1 = 99.5$, $x_3 = 101.5$, and x_2 anywhere in between, a range $R(3) \le 2$ and a $s_x = 1.00 ... 1.15$ result; so far so good. Since the 95% $CL(s_x)$ are 0.52 ... 6.28 for $2 \cdot p - 0.05$ and $\sigma = 1$, the chances of obtaining a $s_x > 2$ on the next try are in the range 22 ... 28% (use Eq. (1.40) or (1.42) with the Excel functions; the approximation in program MSD yields 0.52 ... 6.38)!

4.35 AN EYE-OPENER

A frequently observed sin is that measurements are performed absentmindedly, that is, strictly according to SOP, but with all critical senses turned off,

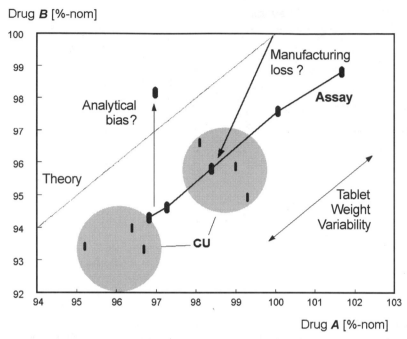

Figure 4.49. Assay and content uniformity (CU) results for six batches of a tablet containing two drugs. One assay result falls completely out of line while there is general loss of component *B* during manufacturing.

and each value is appraised on its own. If the SOP then demands a check against the specification limits, that test might result in a "PASS" even if the data had a story to tell. Assuming a tablet contains two components and the tablet weight varies by 1–2%, Fig. 4.49 might result. The assay values *A* and *B* are strongly correlated, and even the means of the content uniformity tests are on the same trend line, which means the mixture is homogeneous. Three problems are immediately obvious:

1. Drug *B* is about 3% low relative to *A*, and both are about a further 1–2% low relative to the nominal content (if the amount of drug in a tablet is a small fraction of the total tablet weight, the slightest loss of drug to absorption on walls or in the air filters shows up in low assays).

2. One assay falls out of trend by about 4%; this could be a laboratory error (see Section 4.32) that, by chance, does not generate an OOS result and so goes undetected. This out-of-trend (OOT) result could

theoretically be due to a positive (analytical) bias for component B or a negative bias (analytical or manufacturing) for component A; because the corresponding CU results are in a group at left, the first explanation is more likely.

3. The content uniformity values are even further reduced relative to the assays, which is probably due to the sample work-up with very small amounts of drug.

If only one drug had been formulated, then this B vs. A plot would not have been possible; part of the information would have been accessible by plotting A vs. weight, though, see file ASSAY_AB.xls

4.36   BORING BLISS

Those guys from QA are a curious lot: They get paid for taking all the fun out of life, and if something does happen, they see their bonus go up in smoke. Let us look at a thrilling validation report: After the explanations concerning the scope of the work and the particular assumptions, there follows a precise plan on what is to be done and what the success criteria are. The plan is signed off by the author and a clutch of directors from various departments. The next sheet, written at a later date and also signed off by many people, features a summary of the data and a concluding statement like "Revision B of analytical method X-1234 for the assay of component Y in product Z, dated 4/1/1999, has been shown to be adequate for the intended purpose, and accurate, linear, and precise in conformance with the current GMPs." What does the attachment look like that makes all of these people smile?

Example 64: There are long lists of numbers. (See Table 4.42 and 4.43.) There is a discussion that sets the important results in perspective (state of the art, previous methods or revisions, expectations, specification limits).

The HPLC method for which data are given had previously been shown to be linear over a wide range of concentrations; what was of interest here was whether acceptable linearity and accuracy would be obtained over a relatively narrow concentration range around the nominal concentration in the product; the specification limits were 90–110% of nominal. Three concentrations were chosen and three repeat determinations were carried out at each. Two different samples were prepared at each concentration, namely an aqueous calibration solution and a spiked placebo. All samples were worked up according to the method and appropriate aliquots were injected. The area counts are given in the second, respectively the fifth column of Table 4.42.

Table 4.42. Raw Data from Method Validation Tests

Conc.	Calibration Samples			Spiked Samples			
%-Nom	Area	Mean	SD (CV)	Area	Mean	SD (CV)	t-Tests
50	9234	9309	123 (1.3%)	9156	9250	91.2 (1.0%)	$\Delta = -84$
50	9242			9257			$t = 0.65$
50	9450			9338			$p = 0.27$
100	18307	18420	131 (0.7%)	18417	18406	97.0 (0.5%)	$\Delta = -14$
100	18564			18497			$t = 0.15$
100	18389			18304			$p = 0.45$
150	27120	27247	122 (0.4%)	27209	27273	123 (0.4%)	$\Delta = -26$
150	27257			27414			$t = 0.26$
150	27363			27195			$p = 0.40$
n	9	3		9	3		
a	$387.1 \pm 75\%$	$387.3 \pm 580\%$		$287.3 \pm 90\%$	$286.7 \pm 800\%$		
b	$179.4 \pm 1.5\%$	$179.4 \pm 12\%$		$180.2 \pm 1.1\%$	$180.1 \pm 12\%$		
s_{res}	139	116		124	118		
r^2	0.9997	0.9999		0.9998	0.9999		

Table 4.43. Back-Calculated Data from Method Validation Tests

Conc.	Calibration Samples			Spiked Samples			Repeatability
%-Nom	Estim. Conc.	%-Nom	Mean (CV)	Estim. Conc.	%-Nom	SD (CV)	%-nom
50	49.32	98.54	99.47	49.21	98.42	99.43	100.0
50	49.36	98.72	(1.4%)	49.72	99.44	(1.0%)	101.0
50	50.52	101.0	.	50.22	100.4		98.9
							99.1
100	99.90	99.90	100.5	100.6	100.6	100.52	98.8
100	101.3	101.3	(0.71%)	101.0	101.0	(0.52%)	100.5
100	100.4	100.4		99.97	99.97		100.5
							100.2
150	149.0	99.33	99.8	149.4	99.60	99.82	99.6
150	149.8	99.87	(0.47%)	150.5	100.3	(0.44%)	98.2
150	150.4	100.3		149.3	99.53		
Mean		99.94			99.92		99.67
CV		0.93%			0.77%		0.92%

The next two columns contain the means, standard deviations, and CVs. The linear regressions could now be carried out using either the individual points ($n = 9$) or the means ($n = 3$), without any effect on the so obtained slopes and intercepts. A closer examination demonstrates that the $n = 3$ case, while saving some time when typing in the data, has serious drawbacks: A lot of (expensive) information is thrown out and the number of degrees of freedom is drastically reduced. The confidence limits on both the slope and the intercept are much wider (Table 4.42: ±75 $\rightarrow$ ±580%, etc.), and the CL(X) on the back-calculated data degrade from typically ±2% to ±8%, even if the change from $k = 1$ to $k = 3$ repeats is taken into consideration (program LINREG, option $\langle X = f(y)\rangle$). The fact that both s_{res} and r^2 improve can be traced to the elimination of the within-groups variance. (See Table 1.14.) The means are very close, with the spiked samples being a bit smaller on average (-0.9% ($\Delta = 84$), -0.08%, resp. -0.1%); these differences are so small that a t-test between the appropriate items yields $t \leq 0.65$, which corresponds to a probability $p \geq 27\%$ due to chance errors. The F-tests cannot distinguish between the standard deviations at $n = 3$. Even if a worst case is constructed, namely by taking the difference -0.9% and $V_x = (0.4)^2$ (Eq. 1.18) for two groups of $n = 9$ repeats in columns 3 and 6 in Table 4.43), t is only 2.25.

The linearity is tested by comparing the slopes b_{50-100} and $b_{100-150}$, which are (18420 - 9309)/50, etc., that is, 182.22 and 183.12, resp. 176.54 and 177.34, that is, there is a barely perceptible curvature; if the interpolation error $\Delta X = f(y \approx 100|\Delta b)$ this causes it to be no larger than, say, $1 \approx s_x$, the available concentration range is about $100 - 180/(182 - 180) \ldots 100 + 180/(180 - 176) = 10 \ldots 145$, which certainly covers the interesting interval 100 ± 10. A nonlinear model (polynomial, square root, or exponential) could be tried to improve the fit, but one has to consider that the maximal studentized deviations $(\mathbf{y}_{mean} - Y)/s_y$ at $c = 50$ and 100 are of the order $(9309 - 387.1 - 179.4 \cdot 50)/123 \approx 0.4$ to $(18420 - 387.1 - 179.4 \cdot 100)/123 \approx 0.76$, which is already better than the situation schematically depicted in Fig. 2.7.

The conclusions, therefore, are as follows:

- The repeatability (0.92% for $n = 10$; 0.4 … 1.3% for $n = 3$) is acceptable.

- The back-calculated results (99.9 ± 0.9%) for both calibration and spiked samples are acceptable.

- There is no systematic difference between calibration and spiked samples. (Cf. Fig. 3.2.)

- Linearity is acceptable.

4.37 KEEPING TRACK OF DISSOLVING TABLETS

A basic experiment in the pharmaceutical industry is the dissolution rate test in which a solid dosage form such as a tablet or a capsule is immersed in simulated gastric fluid or an appropriate surrogate to determine the speed with which the drug will become available to the body. The basic distinction is between immediate-release and sustained-release or delayed-release forms. The classical tablet, such as many pain-relief medications, are of the first type. The specification will be "80% within 30 minutes," or similar. The test apparatus consists of a glass vessel and a stirrer paddle, the forms of which are tightly defined; the stirrer speed, the dissolution medium, and the tablet-basket are preset. The commercial instrument might pack six such assemblies into one box and provide for flow-through-cell UV photometers, thermostatting, automatic sampling, vessel-rinsing and filling, and tablet-injection for unattended computer-controlled runs. Twelve tablets have to be tested and all have to comply, or the test goes into a mandated extension (Stage II).

For two experimental formulations the data shown in Fig. 4.50 was acquired; for each formulation, there exists a lower and a higher dose. Formulation A obviously rapidly disintegrates and in 20 minutes has set the contained drug free, while Formulation B needs at least an additional hour for the last 5–10% (the two tablets might have contained, on average, 102 and ≈ 97%-of-nominal, respectively).

Example 65: All 30 groups of data were normalized to a mean of 100% and the histogram was calculated for the resulting 347 data points (the instrument rejected 13 readings) to give the distribution shown in Fig. 4.51. The central portion can be approximated by one normal distribution and the tails by another. The form of the curve and the rather small number of data points (10 bins between 90 and 110%) allows for any of a number of parameter combinations that all yield similar goodness-of-fit (GOF)-figures, so there is no clear "best" approximation: The higher of the two dotted curves is for the combination $y = 125 \cdot ND(100, 5.0^2) + 100 \cdot ND(100, 1.6^2)$ while the other is for $y = 70 \cdot ND(100, 5.5^2) + 140 \cdot ND(100, 2.1^2)$; the means μ_1 and μ_2 were fixed at 100; the GOF figures are 670 and 276, respectively. The proportionality coefficients were varied in steps of 5 and the standard deviations in steps of 0.05, for a total of 41 769 tested parameter combinations; 32 400 combinations were found that describe curves with $276 < GOF < 670$; 2200 combinations were within 10% of the best GOF-value! Further refinements might be the modeling of the spike (arrow!) with a third exponential, and adjustments to the assumed means μ_1 and μ_2, but this would be cosmetic and would not add any value to the interpretation, which is that the tablet-to-tablet variation is still larger than the instrumental uncertainty ($\approx 0.75\%$) and so dominates the picture ($s_{\text{tab to tab}} \geq \sqrt{1.6^2 - 0.75^2} = 1.4$).

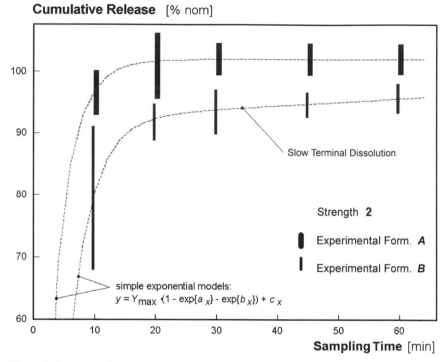

Figure 4.50. Cumulative dissolution results. Two experimental tablet formulations were tested against each other in a dissolution test in which tablets are immersed in a stirred aqueous medium (number of tablets, constructional details and operation of apparatus, and amount of medium are givens). Eighty or more percent of the drug in either formulation is set free within 10 minutes. The slow terminal release displayed by formulation B could point towards an unwanted drug/excipient interaction. The vertical bars indicate $y_{mean} \pm s_y$, with $s_y \approx 3\%$. A simple linear/exponential model was used to approximate the data for the strength 2 formulation. Strengths 1 and 3 are not depicted but look very similar.

A group-by-group analysis of the achieved CVs is given in Fig. 4.52. The three strengths of formulation A are quite similar with the exception of the points marked "c," where instrumental noise sets the tune. This particular picture would also be seen if the solubility of the drug substance was a problem (not the case here), because then saturation would buffer any content-uniformity effects. Formulation B is striking because of the extreme variability seen for the mid- and high-strength dosages at $t = 10$. This is not fully unexpected as at this time the slope in Fig. 4.50 is still very high, but the mechanism should also be in evidence for the low-dose form, and should still play a role at $t = 20$, for a smoother transition from "a" to "b." Over-all, Formulation A is much preferred to B.

Frequency

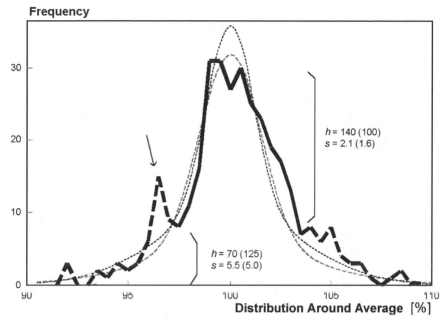

Figure 4.51. Distribution of experimental data. Six experimental formulations (strengths 1, 2, resp. 3 for formulations *A*, respectively *B*) were tested for cumulative release at five sampling times (10, 20, 30, 45, respectively 60 min.). Twelve tablets of each formulation were tested, for a total of 347 measurements (13 data points were lost to equipment malfunction and handling errors). The group means were normalized to 100% and the distribution of all points was calculated (bin width: 0.5%, her depicted as a trace). The central portion is well represented by a combination of two Gaussian distributions centered on $\mu - 100$, one that represents the majority of points, see Fig. 4.52, and another that is essentially due to the 10-minute data for formulation *B*. The data point marked with an arrow and the asymmetry must be ignored if a reasonable model is to be fit. There is room for some variation of the coefficients, as is demonstrated by the two representative curves (gray: coefficients in parentheses, h = peak height, s = SD), that all yield very similar GOF-figures. (See Table 3.4.)

4.38   POKING AROUND IN THE FOG

In a series of examples scattered throughout the first four chapters, conditions outside the laboratory were blamed for difficulties in explaining particular results, or obtaining them at all. It seems that all kinds of things could conspire to frustrate the well-intentioned analyst, and unless all of them were righted simultaneously, there would be no point in trying to measure at all. The GMPs address a wide range of institutions and processes in a factory,

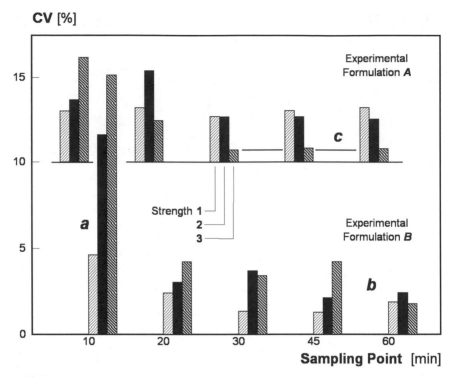

Figure 4.52. Coefficients of variation that reflect both tablet to tablet and analytical variability. For formulation *B*, particularly strengths 2 and 3, the drop in CV with higher cumulative release (*a* → *b*) is marked, cf. Fig. 4.50. When the dissolution rate is high, individual differences dominate, while towards the end analytical uncertainty is all that remains. The very low CVs obtained with strength 3 of formulation *A* (±0.7–0.8%, data offset by +10% for clarity) are indicative of the analytical uncertainty. Because content uniformity is harder to achieve the lower the drug-to-excipient ratio, this pattern is not unexpected.

and if adhered to, instill a large measure of reliability into analytical results.

Even with everything under control, an analyst is well-advised to keep his eyes open so he will have an idea of what artifacts could turn up, and can plan to keep irregularities in check. The list of items in Table 4.44 could turn up in the checklist of any GMP-auditor worth his salt; a corresponding observation would probably trigger his suspicion that there might be further weak spots. The table is given here to provide the reader with an idea of the human and technical factors that can influence the quality of results, and to permit a search for examples that fit a certain category.

Table 4.44. Poor Practices

Examples of Things That Can Go Wrong	Possible Consequences	Refer to Section
(A) PROCESS & ENVIRONMENT • *Outdated technology:* (1) the process is inefficient to start with, (2) the product barely meets today's specifications, (3) there is no margin for improvement, (4) major customers insist on tight specifications, (5) authorities announce shift to higher standards,	Short life cycle. Noncompetitive product. Exclusion from high-yield markets. Licensing authorities and/or customers loose confidence. High cost of goods.	
(6) control methods are imprecise or unspecific. • *Inadequate facilities and/or equipment*	Inefficient production and faulty products due to mismanagement.	4.1
(7) the inability to use controllers or data loggers implies inefficient installation/optimization, qualification, (8) the process cannot be reliably reproduced,	Reduced information flow leads to low-quality decisions, manual data collection introduces transcription errors expensive rework.	4.31
(9) equipment maintenance is compromised and increasingly expensive, (10) the error rate and risk of failed batches is high, (11) there are equipment train incompatibilities or deficiencies. • *Available skill set does not meet requirements:*		4.9
(12) available workers have "muscles & sweat" rather than "fine-tuning" mentality, (13) specialists do not consistently built quality into design, (14) management is not used to handling complex situations involving decentralized decision-making,	Products are not designed and manufactured with a view towards consistently achieved quality, but have "quality" inflicted on them after the fact.	4.7
(15) managers who are far removed from the lab bench are involved in technical decisions • *Organization:* (16) traditional hierarchical structures along specialties instead of teams built around product or process,	A total control and accountability strategy will not compensate the losses incurred through mismanagement, or poor design and execution; ingraining a GMP-mentality takes time,	4.7

Table 4.44. (*Continued*)

Examples of Things That Can Go Wrong	Possible Consequences	Refer to Section
(17) emphasis on seniority rather than merit,	but pays off by keeping a company fit for the future.	
(18) compartimentalization of work flow that reduces individual's involvement, overview, motivation, responsibility, and self-control to zero, individual departments act without coordination,		4.8
(19) internal regulations stress control mechanisms instead of identification and competence,		4.23
(20) Know how is improperly documented or protected,		4.26
(21) nontechnical aspects are given overriding significance, e.g., budgets, headcount issues; unwillingness to admit failure		4.28 4.7

(B) PRODUCT DESIGN		
(1) important design parameters are ignored, management insists on course that courts failure,	The product eventually fails or becomes too expensive to produce.	4.4 4.9 4.33
(2) the product design is not robust: it leaves no margin for error or variability in raw materials or process; production or analytical process require higher-than-standard skills, diligence, or controls; product is insufficiently protected,		2.2.6 4.1 4.20 4.24
(3) specifications are incorrectly set, miss important aspect, or method does not meet requirements; process complexity is too high; customer and/or regulatory demands strain technology to the limit,	Equipment, technology, and skill base do not meet expectations.	4.12 4.31 4.6
(4) an unrecognized combination of hard-to-measure product characteristics results in an analytically intangible quality aspect that affects the process or the customer acceptance.	Loss of control	4.12 4.31

Table 4.44. (*Continued*)

Examples of Things That Can Go Wrong	Possible Consequences	Refer to Section
(C) MANUFACTURING PROCESS		
• *Variations in feed-stock quality:*	Highly variable product quality,	3.1
(1) the supplier does not notify customer of process changes,	unnecessary costs, losses, delays.	4.12
(2) test methods fail to address the real problem,		
(3) specifications are too loose,	The more uncertainties that	
(4) quality control does not detect change,	exist, the more difficult the assignment of cause(s)	4.30
(5) quality assurance does not reject batch.	to observed effects;	
• *Process parameters change from batch to batch:*		
(6) process not properly optimized, or is not robust,	Reliance on "tweaking" and "fire-fighting" inhibits transfer	
(7) there are inadequate parameter limits,	or outsourcing, and creates incessant problems down-	4.31
(8) there is no proper validation,	stream.	4.28
(9) the process is insufficiently controlled,		4.4
(10) wrong or poorly maintained equipment is used.		4.9
• *Management*		
(11) training of operators, engineering staff, supervisors, management is inadequate, procedures could be spruced up,		4.8
(12) leadership qualities do not ensure motivation, compliance, and/or discipline,		3.5
(13) disrupting time schedules, production targets, or management expectations,		
(14) equipment does not support fail-safe operations.		4.8
(D) SAMPLING		3.1
• *Sampling plan:*	Required reliability of results	
(1) wrong number of samples chosen,	cannot be attained.	
(2) sampling points/times do not address pertinent questions or process characteristics,	Sample is modified before it is tested. Results are biased, not repre-	1.1.3

Table 4.44. (*Continued*)

Examples of Things That Can Go Wrong	Possible Consequences	Refer to Section
(3) inadequate sampling tools and/ or processes,	sentative, wrong, or unassignable	
(4) sample storage/transport is not properly organized,		
(5) sample amount is too small or is lost/wasted,		
(6) operator skills/training/instruction are not up to the job.		
(7) unnecessarily extensive sampling plans demotivate and lead to hastily drawn and carelessly handled materials.		
• *Representativity:*		
(8) bias across species/impurities,		
(9) absorption and losses of analyte,		4.30
(10) reaction not under control, sampling point not recorded,		4.33
(11) collection vessels are not properly cleaned,		
(12) mix-ups and mislabelling cannot be excluded with certainty,		
(13) there is chaotic record-keeping,		
(14) blanks are not available or chemically different,		
(15) the spike is not in same compartment in blank as analyte.		3.2
(E) SAMPLE WORK-UP		
• *Technique:*		
(1) does not take variability of matrix into account,	High LOQ, lack of selectivity, analyte-ratios	1.1.3 3.2
(2) allows side reactions, e.g., reequilibration of esters,	distorted, product fails specs.	
(3) native and spiked analyte is in different chemical compartments, recovery is not the same,		4.2 4.29
(4) recovery is too low, saturation effects,		4.1 4.28
(5) handling is too difficult, error prone, method poorly matched to technology,		
(6) sample rapidly deteriorates.		

Table 4.44. (*Continued*)

Examples of Things That Can Go Wrong	Possible Consequences	Refer to Section
• *Equipment:*		
(7) analyte absorption, contamination, or carry-over effects are possible,		4.30
(8) quality of consumables (e.g., paper, filters) changes		
(9) insufficient precision		4.1
• *Reagents:*		
(10) purity or potency of reagent varies		
• *Skills:*		
(11) loss of sample or contamination thereof through sloppy work		

(F) MEASUREMENT		
• *Equipment:*		
(1) Equipment failed installation-, operations-, performance qualification,	Distorted peak shapes, high noise levels, ghost peaks, saturation and base-line	3.2 4.25
(2) maintenance schedule was missed,	effects.	4.23
(3) replacement material or consumables change instrument characteristics,		4.14 1.2.1
(4) components (e.g., HPLC columns) deteriorate with increasing number of analyses, instrument malfunction,	Unexplained bias, compliance problems.	3.1 4.34
(5) unaccounted environmental effects (e.g., diurnal temperature/humidity cycles)		
(6) unrecognized failure of digital digital displays can lead to faulty read-out, e.g., "0" instead of "8", or *vice-versa*.	Required precision is not met, unnecessarily high work-load or turn-around time, analytical	4.32, 4.8 1.8.4
• *Standards:*	uncertainty masks manufacturing problems.	4.14
(7) expired lot or inferior quality material,		
(8) working standard not calibrated against primary standard,		4.32
(9) new lot with different potency or impurity profile,	Lab's GMP license or	
(10) lot from unapproved supplier.	factory's marketing	
(11) no internal standard.	authorization revoked.	4.14
• *Calibration:*		
(12) suboptimal calibration concept/design (number of concentrations, concentration range, number of repeat		4.2 4.5

Table 4.44. (*Continued*)

Examples of Things That Can Go Wrong	Possible Consequences	Refer to Section
determinations), poor preparation and/ or execution, work near LOD.		4.33
(13) crucial measurements not done, of insufficient accuracy or precision.		4.1
• *Procedures:*		4.33
(14) the validation or the systems suitability test failed, outdated methodology,		4.14
(15) there is a lack of in-process controls,		
(16) no positive or negative controls,		
(17) no control charts, no or irregular trend analysis,		4.7
(18) retesting until "right results" are found is neither suppressed nor reported.		4.16
• *Operators:*		
(19) unskilled or improperly trained/ supervised,		
(20) inconsistent habits or lack of concentration lead to higher variability,		1.14 4.7
(21) uncritical acceptance of data "as printed,"		
(22) analyst forgets to note exceptional circumstances/observations, or does not recognize them as being such: "I did not think this was important,"		4.24
(23) raw data and comments get lost, misfiled.		
(24) operator tweaks uncontrolled parameters to keep results within speces, e.g., column pressure or eluent make-up in HPLC.		
(G) DATA PROCESSING		
• *Review:*		3.3
(1) plausibility checks are not applied,	Loss of information,	4.28
(2) supervisor does not immediately check and OK results,	time, and opportunity.	4.35
(3) investigation/reanalysis cannot be performed with original standard solutions,	Data base is undocumented	
• *Audit:*	mixture of good, bad,	
(4) transcription error and tamper-evidence checks are worthless if a scientific	and unreliable numbers.	1.3 4.31

Table 4.44. (*Continued*)

Examples of Things That Can Go Wrong	Possible Consequences	Refer to Section
inconsistency slips through; terminology is contradictory or misleading.	Data processing without analytical or QA back-	
• *Data Clean-Up:*	ground assumes numbers	4.22
(5) the data base is not cleaned up or properly formatted,	to be literally correct and can skew interpre-	4.28
(6) the wrong documents are forwarded to the supervisor/processor	tation.	4.31
• *Processing:*		
(7) raw data is not secured against tampering, loss, or mix-up,	Data base is corrupted or becomes untrustworthy.	1.5.5
(8) data entry is not double-checked,		3.7
(9) automatic data aquisition/ treatment is not validated,		3.8
(10) manual steps in semiautomatic data migration and/or treatment are not audited, e.g., cell formulas and macros in spreadsheet applications,		1.8.4, 2.2.9, 2.3, 4.0, 4.23,
(11) the processing concept/model is inappropriate,		4.34, 1.1.5,
(12) the computer code is not validated,		1.5.2,
(13) insufficient numerical precision,		1.8.1,
(14) inappropriate software or poor presentation of results leads to false conclusions being drawn,		2.1, 2.3, 4.10,
(15) straight statistical interpretation may be wide of mark,		4.13, 4.20,
(16) raw data gets lost, becomes inaccessible because electronic data formats change.		4.29, 4.30

(H) EVALUATION

• *Communication:*

(1) not all parties who need to know or who could influence the interpretation are supplied with the result and comments in a form that both summarizes the essentials and gives detail,	Relevance of result is misjudged.	2.3.2,
(2) the program output is not critically reviewed, data is overinterpreted,		2.4 3.5, 4.17
(3) the formal statistical report is given excessive weight—to the exclusion of other information.		4.21, 4.28, 4.33,

Table 4.44. (*Continued*)

Examples of Things That Can Go Wrong	Possible Consequences	Refer to Section
• *Skills, attitude, experience:*		4.35
(4) participants in the discussion do not represent/present all relevant aspects:		
parts of the available information are		4.7
filtered out; selective, see-no-evil		4.28
data evaluation, e.g., only means, but no SD is given,		
(5) a legalistic reduction to "complies Y/N"	Formalistic aspects	4.14
compresses the many shades of gray	override scientific	
contained in a scientific interpretation	judgment	4.24
to one bit—black or white—and		
sacrifices a large percentage of the		
available, and expensive, information,		
(6) misinterpretation, e.g., "H_0: no significant		1.9
deviation detected" is equated to "There is no effect."		

(I) DECISION		4.7
• *Independence of reviewing party (QA):*	Substandard product is	
(1) partiality (e.g., toward manufacturing,	delivered to customer.	
marketing, finance) allows customer		2.2.6
focus to be reduced to time and money,		
and quality looses out.		

(J) ACTION PLAN		
(1) superficial corrective measures are taken,	Chance to improve product	
but the root causes are not examined	or process is wasted.	
or woven into long-term plan (e.g.,		
training, infrastructure).		
(2) actions are decided upon, but shelved; no follow-up.		
(3) corrective actions are not integrated into SOP; the same mistake is committed the next time around.		

CHAPTER 5

APPENDICES

This chapter contains the algorithms necessary for approximating statistical tables, some program kernels in BASIC, instructions on how to install the VisualBasic programs, and finally, a description of each of the VB programs and the Excel files.

5.1 NUMERICAL APPROXIMATIONS TO
SOME FREQUENTLY USED DISTRIBUTIONS

There are various reasons for replacing tabulated values by numerical approximations, chief among them to be able to automate the table look-up to save time and to present aspects that otherwise would go unnoticed. Commercial programs like Microsoft Excel feature many of the important statistical functions; the file EXCEL_FNC.xls that is provided with this manuscript shows how some functions are applied. The algorithms that are employed are very accurate, but not accessible as such. For the applications demonstrated in this work, appropriate approximations are incorporated into the VisualBasic programs that accompany the book.

To be useful in this context, the algorithms must fulfill the following criteria:

- A wide range of degrees of freedom must be spanned, over which the approximated value changes appreciably.
- Relative accuracies of about 1% or better should be attained, which suffices for practical applications. A decision that rests on paper-thin margins is best reviewed in a nonstatistical context anyway.
- The programmable calculator or PC that is to be used must be able to work with the number of significant digits required by the algorithm; rounding the coefficients can appreciably alter the results of an approximation.
- The use of an algorithm must conserve memory relative to a table-oriented approach. Polynomials or similar functions should be used because recursive functions tend to converge slowly.

The following figures of merit are used:

329

LAR: Largest absolute residual $r_i = |t_{approx} - t_{tabulated}|^{max}$
TAR: Typical absolute residual $r_i = |t_{approx} - t_{tabulated}|^{typ}$
LRR: Largest relative residual $r_i = (|t_{approx} - t_{tabulated}|/t_{tabulated})^{max}$
TRR: Typical relative residual $r_i = (|t_{approx} - t_{tabulated}|/t_{tabulated})^{typ}$

Comment   The sequence of digits in each coefficient depends on the precision (e.g., three decimal places) and number of tabulated values (34, 50, or 64), the form of the optimization software used (Hewlett Packard HP71B Curve Fit Module), and the number of coefficients chosen (3 ... 8). Discrepancies between the approximated and the real table entries of up to $\frac{1}{2}$·LSD could be due either to insufficiencies of the algorithm or the rounding of table entries. The few LRR that are above 1% do not pose a risk for practical applications.

Specific Assignments

Symbol	Explanation
df	degrees of freedom (df was chosen here so as not to cause confusion with coefficient f)
χ^2	Chi square
CP	cumulative probability in the range 0 ... 1(for $z = -\infty ... + \infty$)
F	Fisher's F-value
f_1, f_2	degrees of freedom in F-test associated with the larger, the smaller variance
p	probability of error
s	sign of $z(-1, 0, +1)$, respectively expression $(CP - 0.5)$
t	Student's t
z	normalized deviate

All other variables, such as "u" or "v", are for intermediate results.

5.1.1   The Normal Distribution

The probability density can be calculated by way of Eq. (1.7). Both a forward and an inverse function for the cumulative probability CP are needed:

- For a given deviation $z = (x - \mu)/\sigma$, $CP = f(z)$ calculates the cumulative probability of finding a deviation as large purely by chance; this corresponds to the t-test for very large numbers of degrees of freedom.
- The inverse, $z = f(CP)$, is particularly valuable in connection with the

Monte Carlo method, because normally distributed random numbers can be generated.

Calculation of CP from z

Use: Calculate the cumulative probability (CP) for a given normalized deviate $z = (x_i - x_{mean})/s_x$ or $z = (x - \mu)/\sigma$.

Assumption: Normal distribution.

Reference: Ref. 190, Eq. (26.2.19).

Procedure: Use the algorithm given below; because of the symmetry of the function, only the $0 \leq z \leq \infty$ part is defined.

Accuracy: The algorithm is extremely accurate, no deviation between calculated and tabulated (four decimals) CP being larger than 0.2% relative (LRR, this occurs where both z and CP are very close to zero); typically, the deviations are less than 0.00005 absolute (TAR), and less than 0.03% for z larger than 0.15 (TRR).

Algorithm:

$$u = |z|$$
$$v = a + b \cdot u + c \cdot u^2 + d \cdot u^3 + \ldots + g \cdot u^6$$
$$CP = 0.5 \cdot (1 + s \cdot (1 - v^{-16})) \tag{5.1}$$

with

$$a = 1.0000000000$$
$$b - 0.0498673470$$
$$c = 0.0211410061$$
$$d = 0.0032776263$$
$$e = 0.0000380036$$
$$f = 0.0000488906$$
$$g = 0.0000053830$$

Example:

For $z = -1.56$, CP = 0.0593798 is found (tabulated value: 0.05938).

For $z = +0.80$, CP = 0.788144 is found (tabulated value: 0.78814).

Calculation of z from CP

Use: Inverse of above function; given a cumulative probability CP, the equivalent normalized deviate z is calculated.

Assumption: Normal Distribution.

For the optimization of the coefficients 64 $\log_{10}(1 - CP)$ values (CP: four decimal places) for $z = 0.00 \dots 3.00$ in steps of 0.05, and $z = 3.5, 4.0$, and 4.4 were used; see also the comment under Student's t, Table 5.1.

Reference: None.

Procedure: Use the algorithm given below; because of the symmetry of the function, only the part $0 \le CP \le 0.5$ is defined; cumulative probability values CP lower than 0.5 are transformed to their (decadic) logarithm, the others are first subtracted from 1.00. The sign is appropriately set to -1 or $+1$.

Accuracy: The algorithm is fairly accurate, no calculated z-value being off by more than 0.0166 up to $z = 4.4$, and with most deviations below 0.006 absolute. Monte Carlo simulations ($n = 20\ 000$ events) yield a mean of 0 and a standard deviation of 1.009; this is close enough for most practical purposes. Figures of merit: s_{res}: 0.005; LAR: -0.015 for $z = 0$; TAR: 0.005; LRR: 20% at $z = 0.05$; 2% at $z = 0.15$; 0.3% at $z = 1.5$.

Abbreviations: See Section 5.1. Deviations for $z > 3.5$ (CP < 0.0003 or CP > 0.9997) can be larger than 0.01, but this is irrelevant due to the low probability of having to simulate such a z-value; since empirical evidence points toward wider-than-ND tails, this is actually a step in the right direction.

Algorithm:

$$CP \begin{cases} < 0.5 & u = \log_{10}(CP) & s = -1 \\ \ge 0.5 & u = \log_{10}(1 - CP) & s = +1 \end{cases} \tag{5.2}$$

$$v = a + b \cdot u + c \cdot u^2 + d \cdot u^3 + \dots + g \cdot u^6$$
$$z = s \cdot v$$

with

$$a = -0.9069066$$
$$b = -3.64591$$
$$c = -2.205586$$

$$d = -0.9623506$$
$$e = -0.2366192$$
$$f = -0.02921359$$
$$g = -0.001375013$$

Example:

For CP = 0.05938, $z = -1.5633$ is found (tabulated value: -1.56).

For CP = 0.78814, $z = +0.7982$ is found (tabulated value: $+0.80$).

5.1.2   The Student's *t*-Distributions

This empirical one-line function fits into almost any program, especially if only one significance level is needed:

Calculation of Student's t from df and p

Use: Calculate Student's *t*-values given *p* and df; Student's *t* is used instead of the normal deviate *z* when the number of measurements that go into a mean is relatively small and the assumption of μ and σ being infinitely precise has to be replaced by the assumption of a normally distributed mean and a χ^2-distributed s_x.

Assumptions: Empirical polynomial approximation to *t*-tables. A good overall fit was attempted; relative errors of less than 1% are irrelevant as far as practical consequences are concerned. The number of coefficients is a direct consequence of this approach. Polynomials were chosen *in lieu* of other functions in order to maximize programming flexibility and speed of execution.

Reference: References 191 and 192 came to the authors' attention in the early 1990s; P.M. had independently devised the algorithm in 1974. A somewhat different equation is used for $p = 0.05$, 0.025, and 0.005 in Ref. 193. The degrees of freedom for which data points were taken are (50 points, three decimal places): df = 1, 2, ... , 30, 32, 34, ... , 42, 45, 47, 50, 55, 60, 70, 80, 90, 100, 120, 500; 10000 is used for ∞.

Procedure, accuracy: Use the algorithm given below; the figures of merit are given in Table 5.1.

Algorithm:

$$t = a + b/\text{df} + c/\text{df}^2 + \ldots + j/\text{df}^8 \tag{5.3}$$

Table 5.1. Coefficients for Approximating *t*-Values for Various Confidence Levels
(The coefficients *h* and *j* are necessary only for *p* = 0.0001)

p	a	b	c	d	e	f	g
0.5	0.67447220	0.24667600	0.06681826	0.01190292	0	0	0
0.2	1.281482	0.8489111	0.5543883	0.2734102	0.1197992	0	0
0.1	1.644487	1.539693	1.326133	0.9955611	0.8080881	0	0
0.05	1.959002	2.416196	2.544274	2.583332	2.598259	0.6047031	0
0.02	2.328194	3.608506	7.266717	0.6699166	11.71529	6.192514	0
0.01	2.586279	4.2351680	18.24859	−27.92441	66.51121	0	0
0.002	3.089668	8.178925	19.70815	27.40369	129.5110	−213.5485	343.9671
0.001	3.287494	9.948274	23.103690	52.92142	224.7871	−521.1793	843.7513
0.0001	3.708149	31.50907	−260.1995	2222.611	−4122.848	173.4048	7378.562

$h = 274.0580 \qquad j = 665.3953$

Table 5.2. Figures of Merit for the Approximations Given in Table 5.1

p	s_{res}	LAR	df	TAR	LRR %	df	TRR %
0.5	0.0003	0.0005	4	0.0003	0.072	12	0.05
0.2	0.0008	0.004	2	0.0005	0.22	2	0.03
0.1	0.0014	0.0048	3	0.001	0.20	3	0.06
0.05	0.001	0.0047	3	0.0003	0.15	3	0.02
0.02	0.003	0.02	1	0.0006	0.15	3	0.02
0.01	0.007	0.029	3	0.004	0.49	3	0.1
0.002	0.0006	0.0027	4	0.0004	0.037	4	0.01
0.001	0.002	0.011	4	0.002	0.12	4	0.03
0.0001	0.1	0.40	4	0.06	4.3	1000	1

Use the coefficients $a \ldots g(h, j)$ given in Table 5.1 where df: number of degrees of freedom at which LAR or LRR is found.

Example: for df = 5, $p = 0.05$: $t = 2.5690$ is found (tabulated value: 2.5706)

Calculation of p from Student's t and f

Comment: Instead of calculating a critical t_c and comparing it to the experimental one, the experimental t is converted into an estimated error probability, which is then checked against a preset value, e.g., 0.05. The medical and social science communities prefer using the second approach. This algorithm is theoretically underpinned.

In several programs subroutine PROBAB is used to find the probability p that the result is due to chance alone if a Student's t-factor and the number of measurements is known.

Assumptions: definition of Student's t-factor[192]

Procedure, accuracy: calculate p by either of two algorithms, depending on whether the number of degrees of freedom df is even or odd. The function is very accurate: LAR: -0.0003 at $p = 0.5$; LRR: 0.5% at $p = 0.001$.

Abbreviations: See Section 5.1.

Listing: See Table 5.3.

Example: for $t = 3.182$ and df = 3:

$$p = 0.05002 \text{ is found (tabulated value : 0.05000)}$$

5.1.3   F-Distributions

Use: Calculation of F from f_1 and f_2 for $p = 0.05$ and $p = 0.025$ for the F-test.

Assumption: F-distribution.

Reference: Ref. 194

Table 5.3. BASIC Code for Subroutine PROBAB

Assignments:	
A–D:	scratchpad variables
K:	index
F:	degrees of freedom
T:	Student's t
P:	probability of error

```
      SUB PROBAB(F, T, P)
      RADIANS
      D = ATAN(T/SQR(F))
      A = COS(D)
      B = A*A
      IF INT(F/2) # F/2 THEN 150        if F = odd then goto 150
      A = 1                             F: even-numbered
      IF F = 2 THEN 130 ELSE C = 1
      FOR K = 2 TO F-2 STEP 2
      GOSUB 230
      NEXT K
130   A = A*SIN(D)
      GOTO 260
150   IF F = 1 THEN A = 0 & GOTO 210    F: odd-numbered
      IF F = 3 THEN 200 ELSE C = A
      FOR K = 3 TO F-2 STEP 2
      GOSUB 230
      NEXT K
200   A = A*SIN(D)
210   A = 2*(D+A)/PI                    PI = 3.14...
      GOTO 260
230   C = C*B*(K-1)/K                   subroutine
      A = A + C
      RETURN
260   P = 1 - A                         calculate probability
      SUB END
```

Procedure: The two algorithms FF050 and FF025 permit tabulated F-values for the confidence levels $p = 0.05$ and 0.025 to be approximated with high accuracy. The strong curvature of the $F = f(f_1, f_2, p)$ surface militates against simple and flexible functions like polynomials; only two confidence levels are available.

Accuracy: The two programs very accurately approximate the tabulated values, the relative deviations (LRR) remaining below 0.01%, except when the equations on lines 10 are used, where they remain below 0.6%.

Table 5.4. BASIC Code for the Calculation of F-Values for $p = 0.05$

```
SUB FF050(F1,F2,F)
      IF F2 > 4 OR F1 > 1 THEN 10
      IF F2 = 1 THEN F = (F1 - .09849)/(.0039292*F1 + .0016579) & RETURN
      IF F2 = 2 THEN F = (F1 - .03646)/(.051294*F1 + .000761)   & RETURN
      IF F2 = 3 THEN F = (F1 + 1.094)/(.1173*F1 + .0894)        & RETURN
      IF F2 = 4 THEN F = (F1 + 1.349)/(.1776*F1 + .1271)        & RETURN
      IF F1 = 1 THEN F = 7.71 - (F2 - 4.032)/(.2581*F2 - .4076) & RETURN
10    F = (F1 + 1.288)/(.1751*F1 + .1129)
        - (F2 - 4.119)/(.2511*F2 - .4236)
        - .552
        + 6.53/(F1 + 11.533)
        + 3.993/(F2 + 11.533)
        - 88.889/(F1 + 11.533)/(F2 + 11.533)
      RETURN
```

Examples:

For $f_1 = 6$, $f_2 = 5$, and $p = 0.05$: $F_c = 4.960$ is found (tabulated value: 4.95).

For $f_1 = 6$, $f_2 = 5$, and $p = 0.025$: $F_c = 6.99$ is found (tabulated value: 6.98).

Table 5.5. BASIC Code for the Calculation of F-Values for $p = 0.025$

```
SUB FF025(F1,F2,F)
      IF F2 > 4 OR F1 > 1 THEN 10
      IF F2 = 1 THEN F = (F1   .09582)/(.0009831*F1 + .0004153) & RETURN
      IF F2 = 2 THEN F = (F1 - .00904)/(.025317*F1 + .000416)   & RETURN
      IF F2 = 3 THEN F = (F1 + .9232)/(.07192*F1 + .03836)      & RETURN
      IF F2 = 4 THEN F = (F1 + 1.27)/(.121*F1 + .0648)          & RETURN
      IF F1 = 1 THEN F = 12.22 - (F2 - 4.045)/(.1387*F2 - .2603) & RETURN
10    F = (F1 + 1.739)/(.1197*F1 + .1108)
        - (F2 - 3.986)/(.1414*F2 - .2864)
        - .145
        - .0017*F1
        - 2.706/(F2 + 30)
        + 0.0615*F1/(F2 + 30)
      RETURN
```

Table 5.6. Coefficients for Approximating χ^2-Values for Various Confidence Levels

p	a	b	c	d
0.01	0.8304346	0.4157701	0.09632204	0.05718995
0.025	0.7006128	0.5264558	0.11061850	−0.09982465
0.05	0.5847010	0.6263386	0.03897389	−0.01341259
0.1	0.4359082	0.7156998	0.1145636	−0.1324754
0.9	−1.797326	4.633139	−3.693577	2.017129
0.95	−2.401336	5.986627	−5.246916	3.032807
0.975	−3.004395	7.332854	−6.74877	4.007137
0.99	−3.804041	9.459484	−10.17665	7.618876

p	e	f	g
0.01	−0.05174064	0.0213863	−0.003553602
0.025	0.11871100	−0.05364844	0.008313859
0.05	0.04725319	−0.02531424	0.004102034
0.1	0.0976907	−0.02973045	0.003078732
0.9	−0.4866325	0.001041404	0.01305642
0.95	−0.8467675	0.06028729	0.01034983
0.975	−1.214192	0.134666	0.004115038
0.99	−3.470499	0.8727404	−0.09240059

5.1.4 The χ^2-Distributions

Use: Calculation of χ^2 from df, e.g., for the determination of $CL(s_x)$.

Assumption: χ^2-distribution; the curvature of the χ^2-functions *versus* df is not ideal for polynomial approximations; various transformations on both axes, in different combinations, were tried, the best one by far being a $\log_{10}(\chi^2)$ vs. $\log_{10}(df)$ plot. The 34 χ^2-values used for the optimization of the coefficients (two decimal places) covered degrees of freedom 1–20, 22, 24, 26, 28, 30, 35, 40, 50, 60, 80, 100, 120, 150, and 200.

Reference: None.

Procedure, accuracy: Use the algorithm below with the coefficients given in Table 5.6; the figures of merit are given in Table 5.7. The accuracy of the approximations is sufficient for most applications.

Algorithm:

$$u = \log_{10}(df)$$
$$v = a + b \cdot u + c \cdot u^2 + d \cdot u^3 + \ldots + g \cdot u^6 \tag{5.4}$$
$$\chi^2 = 10^v$$

Table 5.7. Figures of Merit for the Approximations Given in Table 5.6

p	s_{res}	LAR	df	TAR	LRR %	df	TRR %
0.01	0.19	0.9	200	0.05	2	1	0.05
0.025	0.1	0.4	200	0.03	0.19	2	0.05
0.05	0.004	0.16	200	0.05	0.17	2	0.05
0.1	0.04	0.16	200	0.03	0.7	2	0.1
0.9	0.3	1.1	200	0.03	2.3	2	0.5
0.95	0.35	1.4	200	0.05	3.1	2	0.5
0.975	0.3	1.3	200	0.05	3	2	0.5
0.99	0.09	0.3	150	0.05	0.3	150	0.08

Examples:

For df = 5 and $p = 0.025$: $\chi^2 = 12.85$ is found (tabulated value: 12.83).

For df = 5 and $p = 0.975$: $\chi^2 = 0.842$ is found (tabulated value: 0.831).

5.2 CORE INSTRUCTIONS USED IN SEVERAL PROGRAMS

The core instructions used in programs MSD, CORREL, FACTOR8, HISTO, and MULTI are given in a stripped-down version of BASIC in the following tables to allow the reader to follow the ideas from a different vantage point and/or to migrate the program kernels to a programmable calculator. Numerical examples are provided to check the calculations.

The core parts of program MSD are given in Table 5.8; Table 5.9 displays the results of a numerical example, see file QUOTE_RESULT.xls, and compares this to the results obtained using the approximations listed earlier.

Program CORREL is given in Table 5.10; Tables 5.11–13 display the results of a numerical example.

The core part of program FACTOR8 is given in Table 5.14.

The core parts of program HISTO and subroutines NPS and SORT are given in Tables 5.15–17.

Subroutine ANOVA from program MULTI is given in Table 5.18.

5.3 INSTALLATION AND USE OF PROGRAMS

Section numbers are cross-referenced, e.g., Section 2.2.10.

Introduction The intention behind the book and these programs is to provide the user with a number of statistical tools that are applicable to prob-

Table 5.8. BASIC Code for the Core of Program MSD

Assignments	
SX, SXX	statistical sums
XB	x-avg = mean = x-bar
S, VX	standard deviation, variance
R	residual
T0	Student's t for $p = 0.05$
C1...C4	confidence limits for x_{mean} and s_x

Listing
```
SUB MSD(N,X(),M,S)
```

```
SX = 0
FOR I = 1 TO N
SX = SX + X(I)
NEXT I
XB = SX/N
```
calculate mean
Eq. (1.1)

```
SXX = 0
FOR I = 1 TO N
R = X(I) − XB
SXX = SXX + R*R
NEXT I
VX = SXX/(N − 1)
S = SQR(VX)
```
calculate
standard deviation
Eq. (1.2a,d)

```
F = N − 1
A = LGT(F)
T0 = 1.959002 + 2.416196/F
  + 2.544274/F/F + 2.583332/F/F/F
  + 2.598259/F/F/F/F + 0.6047031/F/F/F/F/F
```
degrees of freedom (one estimate
  is involved)
See Section 5.1.2

```
B1 = .7006128 + .5264558*A
  + .1106185*A*A − .09982465*A*A*A
  + .118711*A*A*A*A − .05364844*A*A*A*A*A
  + .008313859*A*A*A*A*A*A
B1 = 10^B1
```
lower bound on χ^2, see
  Section 5.1.4, ($p = 0.025$)

```
B2 = −3.004395 + 7.332854*A
  −6.74877*A*A + 4.007137*A*A*A
  −1.214192*A*A*A*A + .134666*A*A*A*A*A
  + .004115038*A*A*A*A*A*A
B2 = 10^B2
C1 = T0*S
C2 = C1/SQR(N)
C3 = SX*SQR(F/B1)
C4 = SX*SQR(F/B2)
```
upper bound on χ^2 ($p = 0.975$)

t^*s of
the distribution,
the mean,
LCL, UCL of the std. dev.

```
PRINT XB, S, XB − C1, XB + C1, XB − C2, XB
  + C2, C3, "...", C4
SUB END
```

Table 5.9. Rounding a Result Using File QUOTE_RESULT.xls[a]

$n = 15$	$t(14, 0.05) = 2.145$	
$x_{mean} = \mathbf{3.3847}$	$\chi^2(0.975) = 5.629$	$0.3191 \cdot \sqrt{14/5.629} = 0.503$
$s_x = \mathbf{0.3191}$	$\chi^2(0.025) = 26.12$	$0.3191 \cdot \sqrt{14/26.12} = 0.234$
$t \cdot CV(x)$:		
± 0.684	$CL(x)$: 2.70 ... 4.07	(distribution)
$t \cdot CV(x_{mean})$:		
± 0.177	$CL(x_{mean})$: 3.21 ... 3.56	(mean)
$f \cdot s_x$:		
$-0.086, +0.184$	$CL(s_x)$: 0.234 ... 0.503	(standard deviation)
quoted result:	$\mathbf{3.38 \pm 0.32}$	("$n = 15$, confidence interval of the mean for $p = 0.05$")
or	$x_{mean} = 3.4,$	
	$s_x = \pm 0.3,$	
	$n = 15$	

[a]For the approximations given in this chapter, some numbers are marginally different: $t = 2.146$, $\chi^2 = 5.601$ resp. 26.10, $f \cdot s_x$: $-0.085, +0.186$, $CL(s_x)$: 0.234 ... 0.505.

lems encountered in the typical analytical lab, and are instructive. The level is intermediate in statistical complexity, somewhere between the extremes of statistical models that correctly portray only the most simple situations, and others that should only be employed by a specialist armed with years of experience. Programs, too, come in a variety of forms, from those that offer only cryptic prompts and one number results to others that feature the full range of statistical tests, exquisite graphic capabilities, and dozens of options. Here, the attempt was made to come to terms with data sets of up to a few hundred numbers and a few dimensions (variables) using concepts that are easy to explain and are intuitively understood. The authors hope to have come up with a user-interface that emphasizes the didactically important aspects and is an enabler rather than a burden.

5.3.1 Hardware/Configuration

The following minimal hard- and software is necessary:

- A PC with at least 16 MB RAM.
- Windows 95 or NT operating system, or higher.
- A VGA graphics card.
- A Windows-compatible printer.
- At least 5 MB free space on the hard disk.

Table 5.10. BASIC Code for the Core of Program CORREL

Assignments:	
R(,)	data array, N rows (samples), M columns (attributes)
XB()	column averages x_j
SXY(,)	sum of squared residuals per column r_{ii} on diagonal, and sums of products of residuals r_{ij} in cells above diagonal
C(,)	correlation coefficients c_{ij} for each pair of columns i and j
P(,)	probability of the correlation being due to chance

Listing:

```
   SUB CORREL(N, M, R(,))
   DIM SXY(M,M),XB(M),R(N,M),
     C(M,M),N(M,M)
   XB() = 0 & N1 = 0                  set all elements set to zero
   FOR J = 1 TO M
   FOR I = 1 TO N
   IF NOT R(I,J) THEN 80              jump if no value left in list
   XB(J) = XB(J) + R(I,J)
   N1 = N1 + 1
80 NEXT I
   XB(J) = XB(J)/N1                   column means
   NEXT J
   SXY(,) = 0                         clear matrix
   N1 = 0
   FOR J1 = 1 TO M                    calculate sums of squares
   FOR J2 = J1 TO M
   FOR I = 1 TO N
   IF NOT R(I,J1) OR NOT R(I,J2)
     THEN 190
   SXY(J1,J2) = SXY(J1,J2)            calculate correlation sum
     + (R(I,J1) − XB(J1))*(R(I,J2) − XB(J2))
   N1 = N1 + 1
190 NEXT I
   N(J1,J2) = N1
   NEXT J2
   NEXT J1
   FOR J1 = 1 TO M − 1
   FOR J2 = J1 + 1 TO M
```

Table 5.10. (*Continued*)

GOSUB 500	p is calculated from r^2 via t
NEXT J2	
NEXT J1	
PRINT RESULTS IN MATRIX FORM	
END	
500 C(J1,J2) = SXY(J1,J2)/SQR(SXY(J1,J1)	C(J1,J2) contains the correlation
*SXY(J2,J2))	coefficient r_{jk}, which is
T = ABS(C(J1,J2))	then transformed to a Student's t
*SQR((N(J1,J2) − 2)/(1 − C(J1,J2)	and from there to a probability p
*C(J1,J2)))	
F = N(J1,J2) − 1	
CALL PROBAB(R,T,P)	
RETURN	
SUB END	

Table 5.11. Data Table $X(,)$ and Means

Raw Data			Residuals		
Solv. B	Assay Titr.	Assay HPLC	Solv. B	Assay Titr.	Assay HPLC
6.9	98.0	98.0	.24	1.08	1.60
6.2	97.0	94.9	−.46	0.08	−1.50
7.0	94.0	94.0	.34	−2.92	−2.40
6.3	96.4	96.7	−.36	−0.52	0.30
6.9	99.2	98.4	.24	2.28	2.00
6.66	96.92	96.40	Means		

Table 5.12. Sums of Squares

	$K = 1$	$K = 2$	$K = 3$
$J = 1$	0.572	−0.036	0.63
$J = 2$		15.168	13.02
$J = 3$			14.66

Table 5.13. Results of Correlation

Parameters Correlated	Correlation Coefficient	Student's t	Probability of Error	Sign of Slope
B/titr.	−0.0122	0.021	49%	neg
B/HPLC	0.218	0.39	36%	pos
titr./HPLC	0.873	3.10	1.8%	pos

- A resolution of 800 · 600 is necessary (use small font); a resolution of 1024 · 768 is optimal (use either small or large font); higher resolutions can be used with a 17-inch or larger screen (use large font); a resolution of 1280 · 1024 distorts the text.

5.3.2 Software

In this section, instructions for start-up operations and descriptions of programs are given:

Conventions   In the following explanations, square brackets as in [3.56] identify text or numbers to be entered literally, respectively keys that are to be pressed, such as the [F2] key. Options are signalled by angled brackets, e.g. ⟨Open File⟩; "↵" = [ENTER].

The SMAC software is available at ftp://ftp.wiley.com/public/sci_tech_med/SMAC. Follow the instructions posted online for downloading and installing the program.

Starting a Program   The following instructions use program LINREG as an example to guide the user through the options:

1. Double-click the SMAC-icon (Fig. 5.1) on your desktop that features three Gaussian peaks; the main menu screen appears. (See Fig. 5.2.) The full-screen background of this and all screens that follow is light blue and contains about a dozen renditions, in a range of sizes, of the initials "SMAC" the coordinates of which change with every call.
2. Click on a light-blue program description (item B) for a window that lists the main features of the corresponding program; return to the main menu by clicking on the red [EXIT] button (C);
3. Click on the blue cell that contains the program name ⟨LinReg⟩ (A) to load and start the LINREG.exe-file.

Title Screen   Each program starts off with a title screen, see Fig. 5.3, that repeats the major features, displays the "SMAC" initials, lists the program

Table 5.14. BASIC Code for the Core of Program FACTOR

Assignments:

R(,)	low/high levels of factors
Y()	measurements
E()	derived effects
F()	reduced effects
M()	model coefficients

Listing:

PROGRAM FACTOR

INPUT 'measured value for factor 1:'; Y(1)
INPUT 'measured value for factor A:'; Y(2)
INPUT 'measured value for factor B:'; Y(3)
INPUT 'measured value for factor C:'; Y(4)
INPUT 'measured value for factor AB:'; Y(5)
INPUT 'measured value for factor AC:'; Y(6)
INPUT 'measured value for factor BC:'; Y(7)
INPUT 'measured value for factor ABC:'; Y(8)

INPUT levels of factors,
  see scheme:

factor	low	high	
A	40	50	deg. C
B	1	2	ratio
C	6	7	pH

```
E(1) = (-Y(1) + Y(2) - Y(3) - Y(4) + Y(5)
   + Y(6) - Y(7) + Y(8))/4
E(2) = (-Y(1) - Y(2) + Y(3) - Y(4) + Y(5)
   - Y(6) + Y(7) + Y(8))/4
E(3) = (-Y(1) - Y(2) - Y(3) + Y(4) - Y(5)
   + Y(6) + Y(7) + Y(8))/4
E(4) = (+Y(1) - Y(2) - Y(3) + Y(4) + Y(5)
   - Y(6) - Y(7) + Y(8))/4
E(5) = (+Y(1) - Y(2) + Y(3) - Y(4)   Y(5)
   + Y(6) - Y(7) + Y(8))/4
E(6) = (+Y(1) + Y(2) - Y(3) - Y(4) - Y(5)
   - Y(6) + Y(7) + Y(8))/4
E(7) = (-Y(1) + Y(2) + Y(3) + Y(4) - Y(5)
   - Y(6) - Y(7) + Y(8))/4
```

calculate effects

```
F(1) = E(1)/(R(1,2) - R(1,1))
F(2) = E(2)/(R(2,2) - R(2,1))
F(3) = E(3)/(R(3,2) - R(3,1))
F(4) = E(4)/(R(1,2) - R(1,1))/(R(2,2)
   - R(2,1)) * 2
F(5) = E(5)/(R(1,2) - R(1,1))/(R(3,2)
   - R(3,1)) * 2
F(6) = E(6)/(R(2,2) - R(2,1))/(R(3,2)
   - R(3,1)) * 2
F(7) = E(7)/(R(1,2) - R(1,1))/(R(2,2)
   - R(2,1))/(R(3,2) - R(3,1)) * 4
```

calculate factors

Note: F = E/R1/R2*2 means that E
  is first divided by the difference
  R1, then divided by the difference
  R2, and finally multiplied by 2.

Table 5.14. (*Continued*)

ENTER: s_y	Estimated standard deviation s_y
$P = t \cdot s_y/\text{sqr}(8)$	t-Factor $t = 4.604$ for $f = 4$ and
	$p = 0.005$ (one sided) is
	incorporated in program.

```
FOR I = 1 TO 7
IF  F(I) > P THEN M(I) = F(I)          significant effect.
    ELSE M(I) = 0                      insignificant effect.
NEXT I
INPUT 'coordinates A, B, C:'; A, B, C  MODEL: obtain coordinates A, B, C
A(9) = A - (R(1,2) + R(1,1))/2         find coordinates
B(9) = B - (R(2,2) + R(2,1))/2         relative to
C(9) = C - (R(3,2) + R(3,1))/2         of center of cube
Y = 0
FOR I = 1 TO 8
Y = Y + Y(I)
NEXT I
Y = Y/8
Y = Y + M(1)*A + M(2)*B + M(3)*C       extrapolate from center of
Y = Y + M(4)*A*B + M(5)*A*C + M(6)*B*C   cube to point A, B, C
Y = Y + M(7)*A*B*C
```

name, the authors, the ISBN number, the publisher's name, the copyright sign, and the format remark.

Menu Bar, Pull-Down Windows   The first item in the menu bar is the pull-down menu ⟨File⟩, which lists the appropriate selections ⟨Open File⟩, ⟨Close File⟩, and ⟨Close LinReg⟩.

- ⟨Open File⟩ gives access to the standard file-selection window and then branches to ⟨Choose Vector⟩.
- ⟨Close File⟩ discards the data being used and all results and closes the presently open file.
- ⟨Close LinReg⟩ releases the program and returns control to the main menu.

The second item in the menu bar is ⟨Data⟩, which lists the selections ⟨Round, Column Width⟩, respectively ⟨Choose Vectors⟩.

Table 5.15. BASIC Code for the Core of Program HISTO

Assignments:

X() (length = N)	values to be analyzed
U() (length = NB)	number of observations per bin
V() (length = NB)	frequency of observation (%) per bin
W() (length = NB)	cumulative frequencies
XMIN:	smallest x-value
XMAX:	largest x-value
IL:	number of observations x smaller x_{min}
IH:	number of observations x larger x_{max}
W:	width of a bin

Listing:

```
SUB HISTO(N,X())              set up arrays and variables
INPUT "Xmin, Xmax: ";X1,X2    obtain x-range and
INPUT "number of bins:";NB    bin number
DIM U(NB),V(NB),W(NB)

FOR I = 1 TO NB               set counts to zero
U(I) = 0
NEXT I

XMIN = 1E99                   set extremes to "infinity"
XMAX = -XMIN
IL = 0                        set counts to zero
IH = 0
SX = 0                        set statistical sums to zero
SXX = 0
W = (X2 - X1)/NB              bin width
FOR I = 1 TO N                for all measurements in the list X()
X = X(I)
J = INT((X - X1)/W + 1)       assign bin number J, '+1' is necessary
                             so that J > 0
IF J > 0 AND J > NB THEN U(J) increment count in bin J
   = U(J) + 1
IF X < XMIN THEN XMIN = X     update extremes, if necessary
IF X > XMAX THEN XMAX = X
IF X < X1 THEN IL = IL + 1    update counts of extremes
IF X > X2 THEN IH = IH + 1
SX = SX + X                   calculate statistical sum
NEXT I

FOR I = 1 TO NB
V(I) = 100*U(I)/N             calculate frequencies
NEXT I
```

Table 5.15. (*Continued*)

`W(1) = 100*IL/N + V(1)`	cumulative frequencies, starting with
`FOR I = 2 TO NB`	bin 1 that is the sum of bin V(1) and
`W(I) = W(I - 1) + V(I)`	all counts to left of x_{min}
`NEXT I`	
`XB = SX/N`	mean
`FOR I = 1 TO N`	
`R = X(I) - XB`	residual
`SXX = SXX + R*R`	sum of squared residuals
`NEXT I`	
`S = SQR(SXX/(N - 1))`	standard deviation
`H = N*W/S/SQR(2*PI)`	Normalization factor H; PI = 3.14...
`FOR X = XB - 3*S TO XB + 3*S`	calculate and plot normal distribution
`  STEP S/50`	curve of equal area
`Y = H*EXP(-X*X/S)`	
`PLOT (XB - X,Y) and (XB + X,Y)`	
`NEXT X`	
`SUB END`	

- ⟨Round, Column Width⟩ permits the raw data presentation to be adjusted by choosing the number of significant digits to which the numbers are rounded to, see "Presentation of Numbers" below; the column width is automatically adjusted.

- ⟨Choose Vectors⟩ permits two columns (vectors) to be chosen from a data table and assigned to specific roles, i.e., abscissa and ordinate see Fig. 5.4, E, F. The file name and path is confirmed directly under the screen title. The "size-of-file" window gives the number of rows and columns (A); a scroll bar appears automatically if the screen side is insufficient for presenting the whole table. The "rounding" box (D) indicates the presently active rounding scheme; "Not rounded" is the default option. The selection procedure works by clicking on the column in question and then confirming the choice; a yellow window (C) above the table gives the instructions to be followed. In some programs a single vector is all that is needed, and in others a range of columns has to be designated.

The third item in the menu bar is ⟨Options⟩, which lists the program-specific selections, here ⟨Font⟩, ⟨Scale⟩, ⟨Specification Limits⟩, ⟨Select p⟩, ⟨LOD⟩, ⟨Residuals⟩, ⟨Interpolate $Y = f(x)$⟩, ⟨Interpolate $X = f(y)$⟩, ⟨Clear Interpolation⟩, respectively ⟨Weighted Regression⟩.

Table 5.16. BASIC Code for Subroutine NPS

Assignments:	
I, J	indices
X()	data vector
H	scaling factor for ordinate
A, P	intermediate results
FNZ	function to transform %-probability values to NPS-scale; x is the independent variable in the polynomial. (See Section 5.1.1.)

Listing:

```
SUB NPS(N,X())

  CALL SORT(N, X())                            sort x-values in ascending order

  H = -0.5*FNZ(1/N)                            scaling factor; use function FNZ below

  FOR I = 1 TO N - 1
  Y = H*FNZ(I/N)                               calculate coordinate of i-th
  PLOT Y versus X(I)                               point and plot
  NEXT I

  DEF FNZ(P)                                   use function z = f(p) Eq. (5.2) defined
  IF P < 0.5 THEN X = LGT(P) & S = -1            in Section 5.1.1
    ELSE X = LGT(1 - P) & S = 1
  Y = A + B*X + C*X^2 + . . . G*X^6            polynomial only defined for p = CP < 0.5,
  FNZ = S*Y                                      since function is symmetrical about CP
  DEF END                                        = 0.5; S gives sign.

  SUB END
```

Table 5.17. BASIC Code for Subroutine SORT

Assignments:	
X()	data vector

```
SUB SORTX(N, X())

FOR I = 1 TO N - 1
FOR J = 1 TO N   I
IF X(J) > X(J + 1) THEN SWAP X(J), X(J + 1)    The statement "SWAP A, B" can be
NEXT J                                          replaced by 3 statements "C = B";
NEXT I                                          "B = A"; "A = C"

SUB END
```

Table 5.18. BASIC Code for the Core of Subroutine ANOVA

Assignments:
  R(,)   data vector
  S()   sums of values
  N()   number of values in group
  N   number of values in largest group
  M   number of groups
  N1   total number of values
  XB()   group means
  XGM   grand mean

Listing:

```
     SUB ANOVA(N, M, R(,))

     DIM R(N,M), S(M), N(M), XB(M)
     N1 = 0

     FOR J = 1 TO M
     SU = 0
     S(J) = 0
     N(J) = 0

     FOR I = 1 TO N
     IF NOT R(I,J) THEN 120            jump if no value in list
     S(J) = S(J) + R(I,J)
     N(J) = N(J) + 1
120  NEXT I
     N1 = N1 + N(J)
150  SU = SU + S(J)                    grand total
     XB(J) = S(J)/N(J)                 group mean
     NEXT J

     XGM = SU/N1                       grand mean
     SGM = 0
     SXX = 0
     FOR J = 1 TO M                    sum of weighted squared
     SGM = SGM + N(J)*(XB(J) − XGM)*   deviations of group means from
        (XB(J) − XGM)                     grand mean

     FOR I = 1 TO N
     IF NOT X(I,J) THEN 260
     SXX = SXX + (R(I,J) − XB(J))*(R(I,J)   sum of squared residuals. See
        − XB(J))                              Eq. (1.30)

260  NEXT I
     NEXT J

     ST = SGM + SXX                    See Eq. (1.31)
```

Table 5.18. (*Continued*)

F1 = N1 − M	sum of variances = total
F2 = M − 1	variance. See Eq. (1.33)
PRINT SXX, F1, SXX/F1	for example of print-out see
PRINT SGM, F2, SGM/F2	Table 1.9 + 11
PRINT ST, N1 − 1, ST/(N1 − 1)	
SUB END	

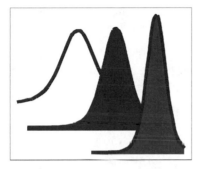

Figure 5.1. The SMAC-Icon.

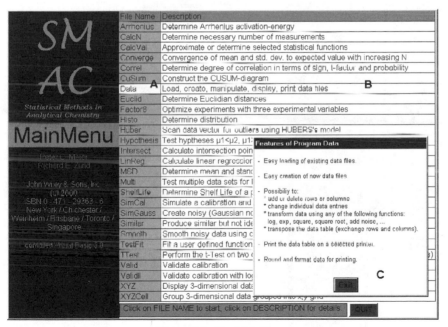

File Name	Description
Arrhenius	Determine Arrhenius activation-energy
CalcN	Determine necessary number of measurements
CalcVal	Approximate or determine selected statistical functions
Converge	Convergence of mean and std. dev. to expected value with increasing N
Correl	Determine degree of correlation in terms of sign, t-factor and probability
CuSum	Construct the CUSUM-diagram
Data	Load, create, manipulate, display, print data files
Euclid	Determine Euclidian distances
Factor8	Optimize experiments with three experimental variables
Histo	Determine distribution
Huber	Scan data vector for outliers using HUBERS's model
Hypothesis	Test hyptheses μ1<μ2, μ1?
Intersect	Calculate intersection point
LinReg	Calculate linear regression
MSD	Determine mean and stand
Multi	Test multiple data sets for
ShelfLife	Determine Shelf Life of a p
SimCal	Simulate a calibration and
SimGauss	Create noisy (Gaussian n
Similar	Produce similar but not ide
Smooth	Smooth noisy data using c
TestFit	Fit a user defined function
TTest	Perform the t-Test on two
Valid	Validate calibration
Validll	Validate calibration with lo
XYZ	Display 3-dimensional dat
XYZCell	Group 3-dimensional data grouped into x,y grid

Features of Program Data

- Easy loading of existing data files.

- Easy creation of new data files

- Possibiliy to:
  * add or delete rows or columns
  * change individual data entries
  * transform data using any of the following functions:
    log, exp, square, square root, add noise, ...
  * transpose the data table (exchange rows and columns).

- Print the data table on a selected printer.

- Round and format data for printing.

Click on FILE NAME to start, click on DESCRIPTION for details

Figure 5.2. The MainMenu.

- ⟨Font⟩ gives access to the installed fonts and text size; a simple font like Arial with a 8 or 10 point size is best; fancy fonts tend to obscure the message and large fonts make it difficult to inspect a large table.

- ⟨Scale⟩ allows the user to select a specific area of the x, y-plane for display. Tic-mark intervals can be tailored to the application; tics are always displayed such that one tic coincides with the origin ($x = 0$ or $y = 0$). The default scale adds 5% to the x- resp. y-ranges on all four sides. (See Figs. 5.5 and 5.6.)

- ⟨Specification Limits⟩ allows (vertical) dashed lines to be added to illustrate the Accept/Reject decision points that the Quality Assurance Department uses in product release.

- ⟨Select p⟩ allows the confidence level to be changed ($p = 0.25, 0.1, 0.05, 0.025, 0.01, 0.005$, or 0.001 (two-sided tests, default $p = 0.05$).

- ⟨LOD⟩ calculation determines the limits of detection and quantitation

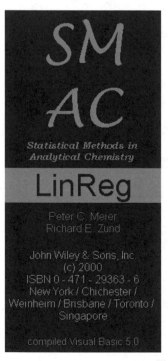

Features

- Choose any 2 vectors and define either as abscissa X or as ordinate Y.

- Calculate linear regression and display graph: points, regression line, upper and lower 95% confidence limits CL for regression line.

- Display key results: number of points N, intercept a, slope b, both with 95 % confidence limits, coefficient of determination r², residual standard deviation.

- Interpolate Y=f(x), ± 95% CL for any x, interpolate X=f(y*,k), ± 95% CL for any mean y* from k determinations.

- Calculate and display the residuals; add specification limits; print all results and table of values.

- Determine the limit of detection LOD and limit of quantitation LOQ according to the interpolation at level y = a + CL of the regression line and its lower CL; this is sensitive to the calibration-point pattern!

Figure 5.3. The Title Page.

Display and Select Data

Linear Regression: Raw Data

A   File D:\smac\Daten\Samples\Cyanide.dat

Size of array	How to proceed	Rounding
33 row(s) B	**Select column for independant** C	Numbers are NOT rounded
8 column(s)	variable x	D

i\j	1	2	3	4	5	6	7	8
	Conc CN- [ug/100ml]	Absorb [AU]	Conc CN- [ug/100ml]	Absorb [AU]	Conc CN- [ug/100ml]	Absorb [AU]	Conc CN- [ug/100ml]	Absorb [AU]
1	0	0	0	0	0	0	0	0
2	10	0.049	20	0.099	2	0.0095	10	0.049
3	30	0.153	40	0.203	4	0.0198	30	0.153
4	50	0.258	60	0.31	6	0.0299	50	0.258
5	70	0.356	80	0.406	8	0.0402	70	0.356
6	90	0.46	100	0.504	10	0.0501	90	0.46
7	110	0.561	120	0.609			110	0.561
8	130	0.671	140	0.708			130	0.671
9	150	0.761	160	0.803			150	0.761
10	170	0.863	180	0.904			170	0.863
11	190	0.956	200	0.997			190	0.956
12	210	1.053	220	1.102			210	1.053
13	230	1.158	240	1.186			230	1.158
14	250	1.245					250	1.245
15							0	0
16	E	F					20	0.099
17							40	0.203
18							60	0.31
19							80	0.408
20							100	0.504
21							120	0.609
22							140	0.708

Figure 5.4. The presentation of the values contained in a data file. A: program and file names; B: size of file; C: instructions on how to proceed; D: currently active rounding option; E: abscissa vector; F: ordinate vector.

according to the calibration-sensitive model described in Section 2.2.7; the associated assumptions and the calculations are displayed in ⟨LOD Explanations⟩.

- ⟨Residuals On/Off⟩ permits the residuals to be superimposed at a chosen ordinate and with a given magnification relative to the ordinate scale.

- ⟨Interpolate $Y = f(x)$⟩ requests the user to either enter a specific x-value into the green box (Fig. 5.6, item D), followed by ↵, or to use the mouse pointer to indicate where the interpolation is to take place (depress left button and slowly pull mouse). The corresponding results are continuously updated in the table. The confidence interval of the result Y is indicated by a bold bar sitting on top of the dashed interpolation line. Clicking on the pale yellow [Print] button sends the numerical results to the selected printer; there is the option of sending a [Form Feed] immediately or after a few interpolations have been done.

- ⟨Interpolate $X = f(y)$⟩ works similarly, only that in addition to the mean measurement y^* the number of repeats k has to be entered, followed by ↵; the default is $k = 2$.

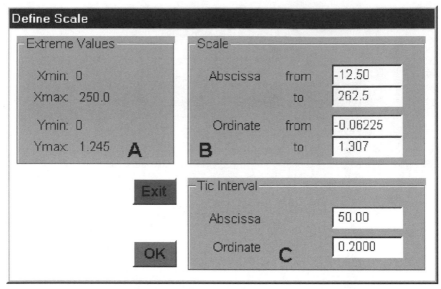

Figure 5.5. The Scale Selection Page. A: extreme values in data set; B: currently active scale (= default scale of extremes $+/- 5\%$ when first opened); C: tic-mark spacing, initially set by default.

- ⟨Clear Interpolation⟩ refreshes the screen without the interpolation results.

- ⟨Weighted Regression⟩ requires the user to define a signal-dependent model of the measurement error, e.g., $s_Y = a + b \cdot x$, which is then used to calculate the weighting factors $1/V_Y$ at every abscissa x_i. For an example on how to enter the model, see "Algebraic Function,"

The item ⟨Display⟩ allows the user to chose between the graph and the corresponding table of data, see Fig. 5.7.

The last item in the menu bar is ⟨Output⟩, which contains the selections ⟨Setup Printer⟩, ⟨Setup Print Job⟩, ⟨Print Graph and Main Results⟩, ⟨Print Table of Detailed Results⟩, and ⟨Copy to Clipboard⟩.

- ⟨Setup Printer⟩ gives access to the standard printer-selection window, which includes the setting of the paper size and orientation.

- ⟨Setup Print Job⟩ permits the following to be defined (See Fig. 5.8.): paper margins, size of rectangular area to which the graph is mapped, units (mm, cm, inch), and line width. In this way, graphics can be reproducibly scaled to be exactly superimposable because the axis and tic

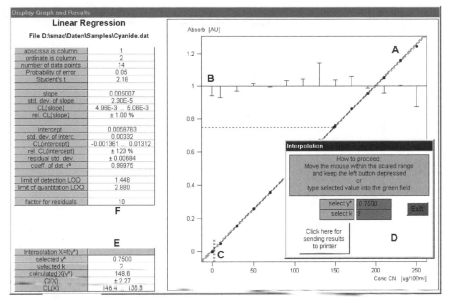

Figure 5.6. The LinReg Graph. A: the regression line with the 95% CL; B: residuals expanded by a factor of 10; C: LOD and LOQ; window D: option for entering specific numerical values for y^* and k, and for sending the interpolation results to the printer; E: numerical results of the specified interpolation; F: other results.

labels are written into a reserved margin between this rectangle and the defined paper margins. If selected too large, the size of the rectangular area defaults to the maximum possible given the size of the paper, the margins, and the space for axis labels (see "Limit" number given below the input box).

- 〈Print Graph〉 sends the graph and the associated results to the selected printer.

- 〈Print Results〉 sends the table of raw data and derived results to the selected printer.

- 〈Copy to Clipboard〉 provides the selection 〈Graph〉, 〈Results〉, and 〈Values〉. (See Fig. 5.9.) One of the three can be selected and sent to the clipboard. After changing to another application, such as Microsoft Word, the contents of the clipboard can be retrieved; use the [Alt] [Tab] keys for rapidly switching between active programs. After returning to the SMAC program, select the next output.

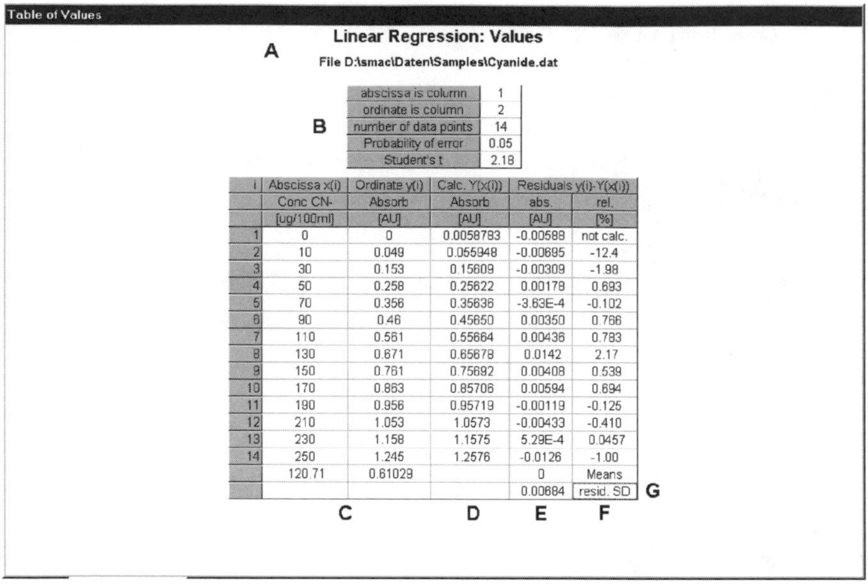

Table of Values

Linear Regression: Values
A
File D:\smac\Daten\Samples\Cyanide.dat

	abscissa is column	1
	ordinate is column	2
B	number of data points	14
	Probability of error	0.05
	Student's t	2.18

i	Abscissa x(i)	Ordinate y(i)	Calc. Y(x(i))	Residuals y(i)-Y(x(i))	
	Conc CN-	Absorb	Absorb	abs.	rel.
	[ug/100ml]	[AU]	[AU]	[AU]	[%]
1	0	0	0.0058783	-0.00588	not calc.
2	10	0.049	0.055948	-0.00695	-12.4
3	30	0.153	0.15609	-0.00309	-1.98
4	50	0.258	0.25622	0.00178	0.693
5	70	0.356	0.35636	-3.63E-4	-0.102
6	90	0.46	0.45650	0.00350	0.766
7	110	0.561	0.55664	0.00436	0.783
8	130	0.671	0.65678	0.0142	2.17
9	150	0.761	0.75692	0.00408	0.539
10	170	0.863	0.85706	0.00594	0.694
11	190	0.956	0.95719	-0.00119	-0.125
12	210	1.053	1.0573	-0.00433	-0.410
13	230	1.158	1.1575	5.29E-4	0.0457
14	250	1.245	1.2576	-0.0126	-1.00
	120.71	0.61029		0	Means
				0.00684	resid. SD	G

C   D   E   F

Figure 5.7. The Output Option ⟨Table of Values⟩. A: option identification and data path; B: data set identity and size, derived Student's *t*, selected *p*; C: abscissa and ordinate values; D: estimated $Y = f(x)$; E: absolute residuals; F: relative residuals; G: mean over absolute residuals and residual standard deviation.

Other options are discussed in the context of the presentation of the individual programs.

Data Input, Data Editor   After each numerical input an editor is called that accepts both Anglo-American and Continental numbers ([NN,NNN.NN], respectively [NN'NNN,NNE-12], with or without an exponent or a separator). Text is not accepted.

Data Storage   All numbers arrived at either through simulation (e.g., program SIMGAUSS) or transformation (e.g., division, square root, logarithm in program DATA) are clipped to five significant digits; this is reasonable in view of the fact that measurement precision is rarely better than the $1 : 10^3$ or $1 : 10^4$ level.

Presentation of Numbers   All numbers are presented in the Anglo-American decimal point format, [NNNN.NNN], that is, without commas to separate groups of three digits. A custom-designed procedure displays a number with the preset number of significant digits as long as the chosen number

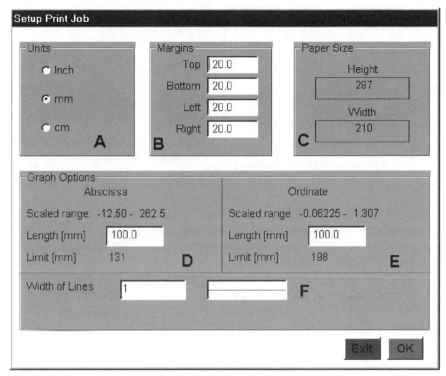

Figure 5.8. The Print Job Page. A: user units; B: paper margins in user units (for axis labels); C: physical size of paper (cannot be changed here, cf. ⟨Set up Printer⟩); D: physical size of graph's abscissa in user units (box around graph area, does not include axis labels); E: idem for ordinate; F: select line width.

of decimal places suffices, otherwise the display switches to scientific format to conserve the number of significant digits. If X significant digits are requested, then the display will ordinarily include up to $X + 5$ symbols (sign, numbers, decimal point, and exponent).

Numerical Accuracy Double precision arithmetic is used throughout; this in principle limits the number of significant digits to 15. (See Table 1.1.) If many operations are involved in obtaining a final result, as in most statistical procedures, truncation will take its toll, and on the order of 8 to 10 digits would remain reliable. Since the measurements that go into a calibration rarely are accurate and precise enough to warrant more than 4–5 significant digits, and the algorithms that are used to eliminate table look ups, see Section 5.1, are accurate to about 2–4 digits, all results are purposely rounded to

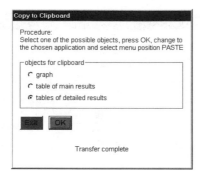

Transfer complete

Figure 5.9. The Copy to Clipboard Page in the ⟨Output⟩ Option.

a final 2–5 digits, which suffices in practical situations. In this way, the user is not misled by superfluous digits and is fored to think about the statements that can be made, and those that cannot. Note that if the selected number of decimal places exceeds that necessary to present a certain number of significant digits, zeros are added. Rounding, as described in "Presentation," affects only the output; all calculations are done using full precision.

Algebraic Function Programs LINREG, SIMCAL, SIMGAUSS, SIMILAR, and TESTFIT provide a formula editor module for entering an algebraic function, much in the way a program is entered in some scientific calculators, see Fig. 5.10. The module permits 24 lines to be entered that consist either of a number, a parenthesis, or an operator. At the end of each line, a ↵ closes that line, the cursor jumps to the next one, and the signs just entered are added to the function window in the lower left corner. The mouse pointer can also be used to advance to the next line. Empty lines are permissible. Errors at this stage can be corrected by clicking on the line in question and editing the contents. An [OK] at the end initiates the parsing of the statement and as a control the value for Y_{AF} $f(x = 2)$ appears in the lowest window. [Continue] returns control to the evaluation program and a line is plotted across the screen that shows the function $Y_{AF} = f(x)$ just entered (the scales are the same as those of the abscissa and ordinate). In the case of a $s_y = f(x)$ function, see programs LINREG and SIMILAR, $Y_{AF} = 0$ is shifted to coincide with the lower edge of the graph. Because standard deviations must be positive, the user is requested to enter the method repeatability (s_y^{min}, short-term stability of signal); the calculated $s_y = Y$ is then limited to be $Y \geq s_y^{min}$ in these programs.

An example of an algebraic function is

$$Y = 0.023 + 0.0034^*(x - 5) + 0.00002^*x^2,$$

Table 5.19. Syntax for the Algebraic Function

Operator	Example	Syntax	Explanation
+	$y = a + b$	+ ↵	addition
−	$y = a - b$	− ↵	subtraction
*	$y = a \cdot b$	* ↵	multiplication
/	$y = a/b$	/ ↵	division
log	$y = \log_{10}(a)$	log ↵ (↵ a ↵) ↵	logarithm to base 10
ln	$y = \ln(a)$	ln ↵ (↵ a ↵) ↵	natural logarithm
exp10	$y = 10^a$	10^ ↵ (↵ a ↵) ↵	exponent to base 10
exp	$y = e^a$	exp ↵ (↵ a ↵) ↵	exponent to base e
sqr	$y = \sqrt{a}$	sqr ↵ (↵ a ↵) ↵	square root
^2	$y = a^2$	^2 ↵ (↵ a ↵) ↵	square

The [ENTER] sign ↵ indicates the end of the line. Note that the first operator cannot be put into row 1, but needs to be preceded by a factor, such as 1↵ * ↵ on the first two lines. After the caret "^" in 10^, a blank must be added.

which would be entered in the following fashion, starting with the top row (empty lines are ignored):

$$[0.023]↵ [+] ↵ [0.0034] ↵ [^*] ↵ [(] ↵ [x] ↵ [-5] ↵ [)] ↵ [+]$$

$$↵ [0.00002] ↵ [^*] ↵ [^2] ↵ [(] ↵ [x] ↵ [)] ↵$$

When the "Function" and the "Result for $x - 2$" windows have been inspected and found to be in order, click on [Continue] to return to the main program. The last few instructions of the function could, of course, have been entered as [0.00002] ↵ [*] ↵ [x] ↵ [*] ↵ [x] ↵.

The available operators are as shown in Table 5.19 and Fig. 5.10.

Graphics Much energy was expended on presenting the situation (location and scatter of data points), and with it the relationship between results and specification limits in a graphical format. Autoscaling is the rule, but there is always the possibility of changing the data-driven scale to one that allows for cross-comparisons between various experiments. The display consists of the box containing the graphical depiction; the tic marks and their labels, the axis labels, a title, and supplimentary information are outside.

- The box is the scalable entity, which is assigned to a defined rectangular area on the printer, see ⟨SetUp Print Job⟩.
- The tic mark intervals can be chosen within certain boundaries (from 1/50 to 100% of the numerical range); tic marks are so arranged that they start from the origin, and not the lower end of the axis range.

Define Weighting Function

Type each element of the empirical formula into a cell.		1	0.02
At the end click OK to accept, then CONTINUE to		2	-
quit/continue. The following elements are accepted		3	exp
(observe the remarks):		4	(
	A	5	0.0043
VALUES:		6	*
numbers fig. 0 … 9, decimal point, pos/neg exponent		7	x
X pos. abscissa Value		8	)
-X neg. abscissa Value		9	**B**
		10	
OPERATIONS: (Remarks)		11	
+ addition		12	
- subtraction	with arguments following	13	
* multiplication	directly or in brackets	14	
/ division		15	
		16	
log logarithm to base 10		17	
ln natural logarithm		18	
exp exponent of e	with arguments following	19	
exp10 exponent of 10	allways in brackets	20	
sqr square root		21	
^2 square		22	
		23	
(opening bracket	nested brackets are	24	
) closing bracket	not allowed		

Function:

$y = 0.02-exp(0.0043*x)$ **C**

Result for $x = 2$:   **D** [OK]   [Exit]   **F**

-.988637086237646 **E**   [Continue]

Figure 5.10. The Algebraic Function. A: instructions and available operators; B: a sample function; C: the parsed function; D: the [OK] button that initiates the parsing operation; E: the evaluation at $x = 2$; F: the [Continue] and [Exit] buttons.

- The axis labels—header and dimension—are given by the data, see program DATA, option ⟨Edit⟩.

- The title identifies the type of graph.

- The supplimentary information comprises the file name and the vector number(s), and important results, but not the data table as such.

Tables In general, a table of the data being used is provided. Depending on the particular program, this may be expanded to include means and standard deviations, interpolations, references, equations, the results of other calculations, and explanations.

Output Formats   Graphics and tables are available in the following formats:

- On the screen.
- As hardprints on a Windows-compatible printer (only little color is used, so that a monochrome printer will also do a good job).
- As clipboard objects (*.bmp format) for export to text and graphics software.
- As *.dat files (some programs, like DATA and SIMILAR, create or add to data files). The data structure is such that a *.dat file can easily be imported into Excel by invoking ⟨Open*.*⟩ and choosing the ⟨Separated by Comma⟩ option.

If the user presses [Print Screen], the contents of the whole screen, including buttons, icons, and status bar will be dumped to the clipboard.

Errors   The authors have spent much time to eliminate the bugs that inevitably creep in whenever someone undertakes a programming task. The data entry and calculation modules were equipped with error-trapping routines that guard against all kinds of mishaps, like letters in number fields and divisions by zero. Colleagues were enlisted to test the result for rugged and fail-safe operation, and user friendliness. Despite best intentions, it is likely that some errors will still be found. Some of these will not be due to carelessness on the authors' part: The sheer complexity of the operating system, uncontrollable configurations (hardware, device drivers, user-supplied data files with really queer numbers, software that is operational in the background), and some residual bugs in VB5.0 can occasionally conspire to drive the system over the brink, even if no programming error is in evidence; control will then revert to the Main Menu. The suggestion is then to shut down all unused applications or even to reboot the PC.

5.4   PROGRAM AND DATA FILE DESCRIPTION

The following programs and associated data files are provided with the book.

5.4.1   Program Flow, User Interface

Most programs are structured along the following sequence of steps:

1. Title Page: display name, version, and features of program.

2. Select the menu item ⟨Open File⟩, or ⟨Start⟩ if no data file is needed.

3. Display the standard Open-File window; load file.

4. Display the data table, with scroll bars, if necessary.

5. Select one or more columns (vectors) for analysis.

6. Select data analysis options.

7. Display a graphical depiction.

8. Display tables of data or results.

9. Modify graph.

10. Select output options.

11. Select other data in same file or other file.

12. Close program.

A consistent style was developed that is apparent across all programs:

- The appearance of the programs is modeled on the familiar Windows surface to reduce the number of hurdles the reader encounters. For instance, loading a data file is accomplished by means of the standard Open-File window.

- The full-screen background is light blue and features the "SMAC" initials in varying positions and sizes so that a uniform non-distracting background is achieved, even if the active window does not fill the full screen. Use of color was restricted to visually distinguishing classes of information (pastel colors), and to prompt the user (yellow explanation and pale yellow input fields (data)).

- The title page is divided into a left side, in shades of blue, that gives bibliographical information and the program name; the right side lists the major features.

- Across the top there is a menu bar with the usual Windows-type pull-down menus arranged from left to right in the order Files, Data Selection, Data Manipulation, Extras/Options, Output, or similar. Those options that are allowed or make sense in a given context are activated.

- Requests for numerical input make use of the standard Windows-type gray box with the question that is to be answered, the white area into which the data is written, and the appropriate confirmatory Yes/No/Cancel buttons.

- Option windows, such as for the selection of scales or printer options, are made up of one or more light blue rectangles with white windows for numerical input or selection of options.

- Numbers can appear outside the white data entry windows for information, but cannot be changed. Examples are the selected paper size in the "Set up Print Job" field, or the default scale in the "Extreme Values" field.

- The raw data table is displayed with the numbers on a white background and the headers and index on a gray one; a yellow How-to-Procede panel gives instructions, respectively confirms choices; two white panels display the file size and the presently selected rounding option.

- Messages, comments, or instructions are displayed wherever appropriate.

5.4.2 Data File Structure

The SMAC data format is identical for all programs: numbers are written with decimal points and are separated by commas; the format can be imported into word-processing and spread-sheet programs under the *.dat or the *.txt option.

Data in Excel can be imported into program DATA; Headers and dimensions can be added afterwards.

Table 5.20. Data File Structure

N, M	(# of rows, # of columns, separated by a comma); two integer numbers.
H$, D$	(M records containing a column heading H$ and a column dimension D$, separated by a comma); the text strings H$ and D$ can be any length, but beyond 10–15 characters the table becomes awkward to read; the square brackets are added automatically when the file is generated or edited in program DATA.
$R(i,j)$	(table entries, $i = 1 \ldots N$ rows, $j = 1 \ldots M$ items/row, separated by commas).
	Note: the array $R(I, J)$ contains the values; "I" is the number, that is the i-th measurement; "J" is the item or dimension number. For a program based on linear regression (LINREG, VALID, SHELFLIFE), since the array $R(,)$ must have $M \geq 2$ columns, it is up to the user to decide which column will be identified with abscissa X (index K), and which with ordinate Y (index L); $R(I, K)$ is the independent variable X, $R(I, L)$ is the dependent variable Y. K (and L, if necessary) are established by clicking on the column(s) after the file has been selected. When any program is started, the available data in the chosen file will be shown for review.

5.4.3 VisualBasic Programs

History The first edition provided the program source code (GW-BASIC for DOS 3.X and higher), which allowed the reader to adapt the programs to his own needs, such as through modification of the input routine. The programs had been conceived for wide distribution and did not make use of any features that were not part of the basic PC configuration. Thus, color was only used if the distinction was also visible on monochrome screens, still a common fixture around the turn of the last decade, and special tricks to speed calculations or to manage proportional fonts were avoided to prevent inexplicable crashes. In the meantime, Windows 95+ has surplanted the older standards, and VisualBasic has become available. The programs have been upgraded and now use the available features for maximal comfort and flexibility:

Because the programs are compiled (*.exe format), they can be run on any PC under Windows, but by the same token, they can no longer be tinkered with. For most users, this will not constitute a disadvantage, as even the authors of very well designed and documented source code occasionally run into problems when attempting to change some "minor" detail. Those who habitually tweak every screw they run into are aware that the old GW- or Q-Basic world was a much simpler place than today's universe of objects, mainly because the barebones syntax was easily memorized.

The practice of using a single data base format compatible with all programs has been retained, as has the possibility to import spreadsheets from Excel.

The following programs were written in VisualBasic 5.0 Beginner's Edition and were compiled. A consistent set of variables was defined. Modules were designed for wide applicability. The algorithms described in Section 5.1 were incorporated wherever necessary.

ARRHENIUS Section 4.21

Purpose From a series of assays done on samples stored at different temperatures over various lengths of time, the assay-vs-time trend is calculated for every temperature. These slopes and the actual storage temperatures [$°K = °C + 273.16$] are used to construct an Arrhenius Activation-Energy diagram, from which the decomposition rate at any temperature within the investigated interval can be estimated, and a shelf-life can be assigned. Note that zero-order decomposition kinetics are assumed (a zero-order reaction proceeds at a constant rate, i.e., independent of the remaining concentration); when this assumption is violated, the activation energy changes with temperature, and the Arrhenius diagram becomes nonlinear. The data format is demonstrated in file ARRHEN1.dat, which was taken from a PhD thesis (Ref. 184).

Features

- Performs linear regressions for every data set (= storage temperature).
- Plots the Arrhenius diagram: slopes ± CL *versus* $1/T$.
- Tabulates assay-vs-LinReg residuals.
- Tabulates LinReg statistics and activation energies.
- Estimates shelf-life for a given temperature.

CALCN Section 1.6

Purpose Estimate the minimal number of determinations that are necessary to assure that the resulting mean and its confidence limits are inside given specification limits.

Features The allowed range is defined by two specification limits, the experimental mean and standard deviation are assumed, and the probability of error α is preset. The calculation is done using Eq. (1.37) in Section 1.6.

- ⟨Select Probability of Error⟩ allows a choice amongst $p = 0.25, 0.1, 0.05$, 0.025, 0.01, 0.005, and 0.001 (one-sided tests, default $p = 0.025$).
- ⟨Set Experimental SD⟩ requests the estimated repeatability.
- ⟨Set Specification Limits⟩ permits the SL to be set at will, or symmetrically relative to a nominal value.
- ⟨Display Values⟩ provides the distance $|x_{mean} - L|$ and the corresponding $x_{mean, min}$ and $x_{mean, max}$ for $n = 2, 3, \ldots, 30$.

CALCVAL Section 5.1

Purpose: Graphically depict the approximations $\chi^2 = f(p, \text{df})$, $F = f(f1, f2, p)$, Student's $t = f(\text{df}, p)$, $p - f(t, \text{df})$, $CP = f(z)$, PD $f(z)$, and $z = f(CP)$.

Features

- ⟨Select⟩ allows the function to be picked; the graph is automatically depicted together with a list of selected values. Drawing the mouse pointer slowly in a horizontal direction with the left button depressed activates an interpolation engine; the coordinates are displayed on the bottom line of the list. Since the selected screen resolution (e.g., 800 ×

600, 1024 × 768, etc.) imposes a raster that prevents many abscissa values from being picked by the cursor, a green input window is provided that allows a specific abscissa value to be entered.

- ⟨Display Values⟩ presents a table of values.

- ⟨Display Accuracy⟩ presents a list of typical and extreme absolute and relative errors incurred when using the approximation; note that the listed errors are in part due to the algebraic approximation as such, and in part to the finite number of digits of the tabulated values.

CONVERGE Section 1.4

Purpose Illustrate what happens when a series of measurements is evaluated for mean and standard deviation each time a new determination becomes available: the mean converges towards zero and the SD towards 1.0. The $CL(x_{mean})$ and the $CL(s_x)$ normally enclose the expected values $E(x) = \mu = 0$, respectively $E(s_x) = \sigma = 1$. Due to the stochastic nature of the measured signal, it can happen that confidence limits do not bracket the expected value; this fact is highlighted by a bold line.

The program also serves as a reminder that under both GMP rules and scientific propriety an investigator has to first set down the experimental design (e.g., number of determinations, evaluation procedure), before starting his measurements. Only in this way can doubts be dispelled that the investigator kept on measuring until he found results that support his preconceived notions.

Features

- Assume a normal distribution ND(0, 1).
- ⟨Select Probability of Error⟩ sets p.
- ⟨Choose Number of Measurements⟩ sets N.
- ⟨Display Values⟩ shows the simulated values and the evaluation.

CORREL Section 4.11

Purpose Find correlations in a data table; the data table is organized into M columns, each of which corresponds to a dimension, e.g., concentrations of impurities, pH, absorbance at various wavelengths, etc. Each row corresponds to a sample, e.g., a batch of material analyzed according to M methods.

Features

- Calculate mean and standard deviation for every column.
- Calculate the correlation coefficient r for every combination of columns, and display the results in a triangular matrix (an absolute value just under 1.00 indicates a strong correlation between the measurements in columns i and j; a minus sign indicates that the slope is negative).
- Using the correlation coefficients, calculate the Student's t factors, and display these (the larger this value, the more probable the correlation; values below about 2 are insignificant).
- Using the Student's t factors and the number of degrees of freedom, calculate the probabilities p that correlations are due to chance alone (error probabilities); these are interpreted as follows:

  — p above about 0.1–0.2: insignificant.
  — p in the range 0.05–0.1: weak.
  — p in the range 0.02–0.05: significant.
  — p below about 0.02: highly significant.

  Negative slopes are flagged: an "N" is to be interpreted in the sense that the slope is negative, and J decreases as I increases; positive slopes remain unmarked.
- Suspected correlations are displayed by calling option ⟨Display Correlation Graph Pairwise⟩; the coordinates can be marked either with a circle (default), or with the appropriate index so that "outliers" can be readily identified.

Any interpretation must take the physical or chemical situation into due consideration:

1. A correlation can indicate a mechanism that links I and J; if I is, say, a concentration (the independent variable), the absorbance J is the dependent variable, but not *vice versa*!
2. A correlation could also be due to a mechanism that links a third, known or unknown, factor to the two observables I and J; an increase in the concentration of a complexing agent, say, could lead to increased solubility (I), and at the same time shift an UV-absorption feature by a few nanometers (J).
3. Spurious correlations are often observed, e.g.,

— When only a small number of observations, N, is available.

— If uncontrolled forces are at work, or important data has not been collected.

— When coincidences apply, for example, when the time between injections is such that the main peak co-elutes with a late peak from a previous injection.

4. Lack of correlation can mean just that, or could be due to the fact that the observations did not span a sufficiently broad range of (controlled) experimental conditions.

Because the results of all correlations are viewed simultaneously, patterns can emerge that are much more powerful indicators of hidden action than any single correlation would be.

CUSUM *Section 1.8.5*

Purpose A technique to detect deviations from random scatter in the residuals (symmetrical about 0, frequent change of sign): Cumulative sum of residuals detects changes in trend or average. Here, an average is subtracted to yield residuals; these residuals are then summed over points $1 \ldots k \ldots N$, with the sum being plotted at every point $x(k)$. Two uses are possible:

1. No averaging has taken place (option 5 in the menu): the individual average is equal to the over-all mean y_{mean}, which is displayed as a horizontal line; this corresponds to the classical use of the Cusum technique. By this means, slight shifts in the average (e.g., when plotting process parameters on control charts) can be detected even when the shift is much smaller than the process dispersion, because the Cusum trace changes slope.

2. If a data set was first appropriately treated in program SMOOTH and the smoothed coordinates were saved, the difference between raw and smoothed values (use subtract function in DATA) can be analyzed; essentially, Cusum now detects how well the smoothed trace represents the measurements. For example, if peak shapes are to be filtered (see data file SIM1.dat) and too wide a filter is used, the smoothed trace might "cut corners"; as a result, the Cusum trace will change slope twice. the Cusum trace can be shifted vertically, and an expansion factor can be chosen. Ordinate rescaling is done automatically.

Features

- $\langle$Select Mean$\rangle$ allows the reference mean to be entered as a number or by way of a range $L \ldots K$, with $\mathbf{y}_{\mathrm{mean,\,range}} = \Sigma(\mathbf{y}_i)/N$, $i = L \ldots K$ and $N = K - L + 1$.

<u>Data</u>

Purpose Create a new SMAC data file or edit/modify an existing one; import (parts of) an existing Excel spread sheet and convert to SMAC format.

Features

- $\langle$File$\rangle$ $\langle$New$\rangle$: A new data file can be generated by defining the number of columns and rows and then filling the table either in a column-by-column or a row-by-row sequence; if the option $\langle$Leave Empty$\rangle$ is chosen, the file is stored "as is." The array size is limited to $n \cdot m \leq 10^5$, but also by the available memory.
- $\langle$Input Data$\rangle$ leads to the same options
- $\langle$Instructions$\rangle$ and $\langle$Import Data from Excel$\rangle$ allow a portion of an Excel spreadsheet to be copied into a SMAC-data file; the necessary steps are as follows: (1) open the Excel file, mark the range to be copied and press [Copy]; (2) open DATA and create a new or open an existing data file; (3) select $\langle$Input Data$\rangle$ $\langle$Import Data from Excel$\rangle$ and position the cursor on an appropriate cell, e.g. cell (1,1). The imported data replaces any data that was in that cell range. If the imported cell range needs more columns or rows than are presently available, the data array is correspondingly increased in size. Headers and dimensions can be added later.
- An existing data file can be loaded for modification.
- $\langle$Modify Table$\rangle$: Once a data table is established, it can be manipulated in various ways: The option offers Add/Delete Row/Column, and Change Entry; choose the option and click on the appropriate item. If many rows or columns need to be added or deleted, it is easier to read the data file into Excel, do the modifications there, and reimport it using the $\langle$Import from Excel$\rangle$ option.
- $\langle$Change Item$\rangle$: the content of individual cells can be edited by clicking on the cell and entering a new number in the yellow box, followed by ⌐ ; empty cells that are ignored by all programs are created by just pressing ⌐ without entering a new value.

- ⟨Delete⟩, ⟨Add⟩: columns or rows can be selectively deleted, or new ones can be added (column M + 1, row N + 1).

- ⟨Transform⟩: the content of a given column (=vector) can be mathematically modified in various ways, the result being deposited in the $(N + 1)^{st}$ column. The available operators are addition of and multiplication with a constant, square and square root, reciprocal, $\log(u)$, $\ln(u)$, 10^u, $\exp(u)$, clipping of digits, adding Gaussian noise, normalization of the column, and transposition of the table. More complicated data work-up is best done in a spreadsheet and then imported.

- ⟨ ⟩: The table can be printed and/or saved.

EUCLID Section 4.12

Purpose, Scheme of Calculation Comparison of two sets of objects (groups *A* and *B*, for a total of $(K_2 - K_1 + 1) + (L_2 - L_1 + 1) \leq M$ objects); each object is profiled by determining its response to measurements in *N* dimensions. These measurements are averaged within each group to establish typical profiles. A "distance" is established between the *A* and the *B* averages by calculating the Euclidian distance in *N*-space: $D = \sqrt{(x_a - x_b)^2 + (y_a - y_b)^2 + \ldots}$, where *x*, *y*, etc. are the means for dimensions $1 \ldots N$, for example a concentration, a pH, an absorbance. The Euclidian distances between every object or sample and the group averages *A* and *B* are calculated analogously; the triangle *A–B–S* is projected onto the screen, the side *A–B* being the base. The scatter and clustering of the upper vertices (*S*) is analyzed: If two distinct groups are formed, then the hypothesis that the objects can be separated into two groups gains credence.

Features

- A table of *M* columns (*M* objects) and *N* rows (dimensions) is assigned to groups *A* and *B* by entering the column numbers *A*1 ... *A*2, respectively *B*1 ... *B*2; the ranges for *A* and *B* may not overlap, but need not be the same size or use all of the available *M* columns.

- The group averages *A* and *B* are formed for every dimension.

- The differences between the group averages are calculated and displayed for every dimension.

- The *N*-space Euclidian distances for every object relative to the group averages *A* and *B* are calculated and plotted.

- The contribution every dimension makes toward the total distance *A–B*, is plotted.

Note: If the N dimensions yield very different numerical values, such as 105 ± 3 mmol/L, 0.0034 ± 0.02 meter, and 13200 ± 600 pg/ml, the Euclidian distances are dominated by the contributions due to those dimensions for which the differences $A–B$, $A–S$, or $B–S$ are numerically large. In such cases it is recommended that the individual results are first normalized, i.e., $x' = (x - \mathbf{x}_{mean})/s_x$, where $\mathbf{x}_{mean}$ and s_x are the mean and standard deviation over all objects for that particular dimension X, by using option ⟨Transform⟩/⟨Normalize⟩ in program DATA. Use option ⟨Transpose⟩ to exchange columns and rows beforehand and afterwards! The case presented in sample file SIEVE1.dat is different: the individual results are wt-% material in a given size class, so that the physical dimension is the same for all rows. Since the question asked is "are there differences in size distribution?," normalization as suggested above would distort the information and statistics-of-small-numbers artifacts in the poorly populated size classes would become overemphasized.

FACTOR8 *Section 2.4.2*

Purpose To determine, from eight initial experiments performed under certain conditions, whether the three controlled parameters have an effect on the measurement, and which model is to be used. This factorial approach to optimization is an alternative to the use of multidimensional simplex algorithms; it has the advantage of remaining transparent to the user.

Features

- The low/high values for the three parameters are entered.

- The eight measurements corresponding to the experimental conditions 1, *a*, *b*, *c*, *ab*, *ac*, and *abc* are entered. "1" means all three parameters are set "low," while "*abc*" connotes the opposite.

- The model is $Y = m(1) + m(2)^* a + m(3)^* b + \ldots m(7)^* a^* b^* c$.

- The model is fit; effects, specific effects, model coefficients, and residuals are displayed.

- The assumed residual standard deviation, i.e., the precision of measurement, can be varied to study its effect.

- By brute-force iteration, the highest Y within the cube spanned in 3-space is located with a resolution of a few percent of the parameter ranges. The cube's center is accordingly moved, and the model is reevaluated; in this fashion a track of steadily higher Y-values in the immediate vicinity of the initial cube is displayed. Thus the direction to be taken for further experiments is indicated.

- The model can be evaluated for any combination of a, b, and c.

HISTO *Section 1.8*

Purpose Determine the distribution of many repeat measurements and compare this distribution to the normal distribution.

Features

- The overall x-range (which scales the plot) and that part of the x-range, which is to be subdivided into B classes (bins) can be individually defined; essentially, this means that the plotted window can be adjusted to be the same for comparing several histograms, while bins need only be defined in that part of the x-axis where the measurements are concentrated. The optimal number of bins is suggested as $B \approx \sqrt{N}$, but can be adjusted.

- No autoscaling is available; that, while convenient, exposes the individual plot limits and bin boundaries to the vagaries of measurement and sampling noise; the user is forced to actively select lower and upper bounds on the subdivided x-range, and the number of bins, to come up with bin boundaries that make sense.

- An area-equivalent Gaussian ("normal") distribution curve can be superimposed.

- Display and/or print tables that contain bin number, number of events, individual, cumulative % population, and the normalized bin boundaries ($z = (bb-\mathbf{x}_{mean})/s_x$), χ^2-components, well as various statistical indicators (extreme values, number of events outside bin-range, mean, SD).

- The cumulative percentage points can be plotted on a distorted %-axis (so-called "normal probability scale") that yields a straight line for perfectly ND data.

- ⟨Stack⟩: a selected range of columns is automatically mapped into a stack of histograms on common x- and y-scales; the vertical offset between histograms corresponds to the largest frequency found in all bins.

HUBER *Section 1.5.5*

Purpose Check a vector for outliers.

Features

- Select a column from a data file and calculate mean, median, and standard deviation. Display the original data together with x_{mean}, x_m, and $x_{mean} \pm s_x$.
- Calculate all deviations $(x(i) - x_m)$, and sort according to absolute size; calculate the median average deviation MAD; calculate cut-off limits for outliers according to Huber[21] by assuming the recommended value for Huber's k (3.5). Different values for this multiplier can be selected.
- Display the cut-off limits and the clipped data set.
- Display x_{mean}, x_m, and s_x before and after elimination.
- The ± 2 and ± 3 sigma cut-off limits of the conventional outlier-detection models are plotted.
- Use the sorted absolute deviations above and plot the critical Huber's k for each $x(i)$ *versus* the percentage of points retained (cumulative number of ordered absolute deviations); this yields insight into the sensitivity of the clipping process to changes in Huber's k (solid line, points). Note that a scale for $k = 1, 2, \ldots$, etc. is given on the left side.
- Plot analogous critical z-values for the mean/SD before resp. after elimination of points (dotted lines). Since the standard deviation will decrease on elimination of suspected outliers, the dotted sensitivity curve for "after elimination" will be higher than the one for "before". Huber's k changes, too, but to a lesser degree. (See Fig. 1.1.)
- The first points to be eliminated are defined by the largest (absolute) deviate from the median, $|x_i - x_m|^{max}$, respectively by the largest (absolute) deviate from x_{mean}, $|x_l - x_{mean}|^{max}$. Since these deviations are measured relative to the median and the mean, which do not always closely coincide, the first point to be eliminated need not be the same for the two rules. Huber's rule can be used with asymmetrical distributions.

HYPOTHESIS : Section 1.9

Purpose Display the type I error (α) and the type II error (β) both as (hatched) areas in the $ND(\mu_{REF}, \sigma_{REF}^2)$ and the $ND(\mu_{TEST}, \sigma_{TEST}^2)$ distribution functions and as lines in the corresponding cumulative probability curves.

Features

- Freely choose μ.
- The means and SDs can be freely chosen to repeat any combination of

reference measurement and test measurement (hypothesis H_0 applies if the REF and TEST measurements are indistinguishable, hypothesis H_1 if the two can be distinguished).

- The following tests can be conducted: REF < TEST, REF <> TEST, and REF > TEST.
- Display and print the graphics.
- Display the estimated error β numerically.

INTERSECT Section 2.2.11

Purpose   Calculate the intersection of two linear regression lines and estimate the 95% confidence limits on the intersection coordinate. (See Fig. 2.19.)

Features

- Select the abscissa and ordinate vectors for the first, then those for the second data set (the two sets can actually be located in separate files); the two data sets need not have the same size.
- The linear regression parameters are calculated independently for each set.
- The $x_{\text{intersect}}/y_{\text{intersect}}$ coordinate is calculated.
- The abscissa interval containing the intersection point is scanned from left to right in 500 steps and the two CL(Y) distributions at each point (resolution: y-range/100) are multiplied to obtain the volume $\iint \text{PD}_{Y1}(x, y) \cdot \text{PD}_{Y2}(x, y) \cdot \delta x \cdot \delta y$ (approximated by $\Sigma\Sigma \, \text{PD}_{Y1}(x, y) \cdot \text{PD}_{Y2}(x, y) \cdot \Delta x \cdot \Delta y$), which is equated with 100%; the points at which the cumulative overlap function reaches 2.5, respectively 97.5% are the results.
- The intersection coordinates and the 95% $\text{CL}(x_{\text{intersect}})$ values are displayed as numbers and in the graph.
- The cumulative-overlap curve is displayed with the 95% $\text{CL}(x_{\text{intersect}})$ values, which need not be symmetrical with respect to $x_{\text{intersect}}$.

LINREG (Standard unweighted and weighted linear regression): Section 2.2

Purpose   Perform a linear regression analysis over the selected data points; display and print results, do interpolations, determine limits of detection.

Features

- ⟨Choose Data⟩ choose any two columns from the *M*-column data file to represent the abscissa *X* respectively the ordinate *Y*; automatically

  — Perform an unweighted linear regression.
  — Approximate Student's *t* for $N - 2$ degrees of freedom and $p = 0.05$.
  — Display the regression line and its confidence limits.

- ⟨Interpolate⟩ provides the options $Y = f(x)$ and $X = f(y)$, including confidence limits on the results; the interpolation result is displayed in the graph and the table.

- ⟨LOD⟩ calculate and display the limits of detection and quantitation LOD, LOQ.[86,87] [*Note*: This form of calculating the LOD or LOQ was chosen because the results are influenced not only by the noise on the baseline, but also by the calibration design; from the educational point of view this is more important than the consideration whether any agency has officially adopted this or that LOD-model. For a comparison, see Figs. 2.14, 2.15, and 4.31].

- Calculate and display the residuals ($y_i - Y$), where $Y = A + B \cdot x_i$.

- Select and display specification limits on the acceptable *X*-range, as when doing assays;

- ⟨Select Probability of Error⟩ gives the choice of $p = 0.0001, 0.001, 0.002,$ 0.01, 0.02, 0.1, 0.2, or 0.5. (See Table 5.1.)

- ⟨Weighted Linear Regression⟩ Section 2.2.10: the user must define a functional dependence of the repeatability (defined as a standard deviation) of the measurements s_y on the independent variable *x*, such as $s_y = a + b \cdot x$ (typical for gas chromatography, where the relative standard deviation (in [%]) of the measured peak area is often constant over a very large concentration range; the constant "a" represents the intrinsic, concentration-independent repeatability of the instrument). Appropriate functions can be fitted using the output of the results table of programs VALID and TESTFIT. See "Algebraic Function."

  — The $s_y = f(x)$ function is depicted across the bottom of the graph.
  — The position of the "center of gravity" $x_{mean, weighted}/y_{mean, weighted}$ is given by crosshairs.

MSD _(Mean, Standard Deviation): Sections 1.1, 1.3.2, 1.7.2_

Purpose   For a given vector, calculate the mean and the standard deviation, and their confidence limits, and compare with specification limits.

Features

- Calculate mean x_{mean} and the two-tailed confidence limits.
- Calculate the standard deviation s_x and the (asymmetrical) confidence limits.
- Display (graphically and as a table) the mean and its confidence limits as a function of p.
- Display (graphically and as a table) the standard deviation and its confidence limits as a function of p.

The following approximations are used:

- Student's t for $N - 1$ degrees of freedom and various p (0.0001, 0.001, 0.002, 0.01, 0.02, 0.05, 0.1, 0.2, and 0.5).
- The χ^2-tables for $N - 1$ degrees of freedom and various p (0.001/0.999, 0.005/0.995, 0.01/0.99, 0.025/0.975, 0.05/0.95, and 0.1/0.9; this corresponds to two-tailed confidence limits for $\alpha = 0.002$, 0.01, 0.02, 0.05, 0.1, resp. 0.2).

MULTI _Sections 1.5.3, 4.4_

Purpose   Test several sets of data for deviations from the null hypothesis H_0 ("all data sets belong to the same population, that is, the individual means do not differ from the grand mean, and the individual standard deviations do not differ from the overall standard deviation"). It is assumed that the individual data sets consist of a number of repeat measurements of one observable, such as (1) repeat moisture or temperature measurements at several locations, or (2) hourly concentration determinations of a given chemical species in several reaction vessels that are being run in parallel, (3) the same reaction in the same vessel on different days, or (4) content uniformity measurements on the, say, 20 buckets of tablets that make up one batch. If the standard deviations do not significantly differ (Bartlett test), continue with the simple ANOVA test to find whether the data set means belong to one or more populations. If more than one population is involved, find which data set means can be grouped into homogenous subpopulations (multiple range test). It is possible that one and the same data set could be grouped into two subpopulations that partially overlap.

Features

- Each data set fills a column; if data sets are of unequal length, N is determined by the largest set (program DATA; nonexistent entries are entered in DATA by just pressing ⏎, and are ignored by MULTI.

- After the chosen data file is read into MULTI, sequential columns K through L are selected for analysis; K and L must be within the bounds $1 \ldots M$, and L must be larger than K. If the columns one wants to analyze are separated by other data, use program DATA to first copy the unwanted column to column $M + 1$ by invoking the option ⟨Add⟩ and selecting the constant to be zero and then deleting the unwanted column.

- The heading, dimension, means x_{mean}, standard deviation s_x, RSDs 100 · s_x/x_{mean}, and number of determinations N_j are displayed for every column $j = K \cdots L$.

- The following three tests can be individually called, but execution is blocked if the preceding test has not been performed.

- ⟨Bartlett Test⟩ yields an uncorrected and a corrected χ^2-value, which are compared to the critical χ^2 for $f1$ degrees of freedom and $p = 0.05$. The interpretation is given. If at least one standard deviation is significantly different from the others, the program stops here. If this happens because one column yields a standard deviation of nearly zero, check whether a signal converter with insufficient digital resolution is the cause. This effect can be studied with program CONVERGE, where a digitizer simulation can be adjusted so that all values collapse on just one or two adjacent levels. In the laboratory, the equipment will have to be modified. To at least get an idea of what results could be extracted from the data set if it were not corrupted, use DATA option ⟨$v = u$ + Noise⟩ to superimpose a ND(0, σ^2) distribution, with σ chosen to correspond to about $\frac{1}{2}$ to 1 LSD.

- ⟨ANOVA⟩ if the standard deviations are indistinguishable, an ANOVA test can be carried out (simple ANOVA, one parameter additivity model) to detect the presence of significant differences in data set means. The interpretation of the F-test is given (the critical F-value for $p = 0.05$, one-sided test, is calculated using the algorithm from Section 5.1.3).

- ⟨Multiple Range Test⟩ yields a triangular matrix of differences $\Delta x_{\text{mean}, ij}$ (difference in x_{mean} for every possible combination of $x_{\text{mean}, i}$ with $x_{\text{mean}, j}$).

- The triangular matrix of differences $\Delta x_{\text{mean}, ij}$ is converted into a triangular matrix of q-values using the values $x_{mean,i}$, $x_{mean,j}$, N_i, and N_j; since tables of q-values are rarely given in statistics textbooks, especially for p-values other than 0.05, the q-values are converted to "reduced" q-values by division through the appropriate Student's t and square root of 2. (See Table 5.1.) This permits a delineation of subpopulations on the basis of a critical reduced q, which MULTI looks up in a table and interpolates, if necessary. The chance of misinterpretation would be small, even if the statistical probability of error p were changed from $p = 0.05$ to some other value (0.02, 0.1, etc.) by multiplying all q_{red}-values by the appropriate factor $t(f, p = 0.05)/t(f, p)$.

- The subpopulations of data sets that can be distinguished are given as lists of x_{mean} values; the number of lines corresponds to the number of subpopulations.

- A graph is displayed that contains the individual data points, and the associated means x_{mean} and standard deviations s_x; the data sets are arranged left to right in the same order as they appear in the data file.

- A second graph, with the data sets ordered according to increasing x_{mean}, depicts the means and standard deviations, and, as stacked horizontal lines, the range of means spanned by the individual subpopulations.

- The tables of differences-of-means respectively of reduced q values are displayed with the corresponding ordered x_{mean} arranged at the top and down the left margin.

SHELFLIFE Section 4.20

Purpose Determine the shelf-life of a product (*viz.* a pharmaceutical) by evaluating analysis results as a function of storage time. The points at which the lower 95% confidence limit of the population and a horizontal at Y% of the nominal content intersect determines the acceptable shelf-life as promulgated by the FDA in their "Guidelines for Submitting Documentation for the Stability of Human Drugs and Biologics, February 1987". The CL is calculated using either the Student's t-factor for 90% (two tailed) or 95% (single tailed). By default, results are given for $y = 90\%$ and 95% of nominal, but can be requested for any other specification limit, too. The program is a modification of the LINREG program:

Features

- The x- and y-ranges are handled as in all other programs. Notice, though, that shelf-lives above 5 years are uncustomary in the pharmaceutical industry.

- An algorithm for calculating the symmetrical (two-tailed) t-factors for $p = 0.1$ is incorporated; its use corresponds to the statement that "the probability that measurements on a future batch, given the linear trend already established, will inadvertently be found to be below the specification limits of $Y\%$ of nominal, at a shelf-life that would lead one to expect a residual content at or above the specification limit, is $p = 0.05$."

- The equation for the confidence limit is $Y = A + B \cdot x \pm t \cdot \sqrt{V_{res} \cdot (1/N + (x - x_{mean})^2/S_{xx})}$, where x is time. Minus sign: for main component; plus sign: for impurity.

SIMCAL Section 2.2.6

Purpose Generate a data set that superimposes normally distributed noise on a linear calibration model to study the effects of the adjustable parameters. A whole calibration—measurement—evaluation sequence can be optimized for quality of the results and total costs

Features

- ⟨Experimental Design⟩.

  — Set the number of calibration solutions (concentrations).
  — Set the number of repeat determinations per calibration standard.
  — Set the number of repeat determinations per unknown.

- ⟨Distribution of Calibration Concentrations⟩:

  — Define the calibration concentrations as being spaced linearly, logarithmically, or arbitrarily within the concentration range.
  — Choose the endpoints of the calibration range; the calibration concentrations are now displayed in the green field.

- ⟨Instrument Characteristics⟩

  — Set the intercept and slope of the instrument's calibration function.
  — Set the number of decimal places displayed to simulate the action of digital signal acquisition; 1 would mean the result of a measurement is clipped to the first decimal.

- ⟨Measurement Noise⟩: Define the measurement noise as being constant, proportional, or an arbitrary function of x; choosing the ⟨Arbitrary⟩ option activates the algebraic function module to define a $SD = f(x)$ function that will also be used in case a weighted regression is selected.

The preceding items must be set; headers and dimensions for the two axes can be defined; if there is information concerning the LOQ, e.g., from a previous validation, that can be entered, too (for display only).

- ⟨Accept⟩: Run interpolations and obtain confidence limits of the estimate; other options as with LINREG.

- ⟨Display Table of Detailed Calibration Results⟩: The simulated calibration points (x_i, y_i), the estimates $Y = f(x_i)$, and both the absolute and relative residuals are given.

- ⟨Display Calibration Check Graph⟩: The calibration points as obtained under ⟨Accept⟩ above remain as is, but a renewed measurement $y_{CS, i}$ of these same samples as "unknowns" is simulated; vertical lines indicate the $CL(X)$ that would be determined for these $X = f(y_{CS, mean})$. The variability so observed mimics the within-calibration repeatability. Use the button [New Check] to repeat the simulation.

- ⟨Display Calibration Check Table⟩: Numerical results for the above item, including the back-calculated values in [% of nominal] format.

- ⟨Cost Factors⟩: Define a cost-accounting model that includes consumables, labor, write-off, workload, and frequency of recalibration to estimate the cost per result.

- ⟨Data⟩ ⟨Change Assumptions⟩: Assumptions can be changed at any time.

- ⟨Data⟩ ⟨Create New File⟩: Define a new file, in which calibration designs can be stored.

- ⟨Data⟩ ⟨Add Data to Existing File⟩: Add the simulated abscissa- and ordinate-values to a file that can be saved for analysis by other programs.

SIMGAUSS Section 1.4

Purpose   Generate data sets using mixed deterministic/stochastic models with $N = 1 \ldots 1000$. These data sets can be used to test programs or to do Monte Carlo studies. Five different models are predefined: sine wave, saw tooth, base line, GC-peaks, and step functions. Data file SIM1.dat was generated for $N = 200$.

Features

- ⟨Select Function⟩ offers the choice among the above-mentioned functions and a user-defined one. (See "Algebraic Function".)
- ⟨Change Waves⟩ allows for one to ten repetitions of the basic wave form within the window 1 ... N.
- ⟨Change Random Noise⟩ asks for the CV.

SIMILAR (SIMulate statistically simILAR data sets): Section 3.5 Monte Carlo Technique

Purpose   Take an existing data file that comprises at least a column X (independent variable) and a column Y (dependent variable). Choose either a function or real data to model statistically similar data sets.

Features

- ⟨Select Type of SD for Noise⟩: After an abscissa and an ordinate are picked, there is the option of a fixed s_y or a user-defined function $s_y = f(x)$. (See "Algebraic Function.")
- ⟨Select Data Calculation⟩ allows to either use the existing ordinate values as basis:

$$y' = y + \text{ND}(0, \sigma^2) \text{ with } i = 1 \ ... \ N$$

or to define an equation $Y = f(x)$ that then will serve as model (cf. "Algebraic Function" and program SIMGAUSS):

$$y' = Y(x) + \text{ND}(0, \sigma^2) \text{ with } i = 1 \ ... \ N$$

- Both algebraic functions, noise and model, can be modified at any time, so the final table can contain simulated data sets for various combinations of noise and model; both functions are displayed.
- ⟨Calculate New Data⟩ generates a statistically similar ordinate value for each x_i by superimposing $\text{ND}(0, s^2)$ noise on the model or previous data; this option can be repeatedly accessed.
- ⟨Add data set to file⟩: The currently active set of simulated ordinate values can be attached to the existing data table as column $M + 1$. Create a number of such columns $(X, Y, Y', Y'', Y''' \ ...)$. Use the modified data files as if several series of measurements had been acquired and test the

intended statistical program and the interpretation scheme for robustness (ruggedness): If, despite the stochastic variability, similar results and the same interpretation are always found, the evaluation procedure is robust. A second set of measured values that cannot be distinguished from the first set or its simulated statistical look-alikes cannot be interpreted to be different.

SMOOTH Section 3.6

Purpose Construct a smoothed trace over a (large) series of observations to improve presentability or to extract a "trend."

Features

- Ordinate values are assigned by choosing a column number in the data table; abscissa values can either be chosen similarly, or can be assigned sequential numbers ("index," which makes them equidistant on the x-scale). The moving average MA and Savitzky-Golay SG filters assume equidistance for correct application. *Note*: If this condition is only mildly violated (Δx changes little, Δy near zero), not much harm is done. The residuals and the derived residual standard deviation are strongly affected if the abscissa distances between successive points (Δx) changes appreciably from one interval to the next and Δy is much larger than the noise level, because the smoothed coordinates (u, v) would then have to be interpolated.

- The coordinates are automatically sorted according to ascending x if x is not an index.

- $\langle$Box-Car$\rangle$ averaging with a box width (filter width, window) in the range $1 \leq W \leq N$; $W = 1$ just connects the points, while $W = N$ calculates the overall x_{mean}. The mean for each box will be displayed as a horizontal line.

- $\langle$Moving Average$\rangle$ a filter width in the range $1 \leq W \leq N$ is possible, W being restricted to odd values. Both the x- and the y-values are averaged in the filter window and the means are assigned to $u(i)$ respectively $v(i)$; the trace is plotted using these smoothed coordinates u/v. The residuals, however, are calculated as $r = y(i) - v(i)$ without interpolation, under the assumption that the filtered $v(i)$ does not appreciably vary locally; if it is suspected that this condition is not met, it is suggested that the filtering be done using the Index $i = 1 \ldots N$ instead of a column from data table (option "column = 0 for abscissa").

- $\langle$Savitzky–Golay$\rangle$ filter widths in the range $5 \leq W \leq 11$ are possible; smoothing can be carried out using polynomial filters of order 2 (quadratic parabola) or 3 (cubic parabola). Differentiating filters and filter widths above 11 were not implemented because such applications only make sense if very large numbers of measurements are available; if such data series are acquired, it is recommended that dedicated programs optimized for speed be used. The SG filter coefficients are generated *in situ* according to Ref. 162 Whereas the usual implementation of the SG-filtering algorithm allows only for a smoothed trace in the range $x(\text{INT}(W/2) + 1) \ldots x(N - \text{INT}(W/2) - 1)$ (the ordinate in the center of the filter is estimated); analogous filters are also calculated here that allow the estimation of the smoothed trace at the ends of the range, i.e., $x(1) \ldots x(\text{INT}(W/2))$. Thus the smoothed trace includes all N points.

- The original (x, y) and the filtered (u, v) data, together with the appropriate residuals $r_i = y_i - v_i$, are displayed and printed.

- The residuals relative to the smoothed trace (to y_{mean}, if no smoothing has been done) are plotted; a vertical shift and an expansion factor can be chosen.

- Interpolation on the smoothed trace can be carried out for any x within the bounds of the trace; the result of the interpolation is displayed and printed in numerical format and is indicated by cross hairs on the screen.

- Interpolation on the smoothed trace can be carried out for any y; multiple intersection with the trace are possible; the results of the interpolation are displayed and printed in numerical format (list) and are indicated by cross hairs on the screen.

TESTFIT Section 4.13

Purpose There are a variety of uses for this program, chief of them being as follows:

- Testing the adequacy of a model with respect to its complexity by visually checking for trends in the residuals, e.g., is a linear regression sufficient, or is a quadratic polynomial necessary?

- The quantitation of the goodness of fit between a model and a data set by calculation of the residual standard deviation.

- Testing the robustness of a (best) model (and second-best contenders) by evaluating sets of statistically similar data created with program SIMILAR; if the derived decisions remain unaffected by measurement noise, the model is adequate.

Features

- ⟨Define Function⟩: Change the model to fit the data. (See "Algebraic Function.")
- ⟨Residuals⟩ plot the residuals $r_i = f \cdot (y_i - Y_i)/Y_i$ at a chosen Y_{res} and with a scale expansion f.

Interpretation   The model can only be improved upon if the residual standard deviation remains significantly larger (F-test!) than the experimental repeatability (standard deviation over many repeat measurements under constant conditions, which usually implies "within a short period of time"). Goodness of fit can also be judged by glancing along the horizontal (residual = 0) and looking for systematic curvature.

TTEST Section 1.5.2

Purpose   For a given set of results (two means, their standard deviations and numbers of determinations: x_1, x_2, s_{x1}, s_{x2}, n_1, n_2) the F-test is performed; depending on the outcome (s_{x1} significantly different from s_{x2}, or indistinguishable) and the identity or nonidentity of n_1 and n_2, different forms of the t-test are carried out. Because cases exist where different authors propose similar but not identical equations[46,49,56] for treating the same data, ambiguous situations can arise; this is due to varying and sometimes unstated assumptions as to what represents a large number ($n > 200$, for example), whereupon some authors simplify equations by replacing ($n - 1$) by (n), on the assumption $n \gg 1$, while others do not, see Eqs. (1.14)–(1.16), Table 1.10. One variation of an equation that is given in more than one textbook is included despite the fact that it probably contains a transcription error ("$n + 1$"[47,49] instead of "$n - 1$"[46]). Equations and references can be displayed and printed.

Features

- ⟨Enter Data Manually⟩ opens an empty table for manual input of the two means, standard deviations and numbers of determinations.
- ⟨Read Data from File⟩ provides a convenient access to data sets that are already in a data file; this file needs to contain at least two columns of data; the mean, the SD, and N are automatically calculated for the two data sets that were selected.
- ⟨New Data Sets⟩ or ⟨Change Data Manually⟩ can be used to modify the means, standard deviations, or numbers of determinations.

- ⟨Choose New Vectors in File⟩ allows to select new·data sets in the open file.
- ⟨References⟩ gives a list of references for the employed equations, cf. Table 1.10.
- ⟨Equations⟩ gives these equations in a common format; because not all t-test procedures return the same number of degrees of freedom or the same t-value, the interpretations can slightly differ. For each of the four combinations of N_1 (=, ≠) N_2 and V_1 (=, ≠) V_2 all equations that apply are used and the corresponding results and interpretations are displayed.

VALID Sections 3.2, 4.23

Purpose   Same as program LINREG; it is assumed that repeat determinations were performed for most concentrations. (See VALID1.dat.)

Features

- The mean, the standard deviation, and the confidence limits of the population at each concentration with multiple measurements are calculated and tabulated.
- Option ⟨Valid⟩ presents a graph of relative standard deviation (c.o.v.) *versus* concentration, with the relative residuals superimposed. This gives a clear overview of the performance to be expected from a linear calibration Signal = $A + B ·$ Concentration, both in terms of (relative) precision and of accuracy, because only a well-behaved analytical method will show most of the residuals to be inside a narrow "trumpet"-like curve; this trumpet is wide at low concentrations and should narrow down to c.o.v. = ±5% and rel. CL = ±10%, or thereabouts, at medium to high concentrations. Residuals that are not randomly distributed about the horizontal axis point either to the presence of outliers, nonlinearity, or errors in the preparation of standards.
- The back-calculation feature in option ⟨Display Values⟩ gives each measurement as the absolute estimate $X(y(i))$ and normalized to the nominal concentration; the normalized results should all be around 100%. The symmetrical limits ±SD are also given.
- The other features are identical to those of program LINREG.

VALIDLL Sections 3.2, 4.23

Purpose   VALIDLL is identical in concept and features to VALID, the difference being that the use of a log-log depiction is assumed. Linear (i.e., nontransformed $x(\ )$: 1, 2, 5, 10, 20; $y(\ )$: 1.3, 2.1, 4.9, 10.7, 18.9) data are read. Thereafter, the data set is logarithmized and the linear regression is calculated.

Features

- interpolations are done by entering transformed values ($y^{\text{lin}} = 16.5 \rightarrow y^{\text{log}} = 1.28$), the results are displayed transformed ($x^{\text{log}} = 1.238 \pm 0.161$) and back-transformed ($X^{\text{lin}} = 17.3 + 7.8/-5.4$); back-transformed confidence limits are nonsymmetrical.
- LOD and LOQ are not calculated.

XYZ Section 2.4

Purpose   Plot pseudo-three-dimensional (isometric) presentations in order that complex relationships can be studied.

Features

- Select any vector to be the left-to-right axis (X-axis).
- Select any vector to be the front-to-back axis (Y-axis).
- Select any vector to be the vertical axis (Z-axis).
- Autonormalize every vector to the range 0 ... 100 so that a cube results.
- Depict every coordinate by a small circle.
- Mark the footprint of every coordinate on the X–Y plane.
- Mark every axis with the corresponding header.
- Rotate about Z-axis in one degree steps in the range -90 ... $+90°$ for better view. (Use the mouse pointer to grab and rotate the bold azimuth line in the plan view in the insert!)
- ⟨Display Values⟩ gives the X-, Y-, and Z-coordinates and the X'- and Y'-projection coordinates for each point.

XYZCELL Section 2.4

Purpose   Condense large amounts of 3-D data into an easy-to-understand presentation.

Features

- The (scaleable) x- and y-ranges are mapped into an n-by-n array of cells, with $n = 10 =$ default.

- ⟨⟩ Each data point is assigned to the appropriate cell, where the following statistics are calculated: number of values, z_{mean}, and s_z.
- The distribution of points on the x,y-plane is displayed, along with the following:

    — The z_{mean} for each occupied cell (dark bar).
    — The s_z for each occupied cell (light bar).
    — The number of data points in each cell.

- The z-range that is displayed in each cell and the SD-scaling factor can be separately adjusted for optimal overview.

5.4.4 DATA FILES for VisualBasic Programs

The following files, among others, are provided as instruction aids and examples:

ARRHEN1.dat: A product was put on stability at $25°$, $30°$, and $40°C$. for 24, 24, resp. 3 months. (See Ref. 184.)

ARRHEN2.dat: A peptide solution was put on stability at $30°$, $40°$, $50°$, $60°$, $73°$ and $80°C$ for between 10 and 298 days. (See Ref. 184.)

ARRHEN3.dat: Another peptide solution was also put on stability at $30°$, $40°$, $50°$, $60°$, $73°$ and $80°C$ for between 10 and 298 days. (See Ref. 184.)

ASSAY1.dat: As part of a cross-validation of a modification of a given analytical method, 20 samples were run on either method. There is at least one result that is out of trend (OOT), and another two to three are indicative of laboratory errors.

ASSAY2.dat: As part of a large validation program, 40 samples were pulled from a production batch. One measurement points to either a gross inhomogeneity or an analytical error; the trend of the others could indicate variations in the manufacturing process.

AUC.dat: Sixty-nine subjects were exposed to three different medications containing the same drug substance in a test of equivalence; each had blood samples withdrawn at defined time points after administration so as to obtain a curve of plasma level of drug vs. time. The area under such a curve is a measure for the amount of medication the subject's body absorbed through the gut wall. These AUC values demonstrate the huge differences between subjects and within a subject (say, day-to-day effect) that is encountered in biological tests ("biovariability"). These three medications were accepted as being "bioequivalent." The contrast to the well-behaved systems a chemist is used to is evident.

***BUILD_UP*.dat: (Fig. 4.37)**   Three batches of an experimental chemical were put on stability for 6 months at 25°C, respectively 40°C and were UV tested at $t = 0, 1, 2, 3$, and 6 months for a key degradation product. The SL is 0.2 AU.

***CALIB*.dat:**   Calibration measurements at eight concentrations (double determinations) using a GC; peak area measurements in [mV*sec] vs. weight in [mg].

***COAT_WEB*.dat:**   During the validation of a coating process a certain length of web was sampled on a 15 (length) × 10 (width) raster; the coating weights are given. The hypothesis was tested that there would be a difference between the left and the right side of the web, and that along the length there would be a periodic change because the huge rollers would not be perfectly cylindrical in shape and parallel to each other. Use MULTI on this and on a transposed version of this array (COAT_WEB˜.dat) to test for differences between groups, and CORREL to test for correlations between any two groups (especially neighboring ones).

***CREAM*.dat:**   A batch of cream containing two drug substances was put on stability and tested for at $t = 0, 3, 7$, and 24 months. Active component 1 remains stable, while AC2 degrades so fast that a shelf-life of only 26 months can be demonstrated for SL = 90%. Use with SHELFLIFE.

***CU_ASSAY1*.dat: (Fig. 4.40)**   A tablet production process was being validated; samples were pulled from the beginning, the middle, and the end of the production run; Components A and B were analyzed. The requirement is that the means do not significantly differ and that the CV remains below 6%.

***CYANIDE*.dat: Section 4.13**   Two calibration series over the same range, and one over a short range (three groups of columns Concentration/Signal), and a fourth group that combines all of the above data; the data can be fitted to a parabola $Y = -0.002125 + 0.005211^*X - 0.0000009126^* x^2$ with a residual standard deviation of ±4.5 mAU. Use with LINREG, TESTFIT.

***EDIT**.dat:*   A 6-row-by-25-column table containing integer numbers in random and not-so-random sequences; a few empty cells are included. Use this file to play with the editing functions contained in program DATA.

***FACTOR**.dat: Fig. 3.5, Table 3.2*   The data used in the calculated example is given.

***FILLTUBE**.dat: Section 4.19*   Tubes must be filled to a nominal 20 g; 10 simulations are provided each for the EU and the Swiss Guidelines; the average fill weights are 19.7 g and higher; $n = 50$ tubes per sample.

***HARDNESS**.dat: Section 4.9*   One-hundred-twenty tablets were taken off a press in the order they emerged, and hardness and weight were individually determined; one column features the stamp number (cf. STAMP.dat); another column gives the weight-corrected hardness values. $N = 120$, $M = 5$; use with programs LINREG, XYZ, and XYZCELL.

***HISTO**.dat: Section 1.8*   19 repeat measurements of a normally distributed (ND) signal to be used for programs HUBER, MSD, and HISTO.

***HPLC1**.dat:*   Eight impurities were measured by the area-% technique; nine batches of a raw material were tested; file can be used with any program except EUCLID.

***HPLC2**.dat:*   Eight impurities were tested over 14 runs; file is to be used with program EUCLID to determine whether the 14 samples belong to two different groups. (See also file SIEVE.dat.)

***HUBER**.dat:*   Two vectors of $n = 9$ each are given for testing program HUBER.

***INTERPOL1**.dat:*   For five different probabilities p the corresponding Student's t for $f = 4$ is listed, together with the log/log coordinates, in which the relationship is nearly linear.

***INTERPOL2**.dat:*   Same as INTERPOL1.dat, but with only three levels of probability.

***INTERSECT1**.dat, **INTERSECT2**.dat: Section 2.2.11*   A titration that changes the conductivity of the electrolyte solution is followed; the measurements are assigned to one of two branches for which the intersection of the linear regression lines defines the equivalence point. Program INTER-

SECT estimates the 95% CL of the intersection to be 79.3 ± 2.1 ml for data set INTERSECT1.dat resp. 10.5 ± 0.2 ml for data set INTERSECT2.dat.

JUNGLE1.dat: Section 2.1   Three parameters (impurity content, HPLC assay, and titration assay) were measured on five batches of a raw material.

JUNGLE2.dat: Section 4.22   Eight parameters are measured per batch; the parameters are partially linked. $N = 48$ batches are simulated. Use with CORREL to find these links.

JUNGLE3.dat: Section 4.22   Same as JUNGLE2.dat, but with very specific artifacts added.

JUNGLE4.dat: Section 4.11, Table 4.15   Nine parameters were determined on a total of 43 batches of a chemical intermediate. Use with CORREL.

LRTEST.dat: Synthetic data for program LINREG.

MOISTURE.dat: Section 4.4   At 10 selected locations inside a dryer samples of eight tablets each were drawn to determine water content by the Karl Fischer method; using MULTI, the hypothesis H_0 is tested that all 10 sample means and standard deviations are indistinguishable.

MSD.dat: Test file for MSD (HISTO, HUBER) that contains four data sets of different size and distribution.

ND_160.dat: Fifteen columns that each contain 160 random numbers. To be used with MSD, HISTO, CORREL, SMOOTH to obtain a baseline, against which to compare real data sets; the ruggedness of evaluations can be checked through comparisons with sets of random numbers.

PACK sort.dat: A set of 3444 weighings of a product coming off a packaging line; the smoothing routine is explained in Fig. 1.29. The second vector contains the residuals $r_i = y_i - Y_{smoothed}$.

PARABOLA.dat: Thirteen EMF vs. temperature measurements that conform to the equation $Y = -20.63 + 0.6395^*X - 0.007295^*(X - 31.09)^2$, to be used with TESTFIT and LINREG.

PKG CLASS.dat: One hundred items' weights are given; in Vectors 2 and 3 the items are classified as either "Hi" or "Lo," to be used with SMOOTH, MSD, HISTO, HUBER, and CUSUM.

***PROFILE*.dat:** The area-% results for the main compound and 10 impurities are given for nine samples of a bulk chemical; use with CORREL to ferret out classes of impurities that presumably are involved in common synthesis pathways.

***QRED_TBL*.dat: Table 1.7** Reduced critical q-values (division of q-values for $p = 0.05$ by the appropriate Student's t-factor and SQR(2)), as used with the multiple range test in MULTI.

***RIA PREC*.dat:** Two hundred thirty-eight calibration data sets were collected and analyzed for repeatability (within group CV) and plotted against the mean concentration. In a double-logarithmic plot the pattern seen in Fig. 4.6 appears.

***RND_1_15*.dat:** A triangular matrix of random numbers that serves the same purpose as ND_160.dat, but introduces the vector length as a factor. Use with MSD or HUBER.

***SHELFLIFE*.dat:** The content (% of nominal) of two active components in a dosage form was assayed at various times (0–60 months) during a pharmaceutical stability trial to determine the acceptable shelf-life of the formulation; the point at which the lower 90% confidence limit of the linear regression model intersects the 90%-of-nominal line gives the answer. Use with SHELFLIFE or LINREG.

***SIEVE1*.dat: Section 4.12** A crystalline raw material is purchased from two different suppliers on the basis of the same specifications; crystal size distribution was relatively loosely defined, so that both vendors' materials passed specs. Production trials with seven batches from each vendor resulted in products of unequal properties: sieve analysis was carried out on retained samples using a laser light-scattering technique, yielding %-content for each of 15 classes. Analysis of the vendor-averaged sets by the conventional χ^2-test yielded no conclusive answer due to the high within-vendor-group variability. Using EUCLID, the 15-point data set for each sample was projected from 15-space into the plane defined by the three Euclidean distances A–B, A–S, and S–B, where A, B, and S are the coordinates in 15-space of the vendor-averages A and B respectively the individual sample S. Two nonoverlapping groups of points could be distinguished that confirmed the impression gained during the casual inspection of the 14-column by 15-row table.

***SIEVE2*.dat: Section 4.12** This file is identical to the first seven rows of file SIEVE1.dat, that is, the eight rows (dimensions, measurement channels) that contribute nearly nothing towards the distance between the group centers have been eliminated.

SIM1.dat: Section 1.4   Five data sets of 200 points each generated by SIM-GAUSS; the deterministic time series sine wave, saw tooth, base line, GC-peak, and step function have stochastic (normally distributed) noise super-imposed; use with SMOOTH to test different filter functions (filer type, window). A comparison between the (residual) standard deviations obtained using SMOOTH respectively HISTO (or MSD) demonstrates that the straight application of the Mean/SD concept to a fundamentally unstable signal gives the wrong impression.

SMOOTH.dat: A 26-point table of values interpolated from a figure in Ref. 162, to demonstrate the capability of the discussed extended Savitzky-Golay filter to provide a smoothed trace from the first to the last point in the time series.

STAMP.dat: Section 4.9   For 120 tablets (see HARDNESS.dat) the individual weights and hardness values are arranged in columns of 10, by stamp ($N = 10$, $M = 24$). Use with MULTI, CORREL, or XYZ.

STEP2.dat: A time series of 200 points consisting of the original signal (a step-function at $t = 0$) and the simulated response. (See. Fig. 1.5.) Use with SMOOTH or TESTFIT.

TABLET C.dat: Section 4.18   Simulated drug content uniformity measurements; 10 different means, starting from 46 mg, with two samples of 10 tablets each at every weight. $N = 10$, $M = 20$. To be used with HUBER, HISTO, but also CORREL to test for spurious correlations in table of random numbers and with MSD to test for conformance with limits.

TABLET W.dat: Section 4.18   Similar to TABLET_C.dat, but with tablet weights, starting at 330 mg, and $N = 20$ tablets per sample. $N = 20$, $M = 20$.

TLC.dat: Section 4.2   The raw data shown in Fig. 4.3.

UV.dat: Section 2.2   A set of five calibration points (Absorbance vs. %-of-nominal Concentration) to be used with LINREG, see example used in Chapter 2, starting with Table 2.2.

UV d.dat, UV t.dat, UV q.dat: The same data as in UV.dat, but in duplicate, triplicate, respectively, quadruplicate, so that the effect of increasing N can be studied for constant s_{res} and S_{xx}/N.

VALID1.dat: Section 4.23   A set of repeat determinations ($m = 10$) at concentrations that are logarithmically spaced (10, 25, 50, 100, 250, 500 ng/ml); to be used with programs LINREG, VALID, and VALIDLL.

VALID2.dat: Section 4.23   A set of duplicate determinations carried out on each of six successive days, with logarithmically spaced concentrations (1, 2, 5, 10, 20, 50, 100, 200, and 500 ng/ml); $M = 7$, $N = 18$.

VALID3.dat: Section 4.23   A set of replicate determinations on each concentration (1, 2, 5, 10, 20, 50, 100, 200, and 500 ng/ml), and multiple determinations at 35 and 350 ng/ml. $N = 46$, $M = 2$; all measurements were carried out on one day.

VAR CV.dat: Section 4.34   Five sets of simulated calibration measurements (in triplicate) at six concentrations under the assumption of a CV of 1, 2, 5, 10, respectively 15%.

VOLUME.dat: Section 1.1.2   A set of five precision weighings of a water-filled 100 ml flask; the weights in grams were converted to milliliters using the standard density-vs.-temperature tables. Use with MSD to test the effect of truncation errors on the calculation of the standard deviation. (See Table 1.1.)

VVV.dat, VWV.dat, WWW.dat: Raw data shown in Fig. 4.1.

WEIGHT.dat: Thirty high-precision calibration measurements carried out on each of two analytical balances (LSD: 0.01 mg) in the course of less than one hour using the calibration weight "30 g", cf. Ref. 25.

WLR.dat: Section 2.2.10   A set of peak area vs. concentration results of a gas chromatography calibration. Use with LINREG to test the effect of a weighting scheme. The originally estimated dependence of the standard deviation of determination vs. concentration is described by the equation SD $= 100 + 5^*x$.

XYZCELL.dat: Section 2.4   Fifty six stability measurements were taken on a series of product batches during a program that included storage at 4°, 25°, 31°, 40°, and 41°C and sampling at 0, 1, 3, resp. 6 months; the assay of the drug substance is given in %-nominal, and the concentrations of two degradants are in % relative to the drug substance. The data table has to be rearranged for 2-D analysis using Excel. The storage conditions "31" and "41"°C were used as isothermal models for "real-life" temperature conditions over the course of a "typical" year that would feature standard controlled storage conditions (e.g., 15°–25°C) for most of the time and short excursions up to 40° or 50°C, as could occur during transport.

5.4.5 EXCEL Files

The following applications were programmed in EXCEL and are being provided in the Excel 97 format to assure maximum compatibility. The intention was to provide accessible examples of how the statistical functions available in Excel can be put to use in demonstration and teaching situations. Most applications make use of the Monte Carlo simulation technique; pressing [F9] recalculates all values. Repeatedly pressing [F9] gives the user an impression of how real measurements can vary from one measurement run to the next without changing their statistical properties. Wherever possible, the basic assumptions were formulated as parametrized models. For instance, analytical drift and bias might have a typical value and a probability of occurrence assigned, so that unique conditions obtain for every simulated run.

The files are password- and write-protected; only the numbers in the colored cells can be changed. The output was optimized for a screen with 800 × 600 resolution. In order that those who use higher resolutions can adapt the graphics to full-screen size, the password is here divulged: "smac."

ASSAY AB.xls: Figs. 4.39 and 4.49   The model assumes a dosage form, e.g., a tablet, the weight of which systematically (effective mean in B2) and stochastically (CV in B3) differs from the nominal weight of 100. The composition of the drug is such that the concentration of the two active components A and B systematically (C2, D2) and stochastically (C3 and D3) deviate from the nominal concentrations (= 100%). Thus the actually measured results for each dosage unit are proportional to B2 · C2, resp. B2 · D2, with a CV approximately by $\sqrt{B3^2 + C3^2}$, respectively $\sqrt{B3^2 + D3^2}$, if all goes well. A systematic error (bias, calibration error) defined by ND(0, (s_x = F3)^2) is superimposed with a probability as given in E3. The six repeat results A, B, and weight will normally be compared against the specification limits, here suggested by the graph boundaries 90 . . . 110. If all results conform, many analysts will not bother to look for correlations or out-of-trend (OOT) values, as can easily be done by plotting each assay value against the tablet weight (boxes at left), or, if two active components are present, one against the other (big box). If the distribution in the big box looks like an elongated ellipse along the diagonal, then the weight variation is where corrective efforts should go. Should an individual value dance out of line, then the corresponding analytical method be looked at. An ellipse that is offset from the diagonal signals either over- or underdosing of a component, or analytical bias. Pressing [F9] initiates a new simulation.

CONV.xls: Fig. 1.20   This Excel spread sheet is similar in concept to program CONVERGE.exe, but shows simulation results for 5 groups of 8 data points. On the left side of the graph the data are depicted for an analog instrument. On the right side the same data is treated as if a digital instrument had been used, with a digitization step that is set as a fraction of the standard deviation. Each CI is given as a vertical bar, with the estimate for x_{mean} respectively s_x appearing as a small gap at the appropriate vertical position.

DECOMPOSITION.xls: Fig. 4.32, 4.39, Sections 2.2.6, 2.4, 3.1   The decomposition of a parent compound to its daughter can occur during processing or storage; if the former is the case, then the reduced assay values will not further diminish during storage, and the daughter product will remain at low levels. If degradation sets in or continues during storage, then this should be picked up in the stability study. The model assumes that an analytical method is used that does not distinguish between parent and daughter compounds except by way of the proportionality factor (slope of the calibration line), as is the case when chemically similar compounds are measured by UV spectrophotometry. The characteristic (and perhaps unobserved) degradation rate is given by the number in $B11$; the analytical error is characterized by $ND(0, (s = C11)^2)$. The daughter's chromophore has an extinction coefficient ϵ that is somewhat smaller than that of the parent compound: $\epsilon_D = D11 \cdot \epsilon_P$. The measurement of the daughter is assumed to be beset by the same error as is the parent compound. The actually observed light blue triangles (regression line is gray) are defined by the linear combination according to Lambert-Beer's Law: $A_{obs} = \text{factor} \cdot (c_P \cdot \epsilon_P + c_D \cdot \epsilon_D) + ND(0, 2 \cdot s_A^2)$.

DEGRAD STABIL.xls: Section 1.8.4   The analysis of stability reports often suffers from the fact that the data for each batch of product is scrutinized in isolation, which then results in a "see-no-evil" attitude if the numerical values are within specifications. The analyst is in a good position to first compare all results gained under one calibration (usually a day's worth of work) irrespective of the products/projects affected, and then also check the performance of the calibration samples against experience, see control charts, Section 1.8.4. In this way, any analytical bias of the day will stand out. For this purpose a change in format from a "Time-on-Stability" to a "Calendar Time" depiction is of help.

The file contains a real-life data set and a simulation model. The model allows for staggering of the on-stability time-points of the four batches (use multiples of 3 or 6 months!, gray cells), an individual degradation rate for each batch (purple cells), a random manufacturing bias (blue cells) and a random analytical error (pale yellow cells). For each analytical run, a calibration bias can be built in (yellow, light green cells; Keep the time shifts in mind!). There is also the possibility of introducing a variation of the analytical method at a certain time, which again brings a method-related bias to bear (white, dark blue, dark pink cells).

ELECTRODE xls: Fig. 4.24 The operation of an ion-specific electrode with a slope of 59.16 mV per decade for mono-valent ions (29.58 mV/dec for di-valent ions) is simulated under the assumption that a digital volt meter with a resolution of, say, 0.1 mV is used. The sample volume and the concentration of the metered titration solution are known. Normally, one would add a few milliliters of the concentrated titration solution and do the calculation spelled out in lines 140–150 in Table 4.22; here, because the sample concentration is known, the result can be normalized to it. The operation of short-cuts (volume correction), unknowns (volume bias, deviation of true slope from theoretical), and equipment shortcomings (digitization) can be studied.

OOS RISK N xls: Section 4.24 A basic model of risk-assessment is presented in three variations: $N = 4$, 5, respectively 6 repeat determinations per result. At the top of the screen the ideal dosage form, e.g., a tablet, is described by the signal (e.g., UV absorbance), the total weight, and the weight due to the active component itself (pale green cells). The calculation of the result is Drug = Absorbance · Weight · Factor, where the Factor is derived from the measurement of an ideal tablet, or, in practice, from the weight-corrected absorbance of a close-to-ideal tablet preparation. The vagaries of real life are entered into the colored cells, i.e.: variability of the drug-to-total-weight ratio (light blue cell), the variability of the tablet weight (light green cell), the measurement noise (pink cell), and a weight bias (yellow cell, due to improper equipment adjustment). The assay result can be rounded to a specific number of decimals (grey cell) before being inserted into the calculation. The symmetrical specification limits are in the blue cell. Due to the weight bias one of the SLs is closer to the calculated mean that the other, and is given in cell H10. The four to six simulated absorbances, tablet weights and assay values are given in the box, together with the respective means, and CVs; the percent-deviation of the mean signals from nominal is given below the box. Each calculated assay value is then compared against the SLs and declared to be either "in compliance" or "OOS." If an OOS result is detected, the percent deviation from the appropriate SL is given at right. The normalized differences $|A - \text{SL}|/s_A$, where A = assay, are used to estimate the cumulative risk of having at least one OOS result among the $N = 4$, 5, or 6 repeats (lower left box). When the CL(mean) are within the specification limits, the box at the lower right displays a green YES, otherwise a red NO ($\text{SL}_L \leq (E20 \pm t \cdot E21/\sqrt{N}) \leq \text{SL}_U$). Hereby it becomes evident that averages, and their confidence limits, obtained from four to six repeats can be within specifications even if a single result is just outside these limits; regulatory-wise, such an average would be regarded as being out of compliance!

PEDIGREE.xls: Figs. 4.44 and 4.45 The cumulation of random and systematic errors in the hierarchy of calibration standards is investigated. The model assumes a Gaussian analytical error coupled with transfer errors and laboratory bias. The light blue cells serve to define the reproducibility of the measurements. Each transfer from one laboratory to another is assumed to also involve an error (yellow cells) due to differences in equipment and/or work practices. In each laboratory it is possible to commit calibration or other bias-inducing errors (pale pink cells); this happens at random under a defined probability of occurrence (pink cells). Each simulation run results in specific, normally distributed errors being assigned to each of the assays and transitions. Below the pale pink cells a random number of rectangular distribution is determined for each of the laboratory tests in the pedigree diagram above; if the random number is smaller than the error probability in the pink cell above it (e.g., 0.02384 in cell H27 is less than 3/100 in cell H23) a chance error of magnitude $ND(0, s^2)$ is assigned to the analysis simulated in the corresponding cell (s = H27, analysis in cell H7, in this case). This chance error then appears in the white cell underneath the affected one. The eight representative samples in column I convey an idea of what kind of true values μ could be declared to be indistinguishable from the standard material. In concrete terms, the problem crops up when one factory produces a product using one standard and another factory tests it using their own standard, which too is tied to the same standard material. If a tested sample result is more than X standard deviations from the mean, three asterisks show up in the cell next to it (the $X \cdot$ SD criterion in cell J25 could be used as an outlier test).

POWER.xls: Fig. 1.34, Sections 1.9, 4.1 A test distribution (e.g., ND(2.8, 0.5^2), light green cells, bold green line for distribution, thin green line for CP) and the reference distribution (bold blue distribution, thin blue CP curve; e.g. ND(0, 1), yellow cell and line for power $= 1 - \beta$) can be defined. The cursor (red cell) marks the cut-off for which the risk alpha (dark blue, to right of cursor) is known. With this information, the corresponding areas (risks) under the green curve to the left of the cursor can be estimated (light green cell I1). Change position and width of the test distribution to evaluate risk β.

PROB REJECT.xls: Section 3.6, Fig. 1.29 In a production environment there are often several superimposed processes that yield measurement series like that depicted in the lower panel: there is drift that unexpectedly changes slope, there is bias and measurement noise, and there are operators who take corrective action. The model includes the probability of drift occurring and a feed-back loop that permits both positive and negative coefficients. The operators can be ordered to react if a single value exceeds a set limit, or only if 2, 3, or more successive values do so. The program calculates the two-sided (asymmetric) total probability of a value being OOS and depicts this in the upper panel on a $\log(p)$ scale. The red horizontal is the upper limit on the total risk as set in cell B20.

QUOTE RESULT.xls: Section 1.1.5 The calculated mean (yellow cell A2) and standard deviation (light blue cell A3) are entered together with the sample size N (blue cell A1) and the risk of error α (light green cell A4). The program calculates the confidence limits for the population, the mean, and the standard deviation. The spread between the upper and lower $CL(s_x)$ is used to determine the number of significant digits that can be displayed. The results given in the gray box should be quoted without terminal zeros. Possible interpretations are given at right.

SHELFLIFE.xls: Section 4.20 The yellow area in the left upper corner can be used to enter up to 10 data points (the unneeded cells should be cleared), for which the regression line is calculated. The thin green lines above and below the trendline indicate the zone within which new measurements must be regarded as belonging to the population; any measurement outside this zone should be investigated. The bold yellow curve represents the lower confidence limit of the red trend line. The interpretation is that at the time point where this curve intersects the specification limit, the probability of the trend being below the SL is 5%. This definition is used by the FDA to set the shelf-life of a medication. The program can also be used to study degradation products, because if a SL is entered that is less than 50%, the sign of the slope (cell B18) changes and not a lower but an upper CL is presented.

SYS SUITAB.xls: Section 1.1.4 A series of 10 measurements is simulated under the assumption ND(0, $(s = B2)^2$); in column 1, only the first two are shown, in column 2 the first three, and so on. In the upper panel the corresponding standard deviations for $N = 2, 3, \ldots$ etc. are given by the bold green line. The horizontal lines are for $E(s_x) = 1$ and the $SL(s_x) = 2 \cdot$ (cell E2). The least significant digit can be set in cell $C2$ in order to simulate the operation of a digital instrument. Besides the calculated SD, four other lines (light blue, dark blue, gold, resp. dark green) give the upper $CL(s_x)$ for $p = 0.2 \ldots 0.025$. This program illustrates the width of the confidence interval for a standard deviation for a low number of repeat measurements, and the large changes induced by the addition of a single additional measurement.

EXCEL FNC.xls   The following functions are shown:

- Generation of random numbers belonging to the ND(0, 1) (Monte Carlo Technique).
- Transformation of CP to z.
- Generation of Student's t.
- Generation of PD and CP curves.

TECHNICAL TIDBITS

The genesis of this book is a fair reflection of the development of electronic tools in the past quarter of a century. The text, the calculations, and the graphics improved in leaps and bounds; the authors take some pride in this, but suffered with each introduction of a new software package or release.

Ancient History   Some of the programs (LinReg, Histogram) go back to our first laboratory exercises in data evaluation with chemistry undergraduates at the ETHZ in 1973: FORTRAN IV was run at the computer center, data entry was on punched cards, and output was in the form of "text graphics" (cf. Table 4.35); the introduction of a HP 9830 desktop computer, with all of 1760 bytes of memory (sic!), BASIC programming, a 1-line display, a printer, and a plotter (total cost twice as high as an assistant professor's annual before-taxes salary) was pure luxury and finally made graph paper obsolete. Later, everything was transferred to a HP-85. The first company PCM worked for did not know computers existed, so he bought a HP-71B BASIC-programmable pocket calculator fitted with 22K RAM, a curve and a math ROM, and a disc drive. A ThinkJet matrix printer was used to graph the majority of the figures at a resolution of 320×320 pixels respectively 0.25 mm/0.01 inch per line pair (after data entry, it took some 3–5 minutes of processor time to generate a single graph, formatted as 32 strings of 40 user-defined characters each). Total cost: one-third of his monthly take-home pay. In no time, the kernels of most of the programs presented here were available and were used on a daily basis. Comfort was at a rock-bottom level and entailed memorizing all promts (one to three letter displays) and the corresponding option lists.

First Edition Soon after PCM's transfer to Cilag A. G. in 1986, work began in earnest. Text was sketched out at night and typed into Quality Control's Wang OIS-140 terminal *in lieu* of going out for lunch. Printouts were available every 2 weeks in the next building. Graphics continued to be done on the HP-71B, with a sprinkling on Symphony and Lotus 1-2-3. The programs provided with the book were written on the GW-BASIC platform contained in MS-DOS 3.0 on a Sharp 286.

Second Edition Shortly after submission of the first edition, the text was migrated to WP5.0/DOS on a 386, and then to WP5.1/Windows 3.11 on a 486. The accompanying programs were continuously improved, first on a Q-BASIC and then on a QuickBasic platform. When—surprise—Wiley asked for a revision, practicality demanded a renewed adaptation of the home PC to company standards, so as not to have to switch back and forth between standards every day: the text was prepared on Microsoft Word 7.0 running under Windows 95 on a 200 MHz Pentium II machine. Graphics were prepared in a more roundabout way: whenever precise *XY* coordinates were involved, an Excel Spreadsheet was prepared, *viz.* Figs. 1.14–17, and the coordinates were saved as a .WK3 file. Some calculations were more easily done in QBASIC, *viz.* Figs. 1.2, 1.5, or 1.12, with results written to .csv files that were then read into Excel. These .WK3 files were then imported into Lotus 1-2-3, and the resulting graphics were saved in .cgm format, which could be imported into VISIO 4.0. VISIO has two advantages, namely the ease of manipulating graphical elements (symbols, text, etc.), and the ImageStream driver to create the final .tif-format images that Wiley prefers. More direct routes from Excel to VISIO or .tif ended in inexplicable losses of control over such things as line widths or positions of text boxes. The only Sigmaplot .jnb file (cf. Fig. 2.21) had to be printed, scanned, stored as a .pcx file, and imported into VISIO for conversion to .tif format.

Copyright/Trademarks The following products are named:

Corel Draw
FORTRAN IV
Hewlett Packard HP-71B BASIC Curve ROM
Hewlett Packard HP-71B BASIC Math ROM

Hewlett Packard HP-71B ThinkJet matrix printer
Hewlett Packard HP-71B BASIC-programmable pocket calculator
Hewlett Packard desktop computer models HP-9830 and HP-85
ImageStream TIF Export 2.0.3
Intel 386, 486, and Pentium II CPUs
Lotus 1-2-3 spread-sheet
Microsoft Excel 7.0
Microsoft GW-Basic 3.23
Microsoft Q-Basic 1.1
Microsoft VisualBasic 5.0
Microsoft Windows 3.11 OS
Microsoft Windows OS 95
Microsoft Word 97
QuickBasic
Sharp PC-5541 286
Sigmaplot
Symphony spreadsheet
VISIO 4.0
Wang OIS-140 word processor
WordPerfect 5.0, 5.1 word-processing program

GLOSSARY

The following terms and abbreviations are repeatedly used:

Symbol	Explanation
a	intercept
AL	action limits
alpha (α)	probability of type I error
assay	determination of content of, for example, active principle (– drug compound)
b	slope
beta (β)	probability of type II error
c.o.v.	coefficient of variation; cf. RSD
Chi^2 (χ^2)	statistical indicator of similarity, χ^2-tables
CHN	elemental analysis for C(arbon), H(ydrogen), and N(itrogen)

CI( )	confidence interval of quantity in parentheses, equal to the distance between CL_L and CL_U
CL( )	confidence limits of the estimate in parentheses
CP	cumulative probability
CV	coefficient of variation; cf. RSD
d (δ)	differential, as in $\delta x/\delta t$
d (Δ)	difference between measured values, e.g., in paired t-test
DVM	digital voltmeter
E( )	expected value
EMF	electromotive force, e.g., as found in pH electrodes
F	test statistic, F-test
f or df	degrees of freedom (in Chapter 5 "df" is used instead of f to avoid confusion with the polynomial coefficient f
$f(\dots)$	algebraic function
FWHM	full width at half maximum, width of a peak
GC	gas chromatography
GLP	good laboratory practices
GMP	good manufacturing practices
GOF	goodness of fit, e.g., χ^2
H_0	null hypothesis
H_1	alternative hypothesis
HPLC	high-pressure liquid chromatography
ICP	inductively coupled plasma spectrometry
IHL	in-house limits
indices$_{L,U}$	"lower and upper, as in SL_L, SL_U, CL_L, or CL_U
Inf (∞)	infinity, either $-\infty$ or $+\infty$
ISE	ion-selective electrode
k	normalization factor in weighted regression
k, m	number of repeat measurements
k, z	safety factors in detection of outliers (Huber's k, classical z) or calculation of LOD (z)
LAR	largest absolute residual
LLS	laser-light scattering
LOD	limit of detection
LOQ	limit of quantitation
LRR	largest relative residual
LSD	least significant digit
m	number of groups
MAD	median absolute deviation, Huber's outlier test
MC	Monte Carlo numerical simulation technique
mu (μ)	true value of mean
n	number of measurements, sample size

n_1	number of samples, first series
n_2	number of samples, second series
$ND(0, 1)$	normal distribution with $\mu = 0$, $\sigma^2 = 1$
PD	probability density
q	test statistic in multiple range test
q_c	critical q-value
QC/QA	quality control/quality assurance, department or function
r	correlation coefficient
$R(n)$	range of n values
r^2	coefficient of determination
r_j	residual
RSD	relative standard deviation
s_d	standard deviation of the mean difference
SI	specification interval
sigma (σ)	true value of standard deviation
SL	specification limits
SOP	standard operating procedure
s_{res}	residual standard deviation
S_T	total sum of squares
s_x	standard deviation of a distribution
$s_{x,\,mean}$	standard deviation of a mean
S_{xx}	sum of squares
$S_{xx,w}$	weighted sum of squares
S_{xy}	sum of squares
$S_{xy,w}$	weighted sum of squares
S_{yy}	sum of squares
$S_{yy,w}$	weighted sum of squares
t	Student's t-value, test statistic in t-test
TAR	typical absolute residual
tau (τ)	time constant
t_c	critical t-value
TRR	typical relative residual
UV	ultraviolet part of spectrum
V_1	variance of first series or within group
V_2	variance of second series or between groups
V_a	variance of the intercept a
V_b	variance of the slope b
V_d	variance of the mean difference
V_p	pooled variance
V_{res}	residual variance
V_x	variance of x
V_X	variance of the estimate $X = (y - a)/b$

V_Y	variance of the estimate $Y = a + b \cdot x$
w_i	weight assigned to individual measurement
X	estimate. as in $X = f(y)$
$x()$	array or vector of values
x_{GM}	grand mean
x_i	ith x-value
x_{ij}	element in array $x(,)$
x_m	median
x_{max}	largest value x_i
x_{mean}	mean or average over several x_i
$x_{mean, w}$	weighted mean
x_{min}	smallest value x_i
XRF	X-ray fluorescence
Y	estimate, as in $Y = f(x)$
y^*	(average) value measured for unknown
y_i	measured value
z	standardized deviate in ND

REFERENCES

1. Hill, H. M., and Brown, R. H., "Statistical Methods in Chemistry," *Anal. Chem.* **40,** 1968, 376R–380R.

2. Green, J. R., and Margerison, D., "Statistical Treatment of Experimental Data," Elsevier Scientific, Amsterdam, Oxford, U.K., and New York, 1978.

3. Kratochvil, B., and Taylor, J. K., "Sampling for Chemical Analysis," *Anal. Chem.* **53,** 1981, 924A–938A.

4. Harris, W. E., "Sampling, Manipulative, Observational, and Evaluative Errors," *International Laboratory*, Jan–Feb 1978, 53–62.

5. Boyer, K. W., Horwitz, W., and Albert, R., "Interlaboratory Variability in Trace Element Analysis," *Anal. Chem.* **57,** 1985, 454–459.

6. Minkkinen, P., "Evaluation of the Fundamental Sampling Error in the Sampling of Particulate Solids," *Analytica Chimica Acta* **196,** 1987, 237–245.

7. Krivan, V., and Haas, H. F., "Prevention of Loss of Mercury (II) During Storage of Dilute Solutions in Various Containers," *Fresenius Z. Anal. Chem.* **332,** 1988, 1–6.

8. Hungerford, J. M., and Christian, G. D., "Statistical Sampling Errors as Intrinsic Limits on Detection in Dilute Solutions," *Anal. Chem.* **58,** 1986, 2567–2568.

9. Dyg, S., Cornelis, R., Griepink, B., and Quevauviller, P., "Development and Interlaboratory Testing of Aqueous And Lyophilized Cr(III) and Cr(IV) Reference Materials," *Analytica Chimica Acta* **286,** 1994, 297–308.

10. Caricchia, A. M., Chiavarini, S., Cremisini, C., Morabito, R., and Scerbo, R., "Influence of Storage Conditions on the Determination of Organotin in Mussels," *Analytica Chimica Acta* **286,** 1994, 329–334.

11. Cobo, M. G., Palacios, M. A., and Cámara, C., "Effect of Physicochemical Parameters on Trace Inorganic Selenium Stability," *Analytica Chimica Acta,* **286,** 1994, 371–379.

12. Thomas, R. P., Ure, A. M., Davidson, C. M., Littlejohn, D., Rauret, G., Rubio, R., and López-Sánchez, J. F., "Three-Stage Sequential Extraction Procedure for the Determination of Metal in River Sediments," *Analytica Chimica Acta* **286,** 1994, 423–429.

13. Nilsson, T., Ferrari, R., and Facchetti, S., "Interlaboratory Studies for the Validation of Solid-phase Microextraction for the Quantitative Analysis of Volatile Organic Compounds in Aqueous Samples," *Analytica Chimica Acta* **356,** 1997, 113–123.

14. Yeung, E. S., and Synovec, R. E., "Detectors for Liquid Chromatography," *Anal. Chem.* **58,** 1986, 1237A–1256A.

15. Cooper, J. W., "Errors in Computer Data Handling," *Anal. Chem.* **50,** 1978, 801A–812A.

16. Horlick, G., "Reduction of Quantization Effects by Time Averaging with Added Random Noise," *Anal. Chem.* **47,** 1975, 352–354.

17. Foley, J. P., "Systematic Errors in the Measurement of Peak Area and Peak Height for Overlapping Peaks," *Journal of Chromatography* **384,** 1987, 301–313.

18. Dose, E. V., and Guiochon, G., "Bias and Nonlinearity of Ultraviolet Calibration Curves Measured Using Diode-Array Detectors," *Anal. Chem.* **61,** 1989, 2571–2579.

19. Horwitz, W., "Evaluation of Analytical Methods Used for Regulation of Foods and Drugs," *Anal. Chem.* **54,** 1982, 67A–76A.

20. Anscombe, F. J., "Rejection of Outliers," *Technometrics* **2,** May 1960, 123–147.

21. Davies, P. L., "Statistical Evaluation of Interlaboratory Tests," *Fresenius Z. Anal. Chem.* **331,** 1988, 513–519.

22. Worley, J. W., Morrell, J. A., Duewer, D. L., and Peterfreund, L. A., "Alternate Indexes of Variation for the Analysis of Experimental Data," *Anal. Chem.* **56,** 1984, 462–466.

23. Shukla, S. S., and Rusling, J. F., "Analyzing Chemical Data with Computers: Errors and Pitfalls," *Anal. Chem.* **56,** 1984, 1347A–1368A.

24. Shatkay, A., and Flavian, S., "Unrecognized Systematic Errors in Quantitative Analysis by Gas-Liquid Chromatography," *Anal. Chem.* **49,** 1977, 2222–2228.

25. Bzik, T. J., Henderson, P. B., and Hobbs, J. P., "Increasing the Precision and Accuracy of Top-Loading Balances: Application of Experimental Design," *Anal. Chem.* **70,** 1998, 58–63.

26. Kucera, J., and Faltejsek, J., "Interactive Programme for Evaluation of Circular Analyses (IPECA)," *Fresenius J. Anal. Chem.* **352,** 1995, 80–86.

27. Wanek, P. M., et al., "Inaccuracies in the Calculation of Standard Deviation with Electronic Calculators," *Anal. Chem.* **54,** 1982, 1877–1878.

28. Thompson, M. R., and Dessy, R. E., "Use and Abuse of Digital Signal Processors," *Anal. Chem.* **56,** 1984, 583–586.

29. Spaink, H. A., and Lub, T. T., "Improvement of the Performance of a Diode-Laser Spectrometer by Using Sweep Averaging and Numerical Data Processing," *Analytica Chimica Acta* **241,** 1990, 83–94.

30. Bauer, C. F., Grant, C. L., and Jenkins, T. F., "Interlaboratory Evaluation of High-Performance Liquid Chromatographic Determination of Nitroorganics in Munition Plant Wastewater," *Anal. Chem.* **58,** 1986, 176–182.

31. Nadkarni, R. A., "The Quest for Quality in the Laboratory," *Anal. Chem.* **63**(13), 1991, 675A.

32. Kateman, G., and Pijpers, *Quality Control in Analytical Chemistry*, John Wiley & Sons, New York, Chichester, U.K., Brisbane, Toronto, and Singapore, 1981.

33. "Guide for Use of Terms in Reporting Data in Analytical Chemistry," *Anal. Chem.* **58,** 1986, 269–270.

34. *Wissenschaftliche Tabellen Geigy*, 8. Aufl., CIBA-GEIGY, Basel, Switzerland, 1980.

35. Taylor, J. K., "Quality Assurance of Chemical Measurements," *Anal. Chem.* **53,** 1981, 1588A–1596A.

36. Keith, H. K., et al., "ACS Committee on Environmental Improvement, Principles of Environmental Analysis," *Anal. Chem.* **55,** 1983, 2210–2218.

37. MacDougal, D., et al. "Guidelines for Data Aquisition and Data Quality Evaluation in Environmental Chemistry," *Anal. Chem.* **52,** 1980, 2240–2249.

38. Wegscheider, W., "Standardization, Quality Control and Education in Analytical Chemistry," *Fresenius J. Anal. Chem.* **349,** 1994, 784–793.

39. Ellison, S., Wegscheider, W., and Williams, A., "Measurement Uncertainty," *Analytical Chemistry News & Features*, October 1, 1997, 607A–613A.

40. Solberg, H. E., "Inaccuracies in Computer Calculation of Standard Deviation," *Anal. Chem.* **55,** 1983, 1611.

41. Bialkowski, S. E., "Data Analysis in the Shot Noise Limit: 1. Single Parameter Estimation with Poisson and Normal Probability Density Functions," *Anal. Chem.* **61,** 1989, 2479–2483.

42. Frazer, J. W., "Computer Experimentation Techniques for the Study of Complex Systems," *Anal. Chem.* **52,** 1980, 1205A–1220A.

43. The United States Pharmacopeia, 1995, U.S. Pharmacopeial Convention, Inc., 12601 Twinbrook Parkway, Rockville, MD 20852.

44. Doerffel, K., "Statistik in der analytischen Chemie", 3. Aufl., *Verlag Chemie*, Weinheim, Germany, Deerfield Beach, Florida, and Basel, Switzerland, 1984.

45. Eckschlager, K., *Errors, Measurement and Results in Chemical Analysis*, Van Nostrand, New York, Toronto, and Melbourne, 1969.

46. Sachs, L., *Angewandte Statistik* (also available in English), Springer-Verlag, Berlin, Heidelberg, Germany, New York, and Tokyo, 1984.

47. Miller, J. C., and Miller, J. N., *Statistics for Analytical Chemistry*, Ellis Horwood, Ltd., Chichester, U.K., 1986.

48. Renner, E., *Mathematisch-statistische Methoden in der Praktischen Anwendung*, 2. Aufl., Verlag Paul Parey, Berlin and Hamburg, 1981.

49. Hays, W. L., *Statistics*, Holt, Rinehart and Winston, London, New York, Sidney, and Toronto, 1969.

50. Miller, R. G., Jr., *Simultaneous Statistical Inference*, Springer-Verlag, New York, Heidelberg, and Berlin, Germany.

51. Harter, H. L., "Critical Values for Duncan's New Multiple Range Test," *Biometrics*, December 1960, 671–685.

52. Hartmann, C., Vankeerberghen, P., Smeyers-Verbeke, J., and Massart, D. L., *Analytica Chimica Acta* **344**, 1997, 17–28.

53. Kelly, P. C., "Outlier Detection in Collaborative Studies," *J. Assoc. Off. Anal. Chem.* **73**(1), 1990.

54. Rorabacher, D. B., "Statistical Treatment for Rejection of Deviant Values: Critical Values of Dixon's "Q" Parameter and Related Subrange Ratios at the 95% Confidence Level," *Anal. Chem.* **63**, 1991, 139–146.

55. *FDA's Summary of Judge Wollin's Interpretations of GMP Issues Contained in the Court's Ruling in USA vs. Barr Laboratories (2-4-93)*; http://www.fda.gov/ora/inspect_ref/igs/pharm.html.

56. Doerffel, K., Herfurth, G., Liebich, V., and Wendlandt, E., "The shape of CUSUM—an Indicator for Tendencies in a Time Series," *Fresenius J. Anal. Chem.* **341**, 1991, 519–523.

57. Marshall, R. A. G., "Cumulative Sum Charts for Monitoring of Radioactivity Background Count Rates," *Anal. Chem.* **49**, 1977, 2193–2196.

58. Davies, O. L., and Hudson, H. E., "Stability of Drugs—Accelerated Storage Tests," Chapter 21B in *Statistics in the Pharmaceutical Industry*, C. R. Buncher and J.-Y. Tsay, Eds., Marcel Dekker, 1994.

59. VanArendonk, M. D., and Skogerboe, R. K., "Correlation Coefficients for Evaluation of Analytical Calibration Curves," *Anal. Chem.* **53**, 1981, 2349–2350.

60. Ripley, B. D., and Tompson, M., "Regression Techniques for the Detection of Analytical Bias," *Analyst* **112**, April 1987, 377–383.

61. Mitchell, D. G., Mills, W. N., Garden, J. S., and Zdeb, M., "Multiple-Curve Procedure for Improving Precision with Calibration-Curve-Based Analyses," *Anal. Chem.* **49**, 1977, 1655–1660.

62. Bialkowski, S. E., "Data Analysis in the Shot Noise Limit: 2. Methods for Data Regression," *Anal. Chem.* **61,** 1989, 2483–2489.

63. Ellerton, R. R. W., and Strong, F. C., III, "Comments on Regression through the Origin," *Anal. Chem.* **52,** 1980, 1151–1153.

64. Schwartz, L. M., "Effect of Constraints on Precision of Calibration Analyses," *Anal. Chem.* **58,** 1986, 246–250.

65. Strong, F. C., "Regression Line That Starts at the Origin," *Anal. Chem.* **51,** 1979, 298–299.

66. Royston, G. C., "Comments on Unrecognized Systematic Errors in Quantitative Analysis in Gas Chromatography," *Anal. Chem.* **50,** 1978, 1005.

67. Wang, C. Y., Bunday, S. D., and Tartar, J. G., "Ion Chromatographic Determination of Fluorine, Chlorine, Bromine, and Iodine with Sequential Electrochemical and Conductometric Detection," *Anal. Chem.* **55,** 1983, 1617–1619.

68. Shatkay, A., "Effect of Concentration on the Internal Standards Method in Gas-Liquid Chromatography," *Anal. Chem.* **50,** 1978, 1423–1429.

69. Schwartz, L. M., "Rejection of a Deviant Point from a Straight-Line Regression," *Analytica Chimica Acta* **178,** 1985, 355–359.

70. Renman, L., and Jagner, D., "Asymmetric Distribution of Results in Calibration Curve and Standard Addition Evaluations," *Analytica Chimica Acta* **357,** 1997, 157–166.

71. Schwartz, L. M., and Gelb, R. I., "Statistical Uncertainties of End Points at Intersecting Straight Lines," *Anal. Chem.* **56,** 1984, 1487–1492.

72. Betti, M., Papoff, P., and Meites, L., "Factors Affecting the Precisions of Analyses, by Potentiometric Titrimetry, of Solutions Containing two Weak Acids," *Analytica Chimica Acta* **182,** 1986, 133–145.

73. Carter, K. N., Scott, D. M., Salomon, J. K., and Zarcone, G. S., "Confidence Limits for the Abscissa of Intersection of Two Least-Squares Lines Such as Linear Segmented Titration Curves," *Anal. Chem.* **63,** 1991, 1270–1278.

74. Cardone, M. J., Palermo, P. J., and Sybrandt, L. B., "Potential Error in Single-Point-Ratio Calculations Based on Linear Calibration Curves with Significant Intercept," *Anal. Chem.* **52,** 1980, 1187–1191.

75. Schwartz, L. M., "Statistical Uncertainties of Analyses by Calibration of Counting Measurements," *Anal. Chem.* **50,** 1978, 980–985.

76. Cousino, M. A., Jarbawi, T. B., Halsall, H. B., and Heineman, W. R., "Pushing Down the Limits of Detection; Molecular Needles in a Haystack," *Anal. Chem.*, *"News and Features,"* September 1, 1997, 545A.

77. Porter, W. R., "Proper Statistical Evaluation of Calibration Data," *Anal. Chem.* **55,** 1983, 1290A (letter).

78. St. John, P. A., McCarthy, W. J., and Winefordner, J. D., "A Statistical Method of Evaluation of Limiting Detectable Sample Concentrations," *Anal. Chem.* **39,** 1967, 1495.

79. S. A. B., "Detection Limits, a Systematic Approach to Detection Limits Is Needed," *Anal. Chem.* **58**, 1986, 986A.

80. Long, G. L., and Winefordner, J. D., "Limit of Detection," *Anal. Chem.* **55**, 1983, 712A–724A.

81. Taylor, J. K., "Limits of Detection," *Anal. Chem.* **56**, 1984, 130A (letter).

82. Vogelgesang, J., "Limit of Detection and Limit of Determination: Application of Different Statistical Approaches to an Illustrative Example of Residue Analysis," *Fresenius Z. Anal. Chem.* **328**, 1987, 213–220.

83. Currie, L. A., "Limits for Qualitative Detection and Quantitative Determination," *Anal. Chem.* **40**, 1968, 586–593.

84. Williams, R. R., "Fundamental Limitations on the Use and Comparison of Signal-to-Noise Ratios," *Anal. Chem.* **63**, 1991, 1638–1643.

85. Williams, T. W., and Salin, E. D., "Hazards of a Naive Approach to Detection Limits with Transient Signals," *Anal. Chem.* **60**, 1988, 725–727.

86. Luthhardt, M., Than, E., and Heckendorff, H., "Nachweis-, Erfassungs- und Bestimmungsgrenze analytischer Verfahren," *Fresenius Z. Anal. Chem.* **326**, 1987, 331–339.

87. Oppenheimer, L., Capizzi, T. P., Weppelman, R. M., and Mehta, H., "Determining the Lowest Limit of Reliable Assay Measurement," *Anal. Chem.* **55**, 1983, 638–643.

88. Clayton, C. A., Hines, J. W., and Elkins, P. D., "Detection Limits with Specified Assurance Probabilities," *Anal. Chem.* **59**, 1987, 2506–2514.

89. Schoonover, R. M., and Jones, F. E., "Air Buoyancy Correction in High-Accuracy Weighing on Analytical Balances," *Anal. Chem.* **53**, 1981, 900–902.

90. Ratzlaff, K. L., and bin Darus, H., "Optimization of Precision in Dual Wavelength Spectrophotometric Measurement," *Anal. Chem.* **51**, 1979, 256–261.

91. Lam, R. B., and Isenhour, T. L., "Minimizing Relative Error in the Preparation of Standard Solutions by Judicious Choice of Volumetric Glassware," *Anal. Chem.* **52**, 1980, 1158–1161.

92. Schwartz, L. M., "Calibration of Pipets: A Statistical View," *Anal. Chem.* **61**, 1989, 1080–1083.

93. Gernand, W., Steckenreuter, K., and Wieland, G., "Greater Analytical Accuracy through Gravimetric Determination of Quantity," *Fresenius Z. Anal. Chem.* **334**, 1989, 534–539.

94. Snyder, L. R., and van der Wal, S. J. "Precision of Assays Based on Liquid Chromatography with Prior Solvent Extraction of the Sample," *Anal. Chem.* **53**, 1981, 877–884.

95. Kaye, W. and Barber, D., "Noise and Digital Resolution in a Microprocessor-Controlled Spectrophotometer," *Anal. Chem.* **53**, 1981, 366–369.

96. Unadkat, J. D., Beal, S. L., and Sheiner, L. B., "Bayesian Calibration," *Analytica Chimica Acta* **181**, 1986, 27–36.

97. Cardone, M. J., "New Technique in Chemical Assay Calculations: 1. A Survey of Calculational Practices on a Model Problem," *Anal. Chem.* **58,** 1986, 433–438.

98. Cardone, M. J., "New Technique in Chemical Assay Calculations: 2. Correct Solution of the Model Problem and Related Concepts," *Anal. Chem.* **58,** 1986, 438–445.

99. Franke, J. P., de Zeeuw, R. A., and Hakkert, R., "Evaluation and Optimization of the Standard Addition Method for Absorption Spectrometry and Anodic Stripping Volametry," *Anal. Chem.* **50,** 1987, 1374–1380.

100. Ratzlaff, K. L., "Optimizing Precision in Standard Addition Measurement," *Anal. Chem.* **51,** 1979, 232–235.

101. Gardner, M. J. and Gunn, A. M., "Optimising Precision in Standard Additions Determinations," *Fresenius Z. Anal. Chem.* **325,** 1986, 263–266.

102. Whang, C. W., Page, J. A., vanLoon, G., and Griffin, M. P., "Modified Standard Additions Calibration for Anodic Stripping Voltammetry," *Anal. Chem.* **56,** 1984, 539–542.

103. Gardner, M. J., and Gunn, A. M., "Approaches to Calibration in GFAAS: Direct or Standard Additions," *Fresenius Z. Anal. Chem.* **330,** 1988, 103–106.

104. Midgley, D., "Systematic and Random Errors in Known Addition Potiometry, A Review," *Analyst* **112,** May 1987, 557–572.

105. Horvai, G., and Pungor, E., "Precision of the Double Known Addition Method in Ion-Selective Electrode Potentiometry," *Anal. Chem.* **55,** 1983, 1988–1990.

106. Haghighi, B., Maleki, N., and Safavi, A., "Standard Additions in Flow-Injection Analysis with Atomic Absorption Spectrometry," *Analytica Chimica Acta* **357,** 1997, 151–156.

107. Schwartz, L. M., "Calibration Curves with Nonuniform Variance," *Anal. Chem.* **51,** 1979, 723–727.

108. Garden, J. S., Mitchell, D. G., and Mills, W. N., "Nonconstant Variance Regression Techniques for Calibration-Curve-Based Analysis," *Anal. Chem.* **52,** 1980, 2310–2315.

109. Watters, R. L., Jr., Carroll, R. J., and Spiegelman, C. H., "Error Modeling and Confidence Interval Estimation for Inductively Coupled Plasma Calibration Curves," *Anal. Chem.* **59,** 1987, 1639–1643.

110. Thompson, M., "Variation of Precision with Concentration in an Analytical System," *Analyst* **113,** October 1988, 1579–1587.

111. Phillips, G. R., and Eyring, E. M., "Comparison of Conventional and Robust Regression in Analysis of Chemical Data," *Anal. Chem.* **55,** 1983, 1134–1138.

112. Thompson, M., "Robust Statistics and Functional Relationship Estimation for Comparing the Bias of Analytical Procedures over Extended Concentration Ranges," *Anal. Chem.* **61,** 1989, 1942–1945.

113. Leary, J. J., and Messick, E. B., "Constrained Calibration Curves: A Novel

Application of Lagrange Multipliers in Analytical Chemistry," *Anal. Chem.* **57,** 1985, 956–957.

114. Schwartz, L. M., "Nonlinear Calibration," *Anal. Chem.* **49,** 1977, 2062–2068.

115. Bysouth, S. R., and Tyson, J. F., "A Comparison of Curve Fitting Algorithms for Flame Atomic Absorption Spectrometry," Journal of Analytical Atomic Spectrometry, **1,** February 1986, 85–87.

116. Kragten, J., "Least-Squares Polynomial Curve-Fitting for Calibration Purposes (STATCAL-CALIBRA)," *Analytica Chimica Acta* **241,** 1990, 1–13.

117. Brubaker, T. A., Tracy, R., and Pomernacki, C. L., "Linear Parameter Estimation," *Anal. Chem.* **50,** 1978, 1017A–1024A.

118. Phillips, G. R., and Eyring, E. M., "Error Estimation Using the Sequential Simplex Method in Nonlinear Least Squares Data Analysis," *Anal. Chem.* **60,** 1988, 738–741.

119. Christensen, M. K., "Determining the Parameters of First-Order Decay with a Nonzero End Point and Unequal Time Intervals," *Anal. Chem.* **55,** 1983, 2324–2327.

120. Brubaker, T. A., and O'Keefe, K. R., "Nonlinear Parameter Estimation," *Anal. Chem.* **51,** 1979, 1385A–1388A.

121. Schwartz, L. M., "Lowest Limit of Reliable Assay Measurement with Nonlinear Calibration," *Anal. Chem.* **55,** 1983, 1424–1426.

122. Beebe, K. R., and Kowalski, B. R., "An Introduction to Multivariate Calibration and Analysis," *Anal. Chem.* **59,** 1987, 1007A–1017A.

123. Keller, H. R. and Massart, D. L., "Artefacts in Evolving Factor Analysis-Based Methods for Purity Control in Liquid Chromatography with Diode-Array Detection," *Analytica Chimica Acta* **263,** 1992, 21–28.

124. Candolfi, A., Massart, D. L., and Heuerding, S., "Investigation of Sources of Variance which Contribute to NIR-Spectroscopic Measurement of Pharmaceutical Formulations," *Analytica Chimica Acta* **345,** 1997, 185–196.

125. Pfeiffer, C. D., Larson, J. R., and Ryder, J. F., "Linearity Testing of Ultraviolet Detectors in Liquid Chromatography," *Anal. Chem.* **55,** 1983, 1622–1624.

126. Dorschel, C. A., Ekmanis, J. L., Oberholtzer, J. E., Warren, F. V., and Bidlingmeyer, B. A., "LC Detectors: Evaluation and Practical Implications of Linearity," *Anal. Chem.* **61,** 1989, 951A–968A.

127. Mestek, O., and Suchánek, M., "Sampling Strategy and Its Effect on the Precision of Results of Ore Analysis," *Fresenius J. Anal. Chem.* **348,** 1994, 188–194.

128. Deming, S. N., and Morgan, S. L., "Simplex Optimization of Variables in Analytical Chemistry," *Anal. Chem.* **45,** 1973, 278A–283A.

129. Moody, J. R., Greenberg, R. R., Pratt, K. W., and Rains, T. C., "Recommended Inorganic Chemicals for Calibration," *Anal. Chem.* **60,** 1988, 1203A–1218A.

130. Proctor, A., and Sherwood, P. M. A., "Smoothing of Digital X-Ray Photoelec-

tron Spectra by an Extended Sliding Least-Squares," *Anal. Chem., Analytical Approach,* **52,** 1980, 2315–2321.

131. Subcommittee on Environmental Analytical Chemistry, ACS Committee on Environmental Improvement, Crummett, W. B., Chairman, "Guidelines for Data Acquisition and Data Quality Evaluation in Environmental Chemistry," *Anal. Chem.* **52,** 1980, 2242–2249.

132. Rodbard, D. and Hutt, D. M., "Statistical Analysis of Radioimmunoassays and Immunoradiometric Assays: A Generalized, Weighted, Iterative Least-Squares Method for Logistic Curve Fitting, in Radioimmunoassay and Related Procedures in Medicine," Int. Atomic Energy Agency, Vienna, 165–192, 1974, Uniput, New York.

133. Hampel, F., "Robuste Schaetzungen: Ein Anwendungsorientierter Ueberblick," *Biom. J.* **22,** 1980, 3–21.

134. Danzer, K., "Robuste Statistik in der analytischen Chemie," *Fresenius Z. Anal. Chem.* **335,** 1989, 869–875. (65 references cited)

135. Koscielniak, P., "Non-linear Robust Regression Procedure for Calibration in Flame Atomic Absorption Spectrometry," *Analytica Chimica Acta* **278,** 1993, 177–187.

136. Gonzalez, A. G., "Two Level Factorial Experimental Designs Based on Multiple Linear Regression Models: A Tutorial Digest Illustrated by Case Studies," *Analytica Chimica Acta* **360,** 1998, 227–241.

137. Vander Heyden, Y., Khots, M. S., and Massart, D. L., "Three-Level Screening Designs for the Optimization or the Ruggedness Testing of Analytical Procedures," *Analytica Chimica Acta* **276,** 1993, 189–195.

138. Rawlins, T. G. R., and Yrjonen, T., "Calculation of RIA Results Using the Spline Function," *International Laboratory,* Nov./Dec 1978, 55–66.

139. Halang, W. A., Langlais, R., and Kugler, E., "Cubic Spline Interpolation for the Calculation of Retention Indices in Temperature-Programmed Gas-Liquid Chromatography," *Anal. Chem.* **50,** 1978, 1829–1832.

140. Kragten, J., "Least-Squares Polynomial Curve-Fitting for Calibration Purposes," *Analytica Chimica Acta* **241,** 1990, 1–13.

141. Schwartz, L. M., and Gelb, R. I., "Statistical Analysis of Titration Data," *Anal. Chem.* **50,** 1978, 1571–1576.

142. de Valle, M., Poch, M., Alonso, J., and Bartroli, J., "Comparison of the Powell and Simplex Methods in the Optimization of Flow-Injection Systems," *Analytica Chimica Acta* **241,** 1990, 31–42.

143. Frazer, J. W., Balaban, D. J., and Wang, J. L., "Simulation as an Aid to Experimental Design," *Anal. Chem.* **55,** 1983, 904–910.

144. Poston, P. E., and Harris, J. M., "Maximum Likelihood Quantitative Estimates for Peaks: Application to Photoacoustic Spectroscopy," *Anal. Chem.* **59,** 1987, 1620–1626.

145. Güell, O. A., and Holcombe, J. A., "Analytical Applications of Monte Carlo Techniques," *Anal. Chem.* **62**, 1990, 529A–542A.

146. Bromba, M. U. A., and Ziegler, H., "Digital Smoothing of Noisy Spectra," *Anal. Chem.* **55**, 1983, 648–653.

147. Bromba, M. U. A., and Ziegler, H., "Digital Filter for Computationally Efficient Smoothing of Noisy Spectra," *Anal. Chem.* **55**, 1983, 1299–1302.

148. Bromba, M. U. A., and Ziegler, H., "Efficient Computation of Polynomial Smoothing Digital Filters," *Anal. Chem.* **51**, 1979, 1760–1762.

149. Bush, I. E., "Fast Algorithms for Digital Smoothing Filters," *Anal. Chem.* **55**, 1983, 2353–2361.

150. Bromba, M. U. A., and Ziegler, H., "Variable Filter for Digital Smoothing and Resolution Enhancement of Noisy Spectra," *Anal. Chem.* **56**, 1984, 2052–2058.

151. Jones, R., "High-Pass and Band-Pass Digital Filtering with Peak to Trough Measurement Applied to Quantitative Ultraviolet Spectrometry," *Analyst* **112**, November 1987, 1495–1498.

152. Steinier, J., Termonia, Y., and Deltour, J., "Comments on Smoothing and Differentiation of Data by Simplified Least Square Procedure," *Anal. Chem.* **44**, 1972, 1906–1909.

153. Lam, R. B., Wieboldt, R. C., and Isenhour, T. L., "Practical Computation with Fourier Transforms for Data Analysis," *Anal. Chem.* **53**, 1981, 889A–901A.

154. Horlick, G., "Digital Data Handling of Spectra Utilizing Fourier Transformations," *Anal. Chem.* **44**, 1972, 943–947.

155. Doerffel, K., Wundrack, A., and Tarigopula, S., "Improving Signal-to-Noise by Evaluation of Correlation Functions," *Fresenius Z. Anal. Chem.* **324**, 1986, 507–510.

156. Bialkowski, S. E., "Real-Time Digital Filters: Finite Impulse Response Filters," *Anal. Chem.* **60**, 1988, 355A–361A.

157. Bialkowski, S. E., "Real-Time Digital Filters: Infinite Impulse Response Filters," *Anal. Chem.* **60**, 1988, 403A–413A.

158. Enke, C. G., and Nieman, T. A., "Signal-to-Noise Ratio Enhancement by Least-Squares Polynomial Smoothing," *Anal. Chem.* **48**, 1976, 705A–712A.

159. Bromba, M. U. A., and Ziegler, H., "Application Hints for Savitzky-Golay Digital Smoothing Filters," *Anal. Chem.* **53**, 1981, 1583–1586.

160. Ziegler, E., *Computer in der Instrumentellen Analytik*, Akademische Verlagsgesellschaft, Frankfurt am Main, 1973.

161. Madden, H. H., "Comments on the Savitzky-Golay Convolution Method for Least-Squares Fit Smoothing and Differentiation of Digital Data," *Anal. Chem.* **50**, 1978, 1383–1386.

162. Gorry, P. A., "General Least-Squares Smoothing and Differentiation by the Convolution (Savitzky–Golay) Method," *Anal. Chem.* **62**, 1990, 570–573.

163. Savitzky, A. and Golay, M. J. E., "Smoothing and Differentiation of Data by Simplified Least-Squares Procedures," *Anal. Chem.* **36,** July 1964, 1627–1639.

164. Leach, R. A., Carter, C. A., and Harris, J. M., "Least-Square Polynomial Filters for Initial Point and Slope Estimation," *Anal. Chem.* **56,** 1984, 2304–2307.

165. Kahn, A., "Procedure for Increasing the Accuracy of the Initial Data Point Slope Estimation by Least-Squares Polynomial Filters," *Anal. Chem.* **60,** 1988, 369–371.

166. Edwards, T. R., "Two-Dimensional Convolute Integers for Analytical Instrumentation," *Anal. Chem.* **54,** 1982, 1519–1524.

167. Ratzlaff, K. L., "Computation of Two-Dimensional Polynomial Least-Squares Convolution Smoothing Integers," *Anal. Chem.* **61,** 1989, 1303–1305.

168. Kuo, J. E., Wang, H., and Pickup, S., "Multidimensional Least-Squares Smoothing Using Orthogonal Polynomials," *Anal. Chem.* **63,** 1991, 630–635.

169. Ratzlaff, K. L., and Natusch, D. F. S., "Theoretical Assessment of Precision in Dual Wavelength Spectrophotometric Measurement," *Anal. Chem.* **49,** 1977, 2170–2176.

170. Ratzlaff, K. L., and Natusch, D. F. S., "Theoretical Assessment of Accuracy in Dual Wavelength Spectrophotometric Measurement," *Anal. Chem.* **51,** 1979, 1209–1217.

171. Stolzberg, R. J., "Uncertainty in Calculated Values of Uncomplexed Metal Ion Concentration," *Anal. Chem.* **53,** 1981, 1286–1291.

172. Cabaniss, S., "Propagation of Uncertainty in Aqueous Equilibrium Calculations: Non-Gaussian Output Distributions," *Anal. Chem.* **69,** 1997, 3658–3664.

173. Lind, B., Elinder, C. G., Nilsson, B., Svartengren, M., and Vahter, M., "Quality Control in the Analysis of Lead and Cadmium in Blood," *Fresenius Z. Anal. Chem.* **326,** 1987, 647–655.

174. Ayers, G., Burnett, D., Griffiths, A., and Richens, A., "Quality Control of Drug Assays," *Clinical Pharmacokinetics* **6,** 1981, 106–117.

175. Moler, G. F., Delongchamp, R. R., and Mitchum, R. K., "Estimation of the Variance of the Area of a Single Chromatographic Peak," *Anal. Chem.* **55,** 1983, 842–847.

176. Johansson, E., and Wold, S., "Minimizing Effects of Closure on Analytical Data," *Anal. Chem.* **56,** 1984, 1685–1688.

177. Hall, P., and Selinger, B., "A Statistical Justification Relating Interlaboratory Coefficients of Variation with Concentration Levels," *Anal. Chem.*, **61,** 1989, 1465–1466.

178. Lin, K. K., Lin. T.-Y. D., and Kelly, R. E., "Stability of Drugs—Room Temperature Tests," Chapter 21A in *Statistics in the Pharmaceutical Industry*, C. R. Buncher and J.-Y. Tsay, Eds., Marcel Dekker, 1994.

179. Kvalheim, O. M., Brakstad, F., and Liang, Y., "Preprocessing of Analytical

Profiles in the Presence of Homoscedastic or Heteroscedastic Noise," *Anal. Chem.* **66,** 1994, 43–51.

180. Helsen, J. A., and Vrebos, B. A. R., "Quantitative X-Ray Fluorescence Analysis: Limits of Precision and Accuracy," *International Laboratory*, December 1986, 66–71.

181. Dunnett, C. W., and Goldsmith, C. H., "When and How to Do Multiple Comparisons," Chapter 22 in *Statistics in the Pharmaceutical Industry*, C. R. Buncher and J.-Y. Tsay, Eds., Marcel Dekker, 1994.

182. Chow, S.-C., and Shao, J., "Estimating Drug Shelf-Life with Random Batches," *Biometrics* **47,** September 1991, 1071–1079.

183. Ruberg, S. J., "Pooling Data for Stability Studies: Testing the Equality of Batch Degradation Slopes," *Biometrics* **47,** September 1991, 1059–1069.

184. Helm, V., *Stabilitätsuntersuchungen an wässrigen Peptidlösungen*, Ph.D. thesis, Hochschulschriften Bd. 13, Lit Verlag, Münster and Hamburg, 1990.

185. Bolton, S., "Outlier Tests and Chemical Assays," Appendix V in *Pharmaceutical Statistics, Practical and Clinical Applications*, Bolton, Sanford, 3rd ed., Marcel Dekker, New York, 1997.

186. *"Investigating Out of Specification (OOS) Test Results for Pharmaceutical Production,"* FDA Draft Guidance for Industry (J:\!GUIDANC\121DFT.WPD) dated 9/4/98.

187. Bolton, S., "Should a Single Unexplained Failing Assay Be Reason to Reject a Batch?," *Clin. Research & Reg. Affairs*, **10**(3), 1993, 159–175.

188. Salit, M. L., and Turk, G. C., "A Drift Correction Procedure," *Anal. Chem.* **70,** 1998, 3184–3190.

189. Davis, A. A., "Analytical Instrument Maintenance and Its Implications," *Analytica Chimica Acta* **238,** (1990), 45–53.

190. Abramowitz, M., and Stegun, I. A., Eds. Chapter 26, "Probability Functions," *Handbook of Mathematical Functions*, Dover Publications, New York, 1970.

191. Gardiner, M. J. and Bombay, B. F., "An Approximation to Student's t," *Technometrics*, **7,** 1965, 71.

192. Dudewicz, E. J., and Dalal, S. R., "On Approximations to the t-Distribution," *J. Quality Technology*, **4,** 1972, 196–198.

193. Chambers, W. F., "Comment on Calculator Program Yielding Confidence Limits for Least Squares Straight Line," *Anal. Chem.* **49,** 1977, 884 (correspondence).

194. Johnson, E. E., "Empirical Equations for Approximating Tabular F Values," *Technometrics*, **15**(2), 1973, 379–384.

INDEX

DATE DUE